AF356216

RNA Viruses and Antibody Response

RNA Viruses and Antibody Response

Editor

Jason Yiu Wing Kam

Basel • Beijing • Wuhan • Barcelona • Belgrade • Novi Sad • Cluj • Manchester

Editor
Jason Yiu Wing Kam
Division of Natural and
Applied Sciences
Duke Kunshan University
Kunshan
China

Editorial Office
MDPI
St. Alban-Anlage 66
4052 Basel, Switzerland

This is a reprint of articles from the Special Issue published online in the open access journal *Viruses* (ISSN 1999-4915) (available at: www.mdpi.com/journal/viruses/special_issues/RNA_antibody).

For citation purposes, cite each article independently as indicated on the article page online and as indicated below:

Lastname, A.A.; Lastname, B.B. Article Title. *Journal Name* **Year**, *Volume Number*, Page Range.

ISBN 978-3-7258-0110-7 (Hbk)
ISBN 978-3-7258-0109-1 (PDF)
doi.org/10.3390/books978-3-7258-0109-1

Contents

About the Editor

Jason Yiu Wing Kam

Jason Yiu Wing Kam's major area of research is adaptive immune response in patients, with a focus on IgG class switching and epitope identification. Additionally, he is interested in studying virus-host interactions, which may help in biomarker identification and immunotherapy development. He hopes to draw on his research knowledge to develop better vaccine candidates against infectious diseases (particularly epitope-specific antibodies and epitope specificity at the single-amino-acid level). Furthermore, he seeks to develop new investigative and data collection methods to explore new areas of research and derive results of publication quality.

He is interested in identifying the use of EdTech tools to improve the learning outcomes of students during self-directed learning and team learning processes.

Jason Yiu Wing Kam has a B.Sc. in Biochemistry from the Hong Kong University of Science and Technology, an M.Phil. in Pharmacology from the Chinese University of Hong Kong and a Ph.D. in Microbiology from the University of Hong Kong. He also served as a postdoctoral researcher between 2008 and 2019 at the Singapore Immunology Network (SIgN), A*STAR. Before joining Duke Kunshan University as an Assistant Professor of Global Health, he was a senior lecturer at Republic Polytechnic, Singapore.

Editorial

Concluding Remarks for the Special Issue on RNA Viruses and Antibody Response

Yiu-Wing KAM

Division of Natural and Applied Sciences, Duke Kunshan University, Kunshan 215316, China; yiuwing.kam@dukekunshan.edu.cn

Infectious diseases represent one of the major public health concerns on the global level. The emergence and re-emergence of different RNA viruses (influenza, SARS-CoV-1, MERS, CHIKV, Zika and SARS-CoV-2) remain a major concern for public health control worldwide [1,2]. In every disease outbreak, valuable knowledge about virus–host interactions can be learnt to better manage and control the spread of RNA virus diseases. Consequently, preventive measures such as the early detection of cases, clinical management, and the development of vaccines can be employed to reduce the socioeconomic impact of RNA-virus-mediated outbreaks. In the context of vaccine responses and immunity, the responses of neutralizing antibodies serve as predictors of protection from infection [3,4]. Understanding antibody kinetics provides important information to improve the accuracy of early detection system development. This Special Issue collects 14 original studies and 2 reviews that contribute to the overall knowledge on antibody responses triggered by RNA virus infections.

Ten original research papers are published in this Special Issue related to SARS-CoV-2, each looking at different aspects. Five studies discuss the relationship between antibody kinetic profiles and protection. These studies provide important information about whether all infected individuals will have the same course of humoral response maturation or evolve with a unique binding and neutralizing capacity which is individual specific. We might be able to use a mathematical model to predict the protection time post-infection according to antibody binding kinetics.

Currently, the detection of viral RNA by qRT-PCR remains the gold standard for the acute phase of infection. Can we use antibody detection as a diagnostic alternative? To address this, two studies in this Special Issue provide additional insights into improving the sensitivity and specificity of a diagnostic system for SARS-CoV-2 infections using either conserved full-length (N protein) or fragmented (RBD) antigens as the early serology detection system. However, more validation from new SARS-CoV-2/RNA-virus-infected cohorts will be important to ascertain the robustness of the identified "immune signature" for pathogen-specific identification. With the availability of better diagnostic alternatives, clinicians can then provide a well-informed disease intervention program for patients.

More importantly, this Special Issue contains four studies and two reviews looking at the treatment strategies for managing COVID-19 patients, particularly in preventing severe clinical outcomes post-infection. We might learn lessons from other infectious diseases (HIV and RSV) or validate different antiviral treatment options in order to improve our understanding of virus infection mechanisms and disease severity development.

RNA viruses apart from SARS-CoV-2 (influenza and CHIKV) remain a major concern for public health control globally and this Special Issue contains three studies that provide additional antibody knowledge in order to expand the development of better prophylactic and therapeutic strategies.

We would like to thank all authors for publishing their high-quality research and reviews in this Special Issue. It has been a great pleasure working with all the talented researchers around the world. This Special Issue will serve as a platform to further improve

Citation: KAM, Y.-W. Concluding Remarks for the Special Issue on RNA Viruses and Antibody Response. *Viruses* **2023**, *15*, 1214. https://doi.org/10.3390/v15051214

Received: 16 May 2023
Accepted: 19 May 2023
Published: 22 May 2023

the current knowledge of antibody responses against RNA virus infections. Hopefully, this platform will continue to facilitate the design and development of prophylactic and therapeutic strategies in preparation for future disease outbreaks.

Funding: This research was supported in part by the Startup Fund from the Division of Natural and Applied Sciences, Duke Kunshan University (00AKUG0130). Y.-W.K. acknowledges funding from Duke Kunshan University.

Institutional Review Board Statement: Not applicable.

Informed Consent Statement: Not applicable.

Data Availability Statement: Not applicable.

Conflicts of Interest: The author declares no conflict of interest.

References

1. Kam, Y.W.; Ong, E.K.; Rénia, L.; Tong, J.C.; Ng, L.F. Immuno-biology of Chikungunya and implications for disease intervention. *Microbes Infect.* **2009**, *11*, 1186–1196. [CrossRef] [PubMed]
2. Kam, Y.W.; Leite, J.A.; Lum, F.M.; Tan, J.J.L.; Lee, B.; Judice, C.C.; Teixeira, D.A.T.; Andreata-Santos, R.; Vinolo, M.A.; Angerami, R.; et al. Specific Biomarkers Associated with Neurological Complications and Congenital Central Nervous System Abnormalities from Zika Virus-Infected Patients in Brazil. *J. Infect. Dis.* **2017**, *216*, 172–181. [CrossRef] [PubMed]
3. Akahata, W.; Yang, Z.Y.; Andersen, H.; Sun, S.; Holdaway, H.A.; Kong, W.P.; Lewis, M.G.; Higgs, S.; Rossmann, M.G.; Nabel, G.J.; et al. A virus-like particle vaccine for epidemic Chikungunya virus protects nonhuman primates against infection. *Nat. Med.* **2010**, *16*, 334–338. [CrossRef] [PubMed]
4. Metz, S.W.; Gardner, J.; Geertsema, C.; Le, T.T.; Goh, L.; Vlak, J.M.; Suhrbier, A.; Pijlman, G.P. Effective chikungunya virus-like particle vaccine produced in insect cells. *PLoS Negl. Trop. Dis.* **2013**, *7*, e2124. [CrossRef] [PubMed]

 *viruses*

Article

Validation of N Protein Antibodies to Diagnose Previous SARS-CoV-2 Infection in a Large Cohort of Healthcare Workers: Use of Roche Elecsys® Immunoassay in the S Protein Vaccination Era

Juan Francisco Delgado [1,*] , Mònica Vidal [1], Germà Julià [1], Gema Navarro [2], Rosa María Serrano [3], Eva van den Eynde [4], Marta Navarro [4], Joan Calvet [5], Jordi Gratacós [5], Mateu Espasa [6] and Pilar Peña [3]

1   Immunology Laboratory, Clinic Laboratories Service, Parc Taulí Hospital Universitari, Institut d'Investigació i Innovació Parc Taulí (I3PT-CERCA), Departament de Medicina, Universitat Autònoma de Barcelona, 8207 Sabadell, Spain
2   Epidemiology Service, Parc Taulí Hospital Universitari, Institut d'Investigació i Innovació Parc Taulí (I3PT-CERCA), Universitat Autònoma de Barcelona, 8207 Sabadell, Spain
3   Occupational Health Department, Parc Taulí Hospital Universitari, Institut d'Investigació i Innovació Parc Taulí (I3PT-CERCA), Universitat Autònoma de Barcelona, 8207 Sabadell, Spain
4   Infection Disease Department, Parc Taulí Hospital Universitari, Institut d'Investigació i Innovació Parc Taulí (I3PT-CERCA), Universitat Autònoma de Barcelona, 8207 Sabadell, Spain
5   Rheumatology Service, Parc Taulí Hospital Universitari, Institut d'Investigació i Innovació Parc Taulí (I3PT-CERCA), Departament de Medicina, Universitat Autònoma de Barcelona, 8207 Sabadell, Spain
6   Microbiology Section, Laboratory Service, Parc Taulí Hospital Universitari, Institut d'Investigació i Innovació Parc Taulí (I3PT-CERCA), Universitat Autònoma de Barcelona, 8207 Sabadell, Spain
*   Correspondence: jdelgado@tauli.cat

Citation: Delgado, J.F.; Vidal, M.; Julià, G.; Navarro, G.; Serrano, R.M.; van den Eynde, E.; Navarro, M.; Calvet, J.; Gratacós, J.; Espasa, M.; et al. Validation of N Protein Antibodies to Diagnose Previous SARS-CoV-2 Infection in a Large Cohort of Healthcare Workers: Use of Roche Elecsys® Immunoassay in the S Protein Vaccination Era. *Viruses* **2023**, *15*, 930. https://doi.org/10.3390/v15040930

Academic Editor: Jason Yiu Wing KAM

Received: 1 March 2023
Revised: 5 April 2023
Accepted: 6 April 2023
Published: 7 April 2023

Abstract: The aim of this study was to validate the detection of anti-nucleocapsid protein (N protein) antibodies for the diagnosis of SARS-CoV-2 infection in light of the fact that most COVID-19 vaccines use the spike (S) protein as the antigen. Here, 3550 healthcare workers (HCWs) were enrolled from May 2020 (when no S protein vaccines were available). We defined SARS-CoV-2 infection if HCWs were found to be positive by RT-PCR or found to be positive in at least two different serological immunoassays. Serum samples from Biobanc I3PT-CERCA were analyzed by Roche Elecsys® (N protein) and Vircell IgG (N and S proteins) immunoassays. Discordant samples were reanalyzed with other commercial immunoassays. Roche Elecsys® showed the positivity of 539 (15.2%) HCWs, 664 (18.7%) were found to be positive by Vircell IgG immunoassays, and 164 samples (4.6%) showed discrepant results. According to our SARS-CoV-2 infection criteria, 563 HCWs had SARS-CoV-2 infection. The Roche Elecsys® immunoassay has a sensitivity, specificity, accuracy, and concordance with the presence of infection of 94.7%, 99.8%, 99.3%, and 0.96, respectively. Similar results were observed in a validation cohort of vaccinated HCWs. We conclude that the Roche Elecsys® SARS-CoV-2 N protein immunoassay demonstrated good performance in diagnosing previous SARS-CoV-2 infection in a large cohort of HCWs.

Keywords: SARS-CoV-2; antibody response; infection; vaccination; nucleocapsid protein; spike protein

1. Introduction

COVID-19 is an acute respiratory syndrome caused by the new coronavirus SARS-CoV-2, first described in Wuhan (Hubei province, China) following an outbreak of pneumonia of unknown origin. It is highly transmissible and has spread throughout the world. The WHO declared its spread a pandemic causing COVID-19 [1].

Clinical manifestations of SARS-CoV-2 infection range from asymptomatic or mild non-specific symptoms to severe pneumonia with organ function damage. Common symptoms are fever, cough, fatigue, dyspnea, myalgia, sputum production, and headache [2,3]. These symptoms are non-specific and cannot be used for an accurate diagnosis; therefore, laboratory testing plays an important role in diagnosing SARS-CoV-2 patients. These tests can also identify those who are asymptomatic.

Laboratory diagnosis of COVID-19 has mainly been based on molecular tests such as real-time reverse-transcription PCR (RT-PCR) [4–6]. Antibody-based techniques are complementary tools for SARS-CoV-2 infection detection. The presence of antibodies is an indirect marker of infection [4,6–10]. The development of an antibody response to COVID-19 occurs between 5 and 14 days after exposure to the virus. As such, serological tests in the market are of little use in the context of acute COVID-19. Sensitivities are less than 50% in the first week of infection [11]. However, the detection of SARS-CoV-2 antibodies is an excellent way to determined past infection with a sensitivity higher than 90% after 7 days [9,12–14]. Serological assays play an essential role in population seroprevalence evaluation and can help to account for asymptomatic cases, symptomatic cases that did not get tested, or patients suspected to have COVID-19 with a negative SARS-CoV-2 RT-PCR.

SARS-CoV-2 has at least four structural proteins: spike (S), envelope (E), membrane (M), and nucleocapsid (N) proteins. Both viral S and N proteins are major structural proteins and highly immunogenic. Therefore, most patients develop antibodies against them. [15]. In addition, the SARS-CoV-2 genome encodes 16 nonstructural proteins [16]. Antibodies against peptides derived from non-structural and accessory proteins are also detectable [17]. Commercial serologic methods target specific antibodies on several SARS-CoV-2 epitopes including the N protein, the S protein, and the receptor-binding domain of the S protein. The tests provide accurate diagnosis if performed on specimens collected 10 to 14 days after symptom onset, but performance varies among methods [9,13]. Different studies relate antibody titers after SARS-CoV-2 infection with age, sex, and severity [18–21]. More than 350 vaccines are currently being investigated for a potential role in mitigating the COVID-19 pandemic (https://www.who.int/publications/m/item/draft-landscape-of-COVID-19-candidate-vaccines, accessed on 27 February 2023). The approved vaccines target the S protein because this is the one that binds to the ACE2 (angiotensin-converting enzyme 2) receptor; thus, developing a humoral immune response against it could generate the formation of neutralizing antibodies to prevent infection.

Several studies have analyzed the antibody response induced by the S protein [22]. In a healthy population, people develop an antibody response from mRNA vaccines (BNT162b2 and mRNA-1273). These antibodies act against the S protein of the original strain [23,24]. mRNA vaccines and other vaccines can induce this response against the S protein [25]. Global vaccination rates range from 40–90% depending on the country [26]. This makes the serological diagnosis of SARS-CoV-2 infection difficult because it is impossible to distinguish antibodies against the S protein by infection vs. those produced by vaccination. It is relevant to study antibodies against other virus proteins for serological diagnosis; the most widely used immunoassay assesses the N protein. [23,24,27].

We present here a study analyzing the performance of anti-SARS-CoV-2 IgM/IgA/IgG Elecsys® (Roche Diagnostics International Ltd., Rotkreuz, Switzerland) on Cobas™ e801 (Roche Diagnostics International Ltd., Rotkreuz, Switzerland) to detect N protein antibodies for diagnosing SARS-CoV-2 infection in 3550 healthcare workers (HCWs) during the first COVID-19 wave. We then compared its performance with the Vircell IgG immunoassay (Vircell, Granada, Spain), which detects peptides from N and S proteins and the RT-PCR.

2. Materials and Methods

This observational retrospective study was approved by the Drug Research Ethics Committee of Parc Tauli University Hospital (code 2020581).

2.1. Study Population

Here, 3550 HCWs were enrolled from 6 to 29 May 2020 when S protein vaccines were not available. The demographic and clinical characteristics of this study cohort are shown in Table 1, and tests performed in this cohort are shown in Figure 1. Inclusion criteria were HCWs from Parc Taulí University Hospital and HCWs who worked at the center during the pandemic. Clinical characteristics were obtained from a survey at the time of enrollment. The HCWs had to record their symptoms from the start of the pandemic in our area (at the end of February 2020) until the time of study participation. The HCWs' serum samples were provided by BioBanc I3PT-CERCA and were processed after standard operating procedures with the approval of the Ethics and Scientific Committees for serological commercial immunoassays.

Table 1. Clinical data of healthcare workers from the study cohort ($n = 3550$).

		HCW SARS-CoV-2 Infection ($n = 563$)	HCW without SARS-CoV-2 Infection ($n = 2987$)	p Value
Clinical	Age in years (median $\pm$ IQR)	39.0 (29.0–50.0)	42.0 (33.0–52.0)	<0.001
	Female/Male ratio	3.8	3.5	0.534
characteristics	Days after onset symptoms (median $\pm$ IQR)	55.0 (44.0–54.0)	57.0 (41.0–70.7)	0.041
	Body mass index (median $\pm$ IQR)	23.5 (21.5–26.6)	23.9 (21.5–26.7)	0.573
	Overweight (%)	27.4	28.7	0.520
	Obese (%)	9.4	10.0	0.665
	Smoker (%)	11.5	25.1	<0.001
	Hospitalization (%)	4.4	0.1	<0.001
Symptoms (%)	Vomits	75.3	11.4	<0.001
	Difficulty breathing	72.4	11.1	<0.001
	Abdominal pain	71.1	8.4	<0.001
	Sore throat	58.4	6.0	<0.001
	Nasal congestion	54.7	8.6	<0.001
	Diarrhea	54.7	7.6	<0.001
	Dry cough	41.9	6.6	<0.001
	Fever	41.0	10.4	<0.001
	Loss of taste	35.6	12.0	<0.001
	Loss of smell	34.0	11.9	<0.001
	Chill	33.4	8.4	<0.001
	Headache	23.6	3.1	<0.001
	Myalgia	25.2	5.4	<0.001
	Comorbidities	16.3	19.0	0.130
	Fatigue	12.5	2.6	<0.001
	Arterial hypertension	6.4	7.5	0.342
	Asymptomatic	30.4	84.5	<0.001

IQR = interquartile range.

A validation cohort was designed using a new cross-sectional study of serologies in HCWs vaccinated with mRNA vaccines (BNT162b2 and mRNA-1273) and HCWs who had had a RT-PCR test. HCWs were classified as infected in the study cohort were excluded. Thus, the validation cohort was made up of 297 infected and 1593 non-infected HCWs according to RT-PCR results. The female/male ratio, age, body mass index and smoking variables were recorded using an online survey at the time of enrollment. These clinical data are shown in Supplementary Table S1.

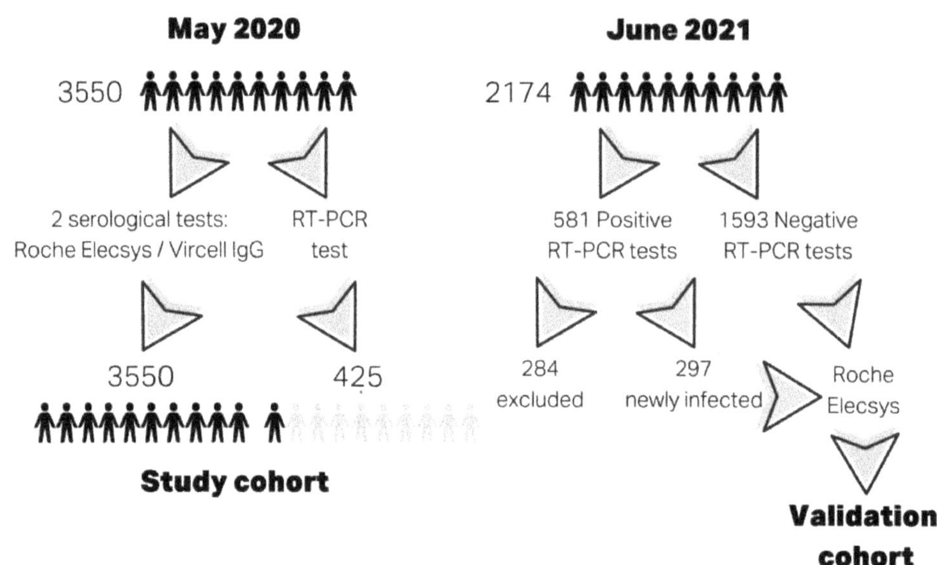

Figure 1. Definition of cohorts and tests performed.

2.2. SARS-CoV-2 Infection Criteria

We confirmed SARS-CoV-2 infection if HCWs were found to be positive by the RT-PCR or were found to be positive in at least two different serological immunoassays. The equivocal results of the immunoassays were interpreted as negative results when assessing the criteria. Serology is a widely known tool for the diagnosis of previous infectious diseases including COVID-19 [13]; however, the use of these infection criteria may have induced a bias when studying the performance of serological immunoassays in the study population, since a large number of HCWs were classified as infected and non-infected using serological tests exclusively, and this variable is part of the infection criteria. We limited the effect of this bias in two ways: by using the existence of two serological tests and not just one to classify a HCW as infected and performing 2 additional serological tests in case of discrepant results.

2.3. RT-PCR

The microbiological diagnosis of SARS-CoV-2 infection was carried out by nucleic acid amplification techniques. All patients had a nasopharyngeal swab and were tested for SARS-CoV-2 infection through a retrotranscriptase PCR (RT-PCR): swabs were processed by the RT-PCR for SARS-CoV-2 with Allplex 2019-nCoV Assay (RP10244Y, Seegene, Seoul, Republic of Korea) or the Simplexa SARS-CoV2 Assay kit (MOL4150, DiaSorin, Gerenzano, Italy) per the manufacturer's instructions for qualitative results.

2.4. Commercial Immunoassays to Detect Antibodies against SARS-CoV-2

All serum samples were thawed to perform the Elecsys® Anti-SARS-CoV-2 IgM/IgA/IgG assay on Cobas™ e801 (09203079190, Roche Diagnostics International Ltd., Rotkreuz, Switzerland) according to manufacturer instructions. Samples were positive if the index was ≥1. The ELISA COVID-19 IgG immunoassay (G1032, Vircell, Granada, Spain) used Triturus® ELISA Instrument (Grifols, Barcelona, Spain) according to the Vircell-adapted protocol for this analyzer; samples were positive if the index was ≥11.2. Antigens and immunoassay characteristics are shown in Supplementary Table S2.

Discordant samples between the two immunoassays were reanalyzed with other commercial immunoassays according to manufacturer instructions or by adapting its protocols to Triturus® ELISA Instrument: ELISA Anti-SARS-CoV-2 (IgG) (EI 2606-9601 G, Euroimmun, Lubeck, Germany); LIAISON® SARS-CoV-2 S1/S2 IgG (311450, DiaSorin, Gerenzano, Italy). Antigens and immunoassay characteristics are shown in Supplementary Table S2.

2.5. Validation Specificity of Immunoassays

In addition, 100 serum samples from healthy donors were collected from Banc de Sang i Teixits in a pre-pandemic period (October 2019) to establish a specificity of > 98% for the Roche Elecsys® and Vircell IgG immunoassays. The healthy donors were aged between 18 and 69 years, including 54 males and 46 women.

2.6. Statistical Analysis

For descriptive purposes, the cohort was characterized with absolute and relative frequencies for categorical variables; medians were used for numerical measurements. Sensitivity, specificity, positive, negative predictive values, and area under the receiver operating characteristic (ROC) curve were calculated for Roche Elecsys® and Vircell IgG immunoassays. Level of agreement with SARS-CoV-2 infection was calculated with Cohen's kappa coefficient. The Kolmogorov–Smirnov test was used to evaluate the suitability of data for normal distribution. We used univariate analysis to test the link between variables with the Chi square test or Fisher's exact test if indicated for categorical variables. A Mann–Whitney U-test was used for continuous quantitative variables. Significant associations were assumed when p was <0.05. Analysis used the statistical software IBM SPSS Statistics v28.0, (Chicago, IL, USA).

3. Results

3.1. Immunoassays' Specificity

Specificity was analyzed with serum samples from 100 healthy donors. None of the processed samples gave a positive result on the Roche Elecsys® immunoassay within the index cutoff established by the manufacturer (1.0). However, five samples were positive for IgG for the Vircell IgG immunoassay within the manufacturer's recommended index cutoff (6.0). The index cutoff for Vircell IgG was 11.2 for a specificity of 98%.

3.2. Immunoassays' Performance

Samples from 3550 HCWs were tested by Roche Elecsys® and Vircell IgG immunoassays: 539 (15.2%) and 664 (18.7%) were found to be positive by each immunoassay, respectively. There were 164 samples (4.6%) with discrepant results between immunoassays (Figure 2). To determine the presence of SARS-CoV-2 antibodies, discrepant samples were tested with Euroimmun IgG and DiaSorin IgG immunoassays. Concordance between immunoassays for 164 discrepant samples are shown in Table 2. Among the 164 discrepant samples, 23 were found to be positive by Roche Elecsys®, and 14 were also found to be positive by both Euroimmun IgG and Diasorin IgG. There were 141 samples found to be positive by Vircell IgG; four of them were found to be positive by Euroimmun IgG, and seven of them were found to be positive by Diasorin IgG. Therefore, in this group, Roche Elecsys® showed a positive concordance with immunoassays, except for Vircell IgG, and with SARS-CoV-2 infection. Vircell IgG showed discordance with the Roche Elecsys® and DiaSorin IgG immunoassays and infection. Therefore, the Roche Elecsys® immunoassay showed better performance on discrepant samples.

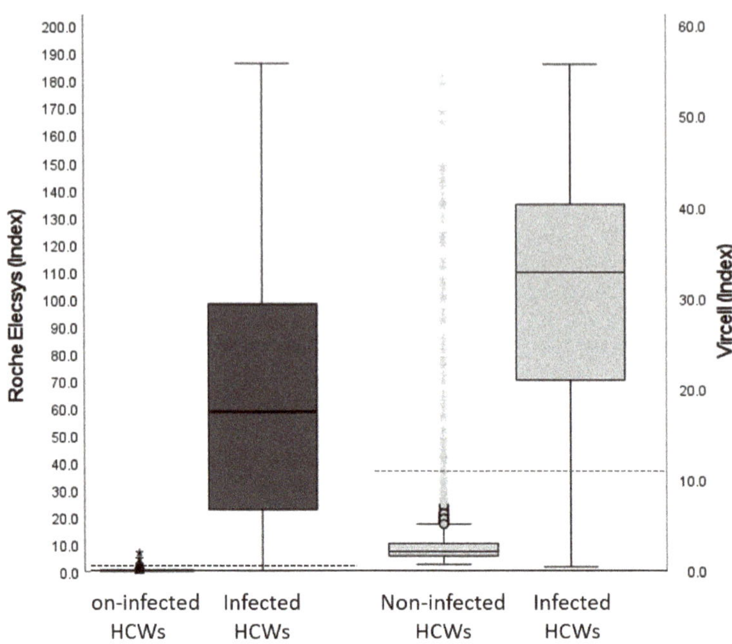

Figure 2. Antibody response against SARS-CoV-2 nucleocapsid protein in the study cohort of health care workers; Roche Elecsys® on the left and Vircell IgG on the right. Subjects were grouped according to their infection status. Boxplots represent the quantification of the distribution of anti-nucleocapsid protein antibodies; upper and lower bounds of the boxes indicate 75th and 25th percentiles, respectively. The dotted horizontal line indicates the cut-off point of each immunoassay.

Table 2. Commercial immunoassay concordance in discrepant samples from the study cohort ($n = 164$).

Manufacturer	Roche Concordance Kappa [95% CI]	Vircell IgG Concordance Kappa [95% CI]	Infection Concordance Kappa [95% CI]
Roche Elecsys®	-	−0.32 [−0.45–(−0.18)]	0.73 [0.58–0.87]
Vircell IgG	−0.32 [−0.45–(−0.18)]	-	−0.27 [−0.38–(−0.15)]
Diasorin IgG	0.58 [0.39–0.76]	−0.18 [−0.28–(−0.08)]	0.81 [0.68–0.94]
Euroimmun IgG	0.61 [0.39–0.82]	0.12 [0.04–0.20]	0.73 [0.56–0.89]

CI = confidence interval.

3.3. RT-PCR Performance

A total of 425 HCWs were tested for SARS-CoV-2 by RT-PCR; 203 (47.8%) gave positive results between the start of the pandemic (February 2020) and the time of enrollment in the study (May 2020). In the group of HCWs with positive RT-PCR results, the Roche Elecsys® and Vircel IgG immunoassay detected antibodies in almost 90% (Table 3). However, concordance between the RT-PCR and serology dropped significantly among HCWs with negative RT-PCR results—antibodies against SARS-CoV-2 were detected in 30% of the HCWs (Table 3). The detection of antibodies using the Roche Elecsys® immunoassay showed a better correlation with SARS-CoV-2 infection than Vircell IgG did among the group of HCWs with RT-PCR results (Table 3).

Table 3. Roche Elecsys® and Vircell IgG performance among HCWs in the RT-PCR group from the study cohort (*n* = 425).

Immunoassay	RT-PCR Positive HCWs Group (*n* = 203)	RT-PCR Negative HCWs Group (*n* = 222)	RT-PCR Kappa [CI] (*n* = 425)	SARS-CoV-2 Infection Kappa [CI] (*n* = 425)
Roche Elecsys®-Positive	182 (89.7%)	67 (30.2%)	0.59 [0.51–0.66]	0.89 [0.85–0.94]
Roche Elecsys®-Negative	21 (10.3%)	155 (69.8%)		
Vircell IgG-Positive	180 (88.7%)	71 (32.0%)	0.56 [0.48–0.64]	0.84 [0.79–0.89]
Vircell IgG-Negative	23 (11.3%)	151 (68.0%)		
SARS-CoV-2 Infection	203 (100.0%)	67 (30.2%)	0.68 [0.62–0.75]	-

CI = confidence interval.

3.4. N protein Antibody Immunoassay Performed to Diagnose SARS-CoV-2 Infection

According to our SARS-CoV-2 infection criteria, 563 among the 3550 HCWs in this study had SARS-CoV-2 infection. Among the infected HCWs, the immunoassay detecting N protein antibodies, Roche Elecsys® immunoassay, was positive in 534 samples (94.8%) and Vircell IgG was positive in 523 (92.9%) samples (Table 4). The sensitivity, specificity, negative predicted value, positive predicted value and accuracy of both immunoassays are shown in Table 4. The N protein antibodies detected by Roche Elecsys® showed excellent correlation and accuracy with SARS-CoV-2 infection—these metrics were higher than those found using the Vircell IgG immunoassay (Table 4).

Table 4. Immunoassay performed to diagnose previous SARS-CoV-2 infection in the study cohort (*n* = 3550).

Manufacturer	Sensitivity	Specificity	Positive Predicted Value	Negative Predicted Value	Accuracy (95% CI)	Infection Concordance (Kappa, 95% CI)
Roche Elecsys®	94.8	99.8	99.1	99.0	99.3 (99.8–99.9)	0.96 (0.95–0.97)
Vircell IgG	92.9	95.3	77.8	98.6	96.9 (96.1–97.6)	0.82 (0.80–0.85)

CI = confidence interval.

3.5. Antibodies against SARS-CoV-2 and RT-PCR Association with Clinical Symptoms

As expected, HCWs with SARS-CoV-2 and hospitalized infected groups had symptoms more frequently associated with infection except for arterial hypertension (Table 1). The infected HCW group had younger workers and fewer smokers than the non-infected HCW group dod (Table 1). Given the strong correlation between antibody detection and infection, the association between antibodies and the presence of symptoms has similar statistical significance (Supplementary Table S3). Of the HCWs with RT-PCR results, the infected HCWs with positive and negative RT-PCR results were compared. Both groups showed similar behavior except HCWs who were overweight or had a dry cough or comorbidities. Differences were observed in the percentage of asymptomatic patients between both groups and the frequency of positive antibodies found by Roche Elecsys® and Vircell IgG. There were more positive antibodies in HCWs with negative RT-PCR results (Supplementary Table S4). Finally, the performance of the immunoassays for the detection of antibodies against SARS-CoV-2 in symptomatic HCWs were compared with that in asymptomatic HCWs: no significant differences were found between Roche Elecsys® and Vircell IgG at a quantitative level (p = 0.243; p = 0.629, respectively) or a qualitative level (p = 0.206; p = 0.687, respectively). One difference between symptomatic and asymptomatic HCWs was seen between smokers (18.1%) in the groups of asymptomatic vs. symptomatic subjects (8.7%), p = 0.001.

3.6. Roche Elecsys® Immunoassay Performance in the Validation Cohort

Finally, we analyzed the performance of the Roche Elecsys® immunoassay in the validation cohort—a cohort of vaccinated HCWs with a positive result found by the RT-PCR who did not belong to the study cohort. Sensitivity, specificity, negative predicted value, positive predicted value, and accuracy are shown in Table 5. The results obtained by the immunoassay in the validation cohort were very similar to those obtained in the study cohort. In this cohort, infected HCWs were younger than non-infected HCWs, at ages of 43.0 (IQR 34.0–52.0) versus 45.0 (37.0–54.0), respectively ($p = 0.007$). Smoking was more frequent in the non-infected HCWs $p < 0.001$ (Supplementary Table S1). There were no significant differences in sex and body mass index between the infected and non-infected HCWs (Supplementary Table S1).

Table 5. Roche Elecsys® immunoassay performed to diagnose SARS-CoV-2 infection in the validation cohort ($n = 1890$).

Manufacturer	Sensitivity	Specificity	Positive Predicted Value	Negative Predicted Value	Accuracy (95% CI)	Infection Concordance (Kappa, 95% CI)
Roche Elecsys	95.3	99.7	98.6	99.1	96.9 (95.3–98.5)	0.96 (0.95–0.98)

CI = confidence interval.

4. Discussion

This study analyzed the performance of the Roche Elecsys® immunoassay directed against the SARS-CoV-2 N protein in a large cohort of HCWs ($n = 3550$). A positive RT-PCR and/or two positive serological tests were the infection criteria. All samples were collected in May 2020 before the approval of S protein vaccines.

The Roche Elecsys® test measures antibodies against the N protein. The results were positive for 539 HCWs. The Vircell IgG test measures antibodies against S protein and N protein peptides and was positive for 664 HCWs. The overall seroprevalence in our cohort was 15.2% (539) for Roche Elecsys® and 18.7% (664) for Vircell IgG in May 2020. Initially, the higher antibody positivity found by Vircell IgG assay could have been due to how it analyzes S and N proteins. The Roche Elecsys® only analyzes N proteins as reported in a meta-analysis [28]. As such, we decided to evaluate the 164 discrepant samples with DiaSorin IgG and Euroimmun IgG immunoassays. Concordance between the Vircell IgG test and the new tests confirmed that the differences were discordant, ranging between −0.18–0.12 (Cohen's kappa coefficient). In Roche's test, however, the concordance was moderate, ranging between 0.58–0.61 (Cohen's kappa coefficient) (Table 2). A possible explanation is the immunoglobulin isotype that each immunoassay detects. Roche Elecsys® measures IgG, IgM, and IgA while Vircel IgG only measures IgG. However, DiaSorin IgG and Euroimmun IgG only detect the IgG isotype and show better performance than Vircell IgG. Thus, the apparent increased sensitivity of the Vircell IgG immunoassay is translated into a loss of specificity and lower positive predictive value (Table 4). In this group of discrepant samples, DiaSorin IgG, Roche Elecsys®, and Euroimmun IgG showed good correlation with SARS-CoV-2 infection (Table 2). However, the Euroimmun and DiaSorin immunoassays use the S protein as an antigen; therefore, they would not be useful for the diagnosis of infection, while the Roche Elecsys® would have this utility given the worldwide COVID-19 vaccination.

May 2020 was a period of the first wave of the pandemic and there was limited access to RT-PCR; thus, not all HCWs with symptoms underwent RT-PCRs during acute infection with SARS-CoV-2. The lack of availability of reagents for the diagnosis of acute infection by RT-PCR causes it to have a low yield with a Cohen's kappa coefficient of 0.68 (Table 3). The main cause was the time between the test's performance and the onset of symptoms. In many cases, this was over two weeks, thus giving negative RT-PCR results; however, 67 of the 202 HCWs found to be negative by RT-PCR were classified as infected due to the positivity of two serological tests. Thus, we evaluated the relevance of serological tests for the diagnosis of past infection.

Among the group of RT-PCR-positive HCWs, around 90% had seroconverted; the infection criterion we used assumes that a positive RT-PCR indicates infection without accounting for the false positive rate found by this technique. The sensitivity of both serologic immunoassays could be slightly underestimated (Table 4) [29]. Among the HCWs, Roche Elecsys® obtained the best Cohen's kappa coefficient (Table 3). Of the RT-PCR-negative HCWs, around 30% had positive antibodies: This could be explained by the lack of reagents—some PCRs were performed outside the period of acute infection, when its sensitivity decreased [30].

The Roche Elecsys® immunoassay was better in terms of sensitivity, specificity, accuracy, and concordance with infection with values of 94.7, 99.8, 99.3, and 0.96, respectively, in the study cohort. The Vircell IgG assay was found to have the same values, 93.0, 95.3, 96.9, and 0.83, respectively. If we used the cutoff point recommended by Vircell for analysis, then specificity would have dropped further and the assay would have had worse results. This is in contrast to the data from Alharbi et al. [31]. The improved performance of the Roche Elecsys® immunoassay agrees with the results of different comparative studies using a small sample size. This research confirms the results obtained by other studies using a large cohort of patients [12,32–34]. Additionally, the Roche Elecsys® immunoassay was analyzed in a validation cohort consisting of vaccinated HCWs who were classified as infected HCWs using RT-PCR test results. In this cohort, the immunoassay showed a sensitivity, specificity, accuracy, and concordance with infection of 95.3, 99.7, 96.9, and 0.96, respectively. The performance of Roche Elecsys® in both study and validation cohorts showed similar results; therefore, the possible bias that the analysis of the performance of the serological test could have as it is also part of the classification criteria was minimized by the correction factors included to reduce this effect.

As expected, symptoms related to SARS-CoV-2 infection were more frequently present in infected HCWs than non-infected HCWs. There are three risk factors associated with severe infection that HCWs in our group did not have: arterial hypertension, obesity, and age (Table 1). This is probably because the population chosen for the study was young with little associated comorbidities compared to the general population. The presence of arterial hypertension was not associated with infection since it is a risk factor for those suffering from more serious diseases than SARS-CoV-2. The mechanism is that treatments with angiotensin-converting enzyme inhibitors induce the upregulation of ACE2, which is the receptor that the virus uses to enter inside cells [35–37]. Obesity was another risk factor evaluated in the survey that was not associated with infection [38,39]. The last risk factor not associated was age—this association was the opposite of what was expected. Infection was more frequent in younger HCWs [40,41]. Another fact is that the smoking rate in non-infected HCWs was higher than that in infected ones. It was reported that infected patients who smoke have a lower antibody response, which is a risk factor for severe SARS-CoV-2 infection; the frequency in hospitalized patients was lower than in the normal population [24,42,43]. In our cohort of HCWs, we saw a paradox of smokers being in the non-infected HCWs group. This fact could be explained due to the bias of the population studied. The clinical variables recorded in the validation cohort showed similar associations between them compared to the study cohort, where infected HCWs were younger and showed a lower smoking rate than non-infected HCWs.

We compared infected HCWs under our infection criteria based on their RT-PCR positivity. We determined if HCWs could be classified as infected exclusively by performing two positive serological tests, which presented a difference with respect to those who had been diagnosed with a positive RT-PCR. In comparison, we observed that most clinical parameters were similar in both groups. Differences were found in terms of the higher frequency of overweight, comorbid, and asymptomatic HCWs between the group diagnosed by serology and the HCW group diagnosed with positive RT-PCRs. There was a higher frequency of dry cough (Table 5). The presence of asymptomatic infection for HCWs diagnosed by serology is explained by the fact that RT-PCR were more likely to be performed during the period of acute infection when symptoms were present than on asymptomatic HCWs, RT-PCR results of whom indicated if there was close contact with a positive case.

Lastly, we analyzed whether asymptomatic HCWs had a different antibody response to infection than HCWs with symptoms; there were no significant differences found between Roche Elecsys® and Vircell IgG ($p = 0.243$ and $p = 0.629$, respectively). There were no differences in sex, age, and weight. We found a higher frequency of HCWs who were smokers in the asymptomatic group vs. the symptomatic group ($p = 0.001$).

5. Conclusions

The Roche Elecsys® SARS-CoV-2 N protein immunoassay was used in a large cohort of HCWs and demonstrated good performance in the diagnosis of previous infection SARS-CoV-2. It is a useful tool to diagnose previous infection in a population vaccinated against SARS-CoV-2 via the S protein.

Supplementary Materials: The following supporting information can be downloaded at: https://www.mdpi.com/article/10.3390/v15040930/s1, Figure S1: Correlation between Roche Elecsys®and Vircell IgG immunoassays; Table S1: Healthcare workers clinical data from the validation cohort ($n = 1890$); Table S2: Commercial immunoassays charasteristics; Table S3: Healthcare workers clinical data according Roche Elecsys®and Vircell IgG status ($n = 3550$); Table S4: Infected healthcare workers (HCWs) clinical data with RT-PCR results ($n = 269$).

Author Contributions: Conceptualization: J.F.D. and P.P.; methodology: P.P., G.N., R.M.S., J.C. and J.G.; data curation: M.N., E.v.d.E. and M.E.; investigation: J.F.D. and M.V.; validation: J.F.D.; formal analysis: J.F.D., G.J. and M.V.; writing—original draft preparation: J.F.D., M.V., M.E. and G.J.; supervision: J.G. and J.C. All authors have read and agreed to the published version of the manuscript.

Funding: This research was funded by the CERCA Programme/Generalitat de Catalunya.

Institutional Review Board Statement: The study was conducted in compliance with data protection regulations and biomedical research laws in Spain. This study was approved by the Drug Research Ethics Committee of Parc Tauli University Hospital (code 2020581).

Informed Consent Statement: Informed consent was obtained from all subjects in the study.

Data Availability Statement: Data are available upon request from the corresponding author.

Acknowledgments: We want to acknowledge the Bank of Sabadell's donation which allowed us to conduct this project at Parc Tauli University Hospital. We want to particularly acknowledge the HCWs and Biobanc I3PT-CERCA for their collaboration.

Conflicts of Interest: The authors declare no conflict of interest.

References

1. Ge, H.; Wang, X.; Yuan, X.; Xiao, G.; Wang, C.; Deng, T.; Yuan, Q.; Xiao, X. The Epidemiology and Clinical Information about COVID-19. *Eur. J. Clin. Microbiol. Infect. Dis.* **2020**, *39*, 1011–1019. [CrossRef]
2. Binnicker, M.J. Challenges and Controversies to Testing for COVID-19. *J. Clin. Microbiol.* **2020**, *58*, e01695-20. [CrossRef] [PubMed]
3. Hu, B.; Guo, H.; Zhou, P.; Shi, Z.-L. Characteristics of SARS-CoV-2 and COVID-19. *Nat. Rev. Microbiol.* **2021**, *19*, 141–154. [CrossRef] [PubMed]
4. Chong, Y.P.; Choy, K.W.; Doerig, C.; Lim, C.X. SARS-CoV-2 Testing Strategies in the Diagnosis and Management of COVID-19 Patients in Low-Income Countries: A Scoping Review. *Mol. Diagn. Ther.* **2023**, 1–18. [CrossRef] [PubMed]
5. Perveen, S.; Negi, A.; Gopalakrishnan, V.; Panda, S.; Sharma, V.; Sharma, R. COVID-19 Diagnostics: Molecular Biology to Nanomaterials. *Clin. Chim. Acta* **2023**, *538*, 139–156. [CrossRef]
6. Huynh, A.; Arnold, D.M.; Smith, J.W.; Moore, J.C.; Zhang, A.; Chagla, Z.; Harvey, B.J.; Stacey, H.D.; Ang, J.C.; Clare, R.; et al. Characteristics of Anti-SARS-CoV-2 Antibodies in Recovered COVID-19 Subjects. *Viruses* **2021**, *13*, 697. [CrossRef]
7. Zhao, J.; Yuan, Q.; Wang, H.; Liu, W.; Liao, X.; Su, Y.; Zhang, Z. Antibody responses to SARS-CoV-2 in patients with novel coronavirus disease 2019. *Clin. Infect. Dis.* **2020**, *71*, 2027–2034. [CrossRef]
8. Guo, L.; Ren, L.; Yang, S.; Xiao, M.; Chang, D.; Yang, F.; Dela Cruz, C.S.; Wang, Y.; Wu, C.; Xiao, Y.; et al. Profiling Early Humoral Response to Diagnose Novel Coronavirus Disease (COVID-19). *Clin. Infect. Dis.* **2020**, *71*, 778–785. [CrossRef]
9. Theel, E.S.; Slev, P.; Wheeler, S.; Couturier, M.R.; Wong, S.J.; Kadkhoda, K. The Role of Antibody Testing for SARS-CoV-2: Is There One? *J. Clin. Microbiol.* **2020**, *58*, e00797-20. [CrossRef]
10. Loeffelholz, M.J. Evaluation of High-Throughput Serological Tests for SARS-CoV-2. *J. Clin. Microbiol.* **2020**, *58*, e02179-20. [CrossRef]
11. Mallano, A.; Ascione, A.; Flego, M. Antibody Response against SARS-CoV-2 Infection: Implications for Diagnosis, Treatment and Vaccine Development. *Int. Rev. Immunol.* **2022**, *41*, 393–413. [CrossRef] [PubMed]

12. Gutiérrez-Cobos, A.; Gómez de Frutos, S.; Domingo García, D.; Navarro Lara, E.; Yarci Carrión, A.; Fontán García-Rodrigo, L.; Fraile Torres, A.M.; Cardeñoso Domingo, L. Evaluation of Diagnostic Accuracy of 10 Serological Assays for Detection of SARS-CoV-2 Antibodies. *Eur. J. Clin. Microbiol. Infect. Dis.* **2021**, *40*, 955–961. [CrossRef] [PubMed]

13. Fox, T.; Geppert, J.; Dinnes, J.; Scandrett, K.; Bigio, J.; Sulis, G.; Hettiarachchi, D.; Mathangasinghe, Y.; Weeratunga, P.; Wickramasinghe, D.; et al. Antibody Tests for Identification of Current and Past Infection with SARS-CoV-2. *Cochrane Database Syst. Rev.* **2022**, *6*, CD013652. [CrossRef]

14. Post, N.; Eddy, D.; Huntley, C.; van Schalkwyk, M.C.I.; Shrotri, M.; Leeman, D.; Rigby, S.; Williams, S.V.; Bermingham, W.H.; Kellam, P.; et al. Antibody Response to SARS-CoV-2 Infection in Humans: A Systematic Review. *PLoS ONE* **2020**, *15*, e0244126. [CrossRef] [PubMed]

15. Liu, D.; Wu, F.; Cen, Y.; Ye, L.; Shi, X.; Huang, Y.; Fang, S.; Ma, L. Comparative Research on Nucleocapsid and Spike Glycoprotein as the Rapid Immunodetection Targets of COVID-19 and Establishment of Immunoassay Strips. *Mol. Immunol.* **2021**, *131*, 6–12. [CrossRef] [PubMed]

16. Wang, M.-Y.; Zhao, R.; Gao, L.-J.; Gao, X.-F.; Wang, D.-P.; Cao, J.-M. SARS-CoV-2: Structure, Biology, and Structure-Based Therapeutics Development. *Front. Cell. Infect. Microbiol.* **2020**, *10*, 587269. [CrossRef] [PubMed]

17. Li, Y.; Xu, Z.; Lei, Q.; Lai, D.; Hou, H.; Jiang, H.; Zheng, Y.; Wang, X.; Wu, J.; Ma, M.; et al. Antibody Landscape against SARS-CoV-2 Reveals Significant Differences between Non-Structural/Accessory and Structural Proteins. *Cell Rep.* **2021**, *36*, 109391. [CrossRef]

18. Kritikos, A.; Gabellon, S.; Pagani, J.-L.; Monti, M.; Bochud, P.-Y.; Manuel, O.; Coste, A.; Greub, G.; Perreau, M.; Pantaleo, G.; et al. Anti-SARS-CoV-2 Titers Predict the Severity of COVID-19. *Viruses* **2022**, *14*, 1089. [CrossRef]

19. Tamizuddin, S.; Cham, J.; Ghiasi, Y.; Borroto, L.; Cao, C.; Orendain, N.; Quigley, M.M.; Nicholson, L.J.; Pandey, A.C. Hospitalization Requiring Intensive Care Unit Due to SARS-CoV-2 Infection Correlated with IgM Depression and IgG Elevation. *Future Sci. OA* **2022**, *8*, FSO783. [CrossRef]

20. Shields, A.M.; Faustini, S.E.; Perez-Toledo, M.; Jossi, S.; Allen, J.D.; Al-Taei, S.; Backhouse, C.; Dunbar, L.A.; Ebanks, D.; Emmanuel, B.; et al. Serological Responses to SARS-CoV-2 Following Non-Hospitalised Infection: Clinical and Ethnodemographic Features Associated with the Magnitude of the Antibody Response. *BMJ Open Respir. Res.* **2021**, *8*, e000872. [CrossRef]

21. Markmann, A.J.; Giallourou, N.; Bhowmik, D.R.; Hou, Y.J.; Lerner, A.; Martinez, D.R.; Premkumar, L.; Root, H.; van Duin, D.; Napravnik, S.; et al. Sex Disparities and Neutralizing-Antibody Durability to SARS-CoV-2 Infection in Convalescent Individuals. *Msphere* **2021**, *6*, e00275-21. [CrossRef] [PubMed]

22. Bayarri-Olmos, R.; Idorn, M.; Rosbjerg, A.; Pérez-Alós, L.; Hansen, C.B.; Johnsen, L.B.; Helgstrand, C.; Zosel, F.; Bjelke, J.R.; Öberg, F.K.; et al. SARS-CoV-2 Neutralizing Antibody Responses towards Full-Length Spike Protein and the Receptor-Binding Domain. *J. Immunol.* **2021**, *207*, 878–887. [CrossRef] [PubMed]

23. Steensels, D.; Pierlet, N.; Penders, J.; Mesotten, D.; Heylen, L. Comparison of SARS-CoV-2 Antibody Response Following Vaccination with BNT162b2 and MRNA-1273. *JAMA* **2021**, *326*, 1533. [CrossRef]

24. Delgado, J.F.; Berenguer-Llergo, A.; Julià, G.; Navarro, G.; Espasa, M.; Rodríguez, S.; Sánchez, N.; Van Den Eynde, E.; Navarro, M.; Calvet, J.; et al. Antibody Response Induced by BNT162b2 and MRNA-1273 Vaccines against the SARS-CoV-2 in a Cohort of Healthcare Workers. *Viruses* **2022**, *14*, 1235. [CrossRef] [PubMed]

25. Mansour Ghanaie, R.; Jamee, M.; Khodaei, H.; Shirvani, A.; Amirali, A.; Karimi, A.; Fallah, F.; Azimi, L.; Armin, S.; Fahimzad, S.A.; et al. Assessment of Early and Post COVID-19 Vaccination Antibody Response in Healthcare Workers: A Multicentre Cross-Sectional Study on Inactivated, MRNA and Vector-Based Vaccines. *Epidemiol. Infect.* **2023**, *151*, e12. [CrossRef]

26. Shah, A.; Coiado, O.C. COVID-19 Vaccine and Booster Hesitation around the World: A Literature Review. *Front. Med.* **2023**, *9*, 1054557. [CrossRef]

27. Sheng, W.-H.; Chang, H.-C.; Chang, S.-Y.; Hsieh, M.-J.; Chen, Y.-C.; Wu, Y.-Y.; Pan, S.-C.; Wang, J.-T.; Chen, Y.-C. SARS-CoV-2 Infection among Healthcare Workers Whom Already Received Booster Vaccination during Epidemic Outbreak of Omicron Variant in Taiwan. *J. Formos. Med. Assoc.* **2022**, S0929-6646(22)00442-9. [CrossRef]

28. Wang, H.; Ai, J.; Loeffelholz, M.J.; Tang, Y.-W.; Zhang, W. Meta-Analysis of Diagnostic Performance of Serology Tests for COVID-19: Impact of Assay Design and Post-Symptom-Onset Intervals. *Emerg. Microbes Infect.* **2020**, *9*, 2200–2211. [CrossRef]

29. Wernike, K.; Keller, M.; Conraths, F.J.; Mettenleiter, T.C.; Groschup, M.H.; Beer, M. Pitfalls in SARS-CoV-2 PCR Diagnostics. *Transbound. Emerg. Dis.* **2021**, *68*, 253–257. [CrossRef]

30. Wikramaratna, P.S.; Paton, R.S.; Ghafari, M.; Lourenço, J. Estimating the False-Negative Test Probability of SARS-CoV-2 by RT-PCR. *Eurosurveillance* **2020**, *25*, 2000568. [CrossRef]

31. Alharbi, S.A.; Almutairi, A.Z.; Jan, A.A.; Alkhalify, A.M. Enzyme-Linked Immunosorbent Assay for the Detection of Severe Acute Respiratory Syndrome Coronavirus 2 (SARS-CoV-2) IgM/IgA and IgG Antibodies Among Healthcare Workers. *Cureus* **2020**, *12*, e10285. [CrossRef] [PubMed]

32. Turbett, S.E.; Anahtar, M.; Dighe, A.S.; Garcia Beltran, W.; Miller, T.; Scott, H.; Durbin, S.M.; Bharadwaj, M.; Thomas, J.; Gogakos, T.S.; et al. Evaluation of Three Commercial SARS-CoV-2 Serologic Assays and Their Performance in Two-Test Algorithms. *J. Clin. Microbiol.* **2020**, *59*, e01892-20. [CrossRef] [PubMed]

33. Dörschug, A.; Schwanbeck, J.; Hahn, A.; Hillebrecht, A.; Blaschke, S.; Mese, K.; Groß, U.; Dierks, S.; Frickmann, H.; Zautner, A.E. Comparison of Five Serological Assays for the Detection of SARS-CoV-2 Antibodies. *Diagnostics* **2021**, *11*, 78. [CrossRef]

34. Krüttgen, A.; Cornelissen, C.G.; Dreher, M.; Hornef, M.W.; Imöhl, M.; Kleines, M. Determination of SARS-CoV-2 Antibodies with Assays from Diasorin, Roche and IDvet. *J. Virol. Methods* **2021**, *287*, 113978. [CrossRef]
35. Kukoč, A.; Mihelčić, A.; Miko, I.; Romić, A.; Pražetina, M.; Tipura, D.; Drmić, Ž.; Čučković, M.; Ćurčić, M.; Blagaj, V.; et al. Clinical and Laboratory Predictors at ICU Admission Affecting Course of Illness and Mortality Rates in a Tertiary COVID-19 Center. *Heart Lung* **2022**, *53*, 1–10. [CrossRef] [PubMed]
36. Savoia, C.; Volpe, M.; Kreutz, R. Hypertension, a Moving Target in COVID-19: Current Views and Perspectives. *Circ. Res.* **2021**, *128*, 1062–1079. [CrossRef]
37. Kreutz, R.; Algharably, E.A.E.-H.; Azizi, M.; Dobrowolski, P.; Guzik, T.; Januszewicz, A.; Persu, A.; Prejbisz, A.; Riemer, T.G.; Wang, J.-G.; et al. Hypertension, the Renin–Angiotensin System, and the Risk of Lower Respiratory Tract Infections and Lung Injury: Implications for COVID-19. *Cardiovasc. Res.* **2020**, *116*, 1688–1699. [CrossRef] [PubMed]
38. Candelli, M.; Pignataro, G.; Saviano, A.; Ojetti, V.; Gabrielli, M.; Piccioni, A.; Gullì, A.; Antonelli, M.; Gasbarrini, A.; Franceschi, F. Is BMI Associated with Covid-19 Severity? A Retrospective Observational Study. *Curr. Med. Chem.* **2023**, *30*. [CrossRef]
39. Hancková, M.; Betáková, T. Pandemics of the 21st Century: The Risk Factor for Obese People. *Viruses* **2021**, *14*, 25. [CrossRef]
40. Minnai, F.; De Bellis, G.; Dragani, T.A.; Colombo, F. COVID-19 Mortality in Italy Varies by Patient Age, Sex and Pandemic Wave. *Sci. Rep.* **2022**, *12*, 4604. [CrossRef]
41. Tong, X.; Huang, Z.; Zhang, X.; Si, G.; Lu, H.; Zhang, W.; Xue, Y.; Xie, W. Old Age Is an Independent Risk Factor for Pneumonia Development in Patients with SARS-CoV-2 Omicron Variant Infection and a History of Inactivated Vaccine Injection. *Infect. Drug Resist.* **2022**, *15*, 5567–5573. [CrossRef] [PubMed]
42. Mori, Y.; Tanaka, M.; Kozai, H.; Hotta, K.; Aoyama, Y.; Shigeno, Y.; Aoike, M.; Kawamura, H.; Tsurudome, M.; Ito, M. Antibody Response of Smokers to the COVID-19 Vaccination: Evaluation Based on Cigarette Dependence. *Drug Discov. Ther.* **2022**, *16*, 78–84. [CrossRef] [PubMed]
43. Watanabe, M.; Balena, A.; Tuccinardi, D.; Tozzi, R.; Risi, R.; Masi, D.; Caputi, A.; Rossetti, R.; Spoltore, M.E.; Filippi, V.; et al. Central Obesity, Smoking Habit, and Hypertension Are Associated with Lower Antibody Titres in Response to COVID-19 MRNA Vaccine. *Diabetes Metab. Res. Rev.* **2022**, *38*, e3465. [CrossRef] [PubMed]

Communication

Correlation of SARS-CoV-2 Neutralization with Antibody Levels in Vaccinated Individuals

Shazeda Haque Chowdhury [1], Sean Riley [1], Riley Mikolajczyk [1], Lauren Smith [1], Lakshmanan Suresh [2,3] and Amy Jacobs [1,*]

1   Department of Microbiology and Immunology, State University of New York at Buffalo, Buffalo, NY 14213, USA
2   Department of Oral Diagnostic Sciences, State University of New York at Buffalo, Buffalo, NY 14215, USA
3   KSL Diagnostics, Inc., Buffalo, NY 14225, USA
*   Correspondence: ajacobs2@buffalo.edu

Abstract: Neutralizing antibody titers are an important measurement of the effectiveness of vaccination against SARS-CoV-2. Our laboratory has set out to further verify the functionality of these antibodies by measuring the neutralization capacity of patient samples against infectious SARS-CoV-2. Samples from patients from Western New York who had been vaccinated with the original Moderna and Pfizer vaccines (two doses) were tested for neutralization of both Delta (B.1.617.2) and Omicron (BA.5). Strong correlations between antibody levels and neutralization of the delta variant were attained; however, antibodies from the first two doses of the vaccines did not have good neutralization coverage of the subvariant omicron BA.5. Further studies are ongoing with local patient samples to determine correlation following updated booster administration.

Keywords: SARS-CoV-2; COVID-19; antibody response; RNA viruses

Citation: Chowdhury, S.H.; Riley, S.; Mikolajczyk, R.; Smith, L.; Suresh, L.; Jacobs, A. Correlation of SARS-CoV-2 Neutralization with Antibody Levels in Vaccinated Individuals. *Viruses* **2023**, *15*, 793. https://doi.org/10.3390/v15030793

Academic Editor: Jason Yiu Wing KAM

Received: 10 March 2023
Revised: 19 March 2023
Accepted: 20 March 2023
Published: 21 March 2023

1. Introduction

With over 102 million cases of and over 1 million deaths from COVID-19 worldwide since its identification in December 2019 (COVID Data Tracker, CDC), SARS-CoV-2 has become an increasingly persistent and strenuous challenge for healthcare professionals and the public alike. While COVID-19 is often characterized by upper respiratory symptoms including generalized malaise, fevers, nasal congestion, and cough, disease presentation is heterogeneous and can range from asymptomatic infection to multiorgan failure and death. Gastrointestinal symptoms include diarrhea, abdominal pain, nausea and vomiting, and anorexia [1], while acute myocardial injury, infarctions and other acute cardiac compromise can be presented by the patients [2]. Another complication of SARS-CoV-2 infection is a syndrome termed 'Long COVID', which is characterized by the persistence of diverse symptoms due to unidentifiable causes 12 weeks post-SARS-CoV-2 infection, and these have also proven to be a significant healthcare and economic burden [3].

As a beta-coronavirus, SARS-CoV-2 is closely related to SARS-CoV, as demonstrated through their structural proteins (envelope [E], membrane [M], and spike [S]), receptor (angiotensin-converting enzyme 2 [ACE2]), and receptor binding domains (RBD) on the S1 subunit of the spike protein [4]. Despite these similarities, the RBD in SARS-CoV-2 tends to be in the 'lying-down' conformation rather than the 'standing-up' conformation of the RBD in SARS-CoV, which may facilitate immune surveillance evasion by SARS-CoV-2, as demonstrated by the decreased development of neutralizing antibodies in patients with SARS-CoV-2 compared to SARS-CoV.

The patient-specific factors that have been demonstrated to increase risk for severe illness include increased age, obesity, tobacco use, and pre-existing comorbid conditions, such as hypertension, cardiovascular disease, and diabetes mellitus, though many of these are interrelated [5]. In addition to patient-specific factors associated with severe COVID-19 illness, disease manifestation is also dependent upon the molecular mechanisms of infection, such as the localization of the receptor, ACE2 and TMPRSS2, a serine protease which cleaves the spike protein into S1 and S2; co-expression in the gut, brain, kidney and cardiovascular system leads to viral entry and disease exacerbation [1].

The first vaccines were approved for emergency use in December 2019 by the U.S. Food and Drug Administration (FDA). These were mRNA-based vaccines, mRNA-1273 (Moderna) and BNT162b2 (Pfizer/BioNTech), which were also approved for use in Europe by the European Medicines Agency (EMA). Apart from these, non-replicating viral vector vaccines have also been in use; AZD 1222 (Oxford/AstraZeneca) has been in use in the UK, EU, and some Asian countries, and the AD26.COV2-S (Johnson & Johnson) has been approved for use in the US [6].

It has been widely demonstrated that low titers or neutralization potency of anti-RBD IgG antibodies are correlated with worsened morbidity and mortality outcomes [7]. This has been applied through the investigational use of convalescent plasma therapy in patients with impaired humoral immunity or in non-hospitalized patients who are at high risk for progression to severe disease under the Emergency Use Authorization (EUA) by the U.S. Food and Drug Administration (FDA). As the efficacy of this therapy is likely dependent on the infusion containing sufficient antibody levels, the EUA is for 'high-titer plasma', which is defined according to a qualifying result that is specific for each of the bioassays and serologic binding assays accepted by the U.S. Food and Drug Administration (FDA). These bioassays are viral or pseudoviral neutralization assays and the serologic binding assays are either enzyme-linked immunosorbent assays (ELISA) or chemiluminescence assays (CLIA).

While previous studies have demonstrated a correlation between anti-RBD IgG levels and neutralizing antibody titers, these were often conducted using ELISA and microneutralization assays. To our knowledge, there are no prior studies that use CLIA and plaque reduction neutralization (PRNT) assays, both of which have been shown to be more sensitive than their respective counterparts, to elucidate this relationship. Herein, we show the correlation between CLIA and PRNT assay results for patients who received the primary doses of the mRNA-1273 (Moderna) and BNT162b2 (Pfizer/BioNTech) vaccines. All samples were tested for the neutralization capability of both Delta (B.1.617.2) and Omicron (BA.5).

2. Materials and Methods

Chemiluminescence Immunoassays—The detection of antibodies to the RBD on S1 was performed on serum at the KSL Diagnostics Laboratories using KSL chemiluminescence immunoassays (CLIA). In this technique, electromagnetic radiation caused by a chemical reaction produces light. The analytic reaction produces visible or near-visible radiation generated when an electron transitions from an excited state to ground state. The luminophore markers used in KSL CLIA assays are acridinium esters. The CLIA assay provides qualitative detection of human IgA, IgG, or IgM antibodies to SARS-CoV-2. The test was authorized by NY State and EUA on 25 November 2020, under Project ID No. 84341. The samples were incubated with a magnetic bead coated with the S1 subdomain from SARS-CoV-2. Once the unbound materials were washed away by magnetic separation, the acridinium ester marker is added for incubation. After a wash step, SARS-CoV-2 antibodies are detected with a substrate that produces a luminescence reaction with the acridinium ester. The luminescence intensity of acridinium ester is proportionate to the amount of antibody against novel coronavirus and yields a test result expressed by cut-off index (COI). If the sample value is less than 0.8 COI, no SARS-CoV-2 antibody is detected. If the value is within 0.8–1.0 COI, the SARS-CoV-2 antibody is indeterminate. If the value is greater than 1.0 COI, then SARS-CoV-2 antibody is detected.

Virus Stock Generation—Inhibition of infection by the test sera was studied by infecting the sera with the viral variants, SARS-CoV-2 Delta (B.1.617.2) and Omicron (BA.5).

Virus was obtained from BEI Resources were propagated in Calu-3 (ATCC, VA) for 3 days (Delta) or 7 days (Omicron). Infection was confirmed by the observance of cytopathic effect (CPE), such as the rounding and dislodgement of cells. Harvested virus was then titered using a PRNT assay so that a viral load of 6×10^3 PFU/mL (30 PFU/well) could be added to the test samples.

Plaque Reduction Neutralization Assays—The PRNT50 protocol used was adapted from Bewley et al [8]. Briefly, serum samples were diluted in a 96-well plate, before SARS-CoV-2 virus was added to the diluted serum at BSL-3 and neutralization was allowed to occur. The neutralized or control virus was then transferred onto Vero-E6 cells (ATCC, VA), allowed to adsorb, overlaid with 1% agar in DMEM and incubated at 37 °C and 5% CO_2. The incubation time for viral propagation and infection was increased from 3 days to 4 days when using the Omicron variant as compared to the Delta strain, since plaques formed by the omicron variant were smaller and took longer to develop. Plates were then fixed with 4% paraformaldehyde, and agar plugs were removed so that plaques could be counted and scored (Figure 1). PRNT50 values were then calculated using Prism GraphPad (Version 9.5.0) according to the instructions outlined in Bewley et al. (Figure 2). Samples were tested in two biological and two technical replicates.

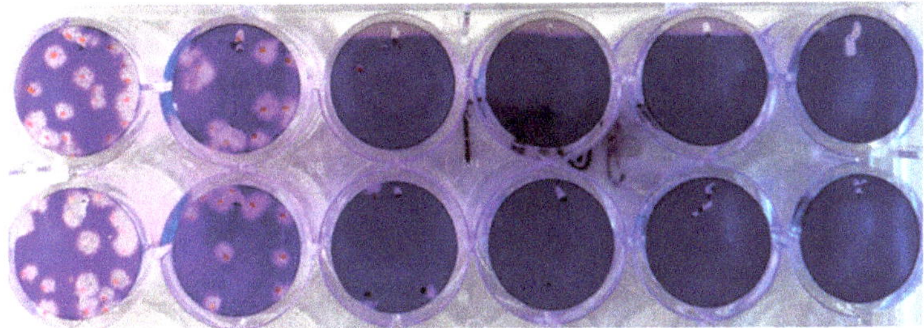

Figure 1. Example of plaque reduction neutralization performed in duplicate. Overlay fixed with 4% paraformaldehyde and stained with crystal violet before plaques were counted.

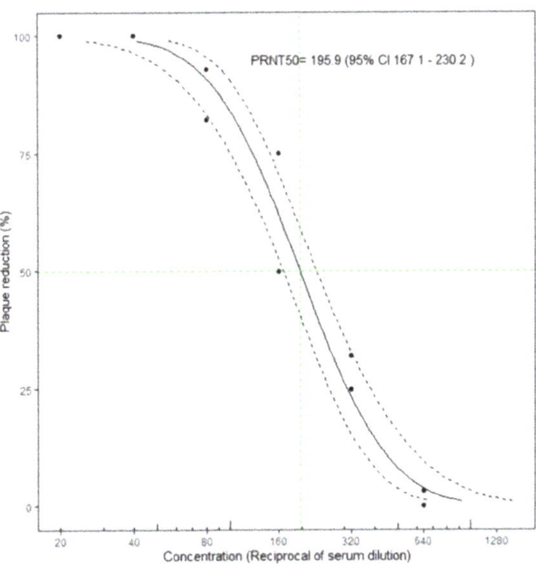

Figure 2. Representative PRNT50 analysis performed according to Bewley et al. [8].

Sample characteristics—In this study, 110 participants (males or females) between the ages of 18 and 85 with no known prior SARS-CoV-2 infection as confirmed by RT-PCR were included. The presence of SARS-CoV-2 nucleocapsid antibodies was considered an exclusion criterion. Participant inclusion and exclusion criteria are listed in Table 1. Participants had received the first 2 doses of the mRNA-1273 (Moderna) (55 patients) and BNT162b2 (Pfizer/BioNTech) (55 patients) vaccine as per the provider's protocols. Fifteen days post vaccination, sera were tested for IgG levels using KSL Chemiluminescence Immunoassays (CLIA), and samples were grouped according to the amount of IgG measured (Table 2). Following that, a total of 17 samples from all the groups were tested for viral neutralization using live virus in BSL-3.

Table 1. Inclusion/Exclusion Criteria of participants enrolled in study.

Inclusion Criteria	
1.	Age and Sex: Male or female participants between the ages of 18 and 85 at the time of enrollment
2.	Participants with no known exposure to COVID-19 infection
3.	Had FDA approved vaccines administered as per the vaccine company recommended protocols
4.	Participants who are willing and able to comply with all scheduled visits and laboratory tests.
Exclusion Criteria	
1.	Age below 18 years of age
2.	History of COVID-19 infections
3.	History of antibodies to SARS-CoV2 nucleocapsid antibodies
4.	People with non-adherence to vaccine administration protocols

Table 2. Sample groups according to IgG COI.

Group	COI * Value Range
1	<0.5
2	0.5–<1.0
3	1.0–2.99
4	3.0–5.99
5	6.0–9.99
6	10.0–14.99
7	15.0–19.99
8	20.0–24.99
9	25.0–29.99
10	30.0–39.99
11	40.0+

*—Cut-off Index.

Statistical analysis—Pearson's correlation test was used to assess the correlation between calculated PRNT50 titers and IgG COIs or age. p-values < 0.05 were considered statistically significant. The neutralization induced by either vaccine was compared using a boxplot displaying median and 95% CI. Data analysis and visualization was conducted in Prism GraphPad (Version 9.5.0).

3. Results

3.1. IgG COI Values Correlate with PRNT50 Titers

It was observed that the calculated PRNT50 values correlated with the measured IgG COI (Figure 3), with the delta variant showing a stronger statistically significant correlation ($r^2 = 0.6908$, p-value < 0.0001) compared to the omicron variant ($r^2 = 0.165$, p-value = 0.0756 n.s.). At COIs < 20.0, no neutralization was observed, with the number of plaques counted being the same as the virus only control (VOC) of both viral strains tested. As summarized in Table 3, when tested with delta variant, robust neutralization was observed at COIs >20.0, as high titers were obtained, while suboptimal neutralization and low titers were obtained for COI between 10.0 and 20.0. In contrast, when tested with the omicron BA.5 variant, suboptimal neutralization indicated by low titers was observed at COIs >10.0.

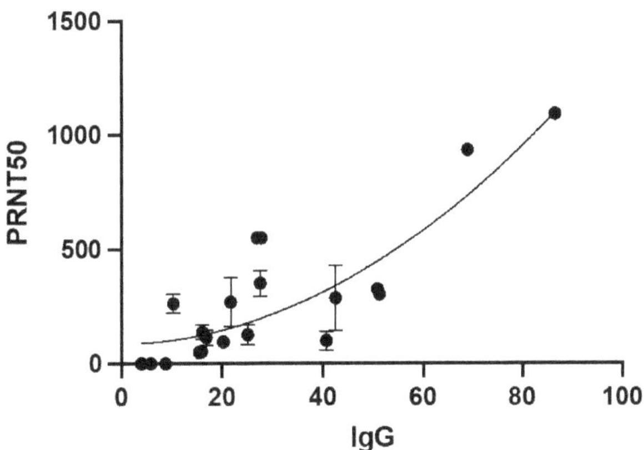

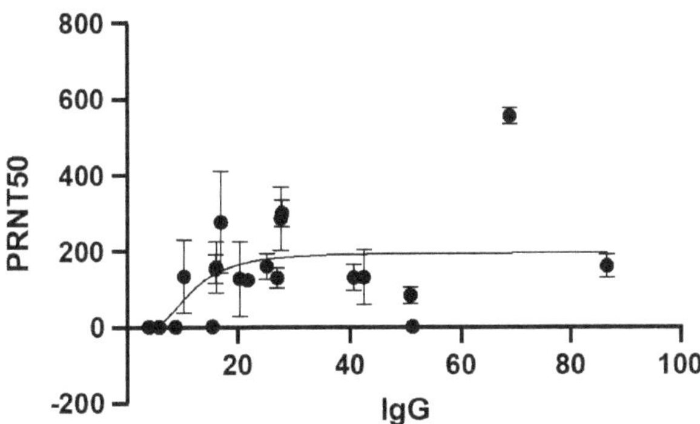

Figure 3. Correlation of PRNT50 titers with IgG COI when challenged with Delta and Omicron variants of SARS-CoV-2. Viral load of 6×10^3 PFU/mL (30 plaques/well) were added to Vero-E6 cells. Data represented mean ± SD of two biological replicates. $\mu = 0.05$.

Table 3. Average PRNT50 values obtained for different COI groups when challenged with the delta or omicron variant.

Group Number	COI Value Range	PRNT50 Average (Delta)	PRNT50 Average (Omicron)	Conclusion
1–5	0–9.99	Same as VOC	Same as VOC	No neutralization of both strains
6–7	10–19.99	123.14	144.84	Similar neutralization of both strains
8–10	20–39.99	322.92	187.81	Delta neutralized more effectively than Omicron
11	>40	235.06	176.58	Delta neutralized more effectively than Omicron

3.2. Age Does Not Correlate with PRNT50 Titers

Analysis showed that the age of the patient during vaccination did not have any effect on the PRNT50 titers obtained (Delta $r^2 = 0.015$, p-value = 0.6332 n.s.; Omicron $r^2 = 0.041$, p-value = 0.4348 n.s.) (Figure 4). This was in tandem with the result that the measured IgG COIs also did not correlate with age ($r^2 = 0.0343$, p-value = 0.4620 n.s.).

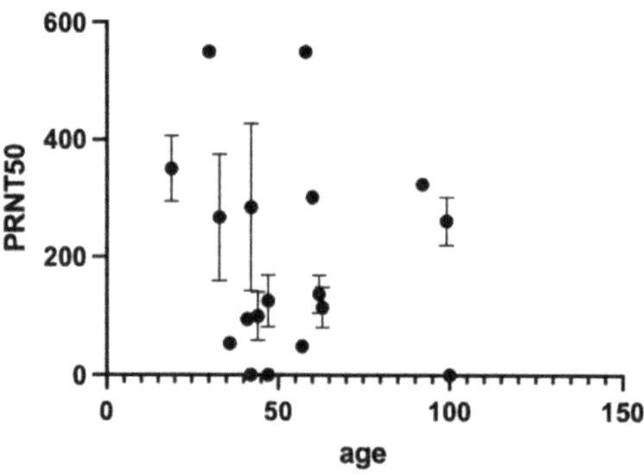

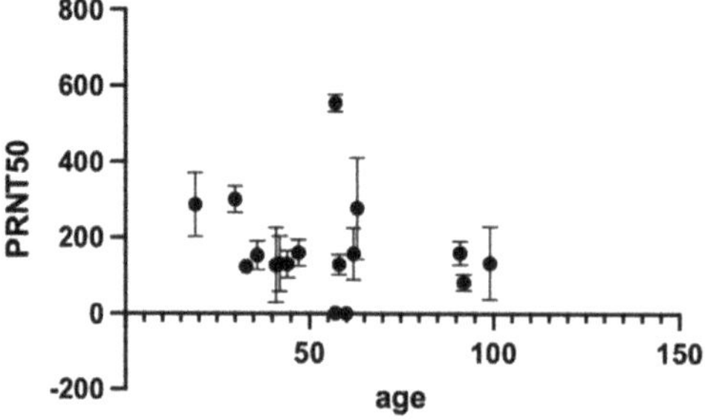

Figure 4. Correlation of PRNT50 titers with age when challenged with Delta and Omicron variants of SARS-CoV-2. Viral load of 6×10^3 PFU/mL (30 plaques/well) were added to Vero-E6 cells. Data represented mean ± SD of two biological replicates. $\mu = 0.05$.

3.3. Comparison of Neutralization following Primary Doses of Either the Moderna or Pfizer/BioNTech Vaccines

The average median PRNT50 titer obtained following vaccination with the primary doses of the Pfizer vaccine was 242.2 when challenged with the Delta strain, and 146.1 when challenged with the Omicron strain. In comparison, following primary doses of the Moderna vaccine, the average median titer when challenged with the Delta strain was 78.31, while it was 118.2 when challenged with the omicron strain (Figure 5). This indicates that neutralization capacity by antibodies produced following vaccination with the Pfizer mRNA vaccine was more effective when compared to the neutralization by antibodies produced by the Moderna vaccine—a result that held true when challenged with both the delta and omicron strain.

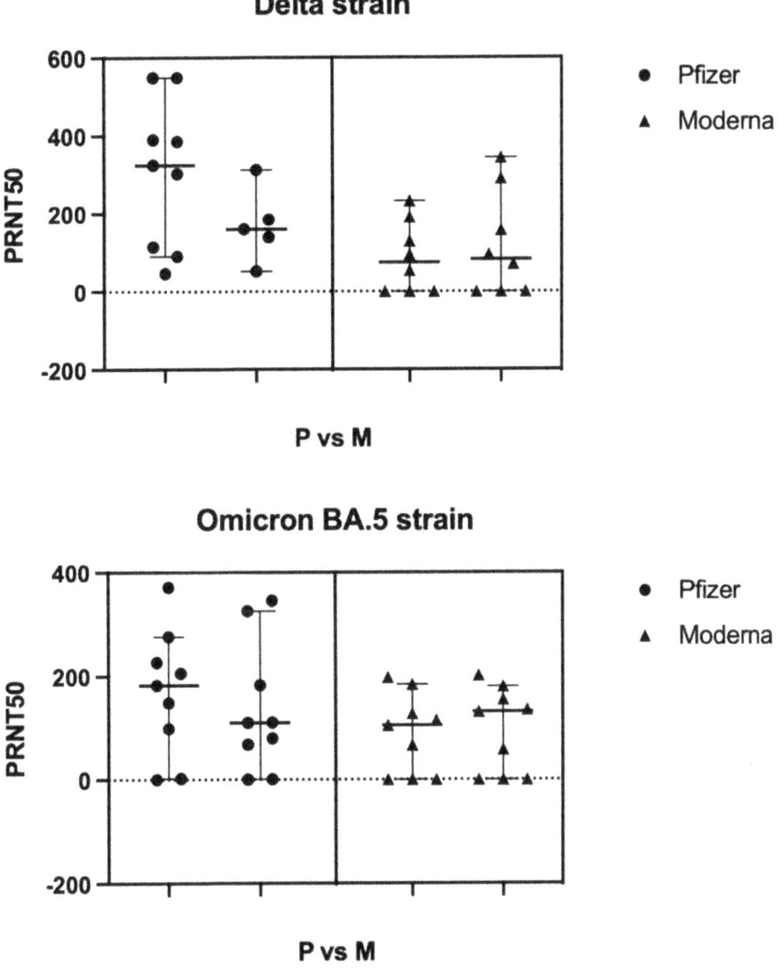

Figure 5. Comparison of PRNT50 titers obtained following vaccination with the primary doses of the Moderna or Pfizer/BioNTech mRNA vaccines. Viral load of 6×10^3 PFU/mL (30 plaques/well) were added to Vero-E6 cells. Data represented median $\pm$ 95% CI of two biological replicates.

4. Discussion

Humoral immunity is characterized by the production of antibodies by B cells as a response to antigens. Although both IgM and IgA appear within the first week of symptom onset, IgG is the most abundant antibody type and provides longer-lasting immunity. IgG is seen in circulation from about 7 days onwards [9]. This response of immunity is also typical for SARS-CoV-2. IgG titers remain stable for at least 4 to 6 months following diagnosis among PCR-confirmed individuals, whereas IgA and IgM titers rapidly decay. Antibodies targeting the spike glycoprotein of the SARS-CoV-2, especially the receptor binding domain (RBD) within the S1 subunit, show the highest neutralizing capacity. The presence of neutralizing antibodies is considered a functional correlate of immunity and provides at least partial resistance to subsequent. Although some serological assays showed a high correlation between IgG and neutralizing antibodies [10], others have poor correlation [11]. Therefore, comparison with virus-neutralization experiments is important as part of the validation of new serological assays.

Several laboratory-developed and commercially available assays utilizing various technology platforms are available to detect anti-SARS-CoV-2 spike antibodies. While these platforms provide a high-throughput means of detecting antibodies against SARS-CoV-2, they are unable to measure the immunological function of SARS-CoV-2-specific antibodies. In contrast, the plaque-reduction neutralization test (PRNT) quantifies levels of neutralizing antibodies capable of blocking the interaction that mediates virus entry into susceptible host cells and subsequent virus replication [12]. For SARS CoV-2, this interaction involves binding of the RBD of the spike glycoprotein with the ACE2 on host cells. This makes the conventional PRNT the reference standard for the evaluation of virus-neutralizing antibodies. Prior to this study, to our knowledge, no other study has compared the effect of neutralizing antibodies on the omicron or delta strain using the plaque reduction neutralization test, although a few have studied the effects of the USA-WA1/2020 isolate [13].

In our lab, testing the neutralizing capabilities of vaccine-induced IgG antibodies showed that at very low IgG levels, no neutralization is achieved. Further, there is a positive correlation between the measured IgG using CLIA and PRNT50 values obtained. As expected, neutralization of virus by vaccine-induced antibodies is more robust when challenged with the delta strain when compared to the omicron BA.5 strain. This is similar to what is reported in the literature [14]. Since the samples obtained are from individuals vaccinated with only the first doses of the vaccine, but not the boosters, the antibodies produced are not effective against the omicron variant. Studies have shown that neutralizing antibodies (nAbs) are progressively less effective against each new variant of concern (VOC) or variant of interest (VOI). This is especially true for omicron BA.5 subvariant, used in this study, which escapes nAbs due to extensive mutations and antigenic remodeling of its spike trimer, and predominance of the closed state of the spike protein [7]. Compared to the original strain, the omicron BA.1 variant has 35 mutations, 15 of them in the RBD which is the region that binds to the host cell receptors and is a target of nAbs. Of the 15 mutations, nine fall in the region specific for binding to the ACE2 receptor on host cells, thereby allowing stronger binding and nAb escape [15]. The subsequent variants that emerged, BA.2, BA.2.12.1, BA.4 and BA.5, each have more numerous and unique mutations, which make the circulating omicron variants more transmissible and less susceptible to vaccine induced immunity. In fact, it has been determined that efficiency of the Moderna vaccine following the first booster (third dose) is only high against the omicron BA.1 variant, and the second booster (fourth dose) is required to achieve a high vaccine efficiency against BA.2, BA.2.12.1, and BA.4, while vaccine efficiency against BA.5 is low even after the fourth dose [16]. The same was also observed when testing vaccine efficiency of the Pfizer/BioNTech vaccine [17]. Further, even though the fourth dose conferred high vaccine efficiency against the BA.5 variant, the effect waned within six months in both studies.

However, it has been reported that some classes of neutralizing antibodies which exert their function by binding to the less immunologically challenged and therefore more conserved stem-helix region of the S2 domain show neutralization at very high concentrations (IC50 > 1 ug/mL) in vitro while in in vivo studies, they show the same effect at lower concentrations [18,19]. Further, it is important to note that in the in vitro studies conducted in our lab, it was observed that infectivity of omicron was to a lesser degree when compared to the delta strain. This was concluded due to the longer infection cycles required to obtain plaques when infecting with omicron. Further, titering of viral particles showed lower numbers of plaques for dilutions of omicron when compared to the same dilution series of the delta variant. This might attribute to the less severe disease manifestation during omicron infections [20]. However, further studies are required to assess this observation.

It was also observed that the age of the patient did not correlate with either the IgG COIs measured, or the PRNT50 titers obtained. However, it has been previously shown that following vaccination, age plays a significant role in neutralization of SARS-CoV-2 by vaccine-induced antibodies, wherein an increase in age correlates with decreased immunity [21]. The results obtained in our study can be attributed to a small sample size and a limited number of participants of different ages in each IgG COI group. It must also be noted that certain individuals who received the primary doses of the vaccine did not produce any measurable IgG, and this may be attributed to an impaired immune system of the patient.

In our study, we observed that vaccination with the primary doses of the BNT162b2 (Pfizer/BioNTech) vaccine induced antibodies that produced higher PRNT50 titers and thereby caused more effective neutralization when compared to the effect of the mRNA-1273 (Moderna) vaccine, which led to lower PRNT50 titers overall, specifically against the delta strain; the effects on omicron were substantially low and similar. It was previously shown that the Moderna vaccine induced a higher amount of functional antibodies; the difference observed in our study is relatively small, and can be attributed to a small sample size [22].

While assays such as ELISA measure the amount of antibodies present, the PRNT provides an insight into the neutralizing functionality of the circulating antibodies. This makes it the gold standard for assessing the neutralizing capability of antibodies. However, even though the PRNT is often used as the reference standard for the evaluation of virus-neutralizing antibodies, this assay is time-consuming, laborious, and requires biosafety containment level 3 (BSL-3) facilities to work with the high-risk group-3 pathogen. As such, this is not practical for large-scale community testing, due to low turnaround time and high manual input. In this study, due to the high labor demand and low efficiency of testing method, a small sample size was studied, which is a limitation of the study, and which may attribute to the differences observed between existing literature and our observations. However, using the tested samples as a reference, high throughput testing techniques can be designed and validated. One such technique is a microneutralization assay (currently being validated in-house) which uses labeled antibodies to count plaques in a 96-well or 384-well format [8]. This provides an added advantage over other immunofluorescence techniques such as ELISA because it counts the foci of infected cells instead of just the absorbance. Future studies will be aimed at assessing the neutralizing capabilities following vaccinations with the booster shots, and also characterize the types of IgG and their effect on neutralization. Trends in variant mutation of SARS-CoV-2 and their nAb escape make it important to develop efficient and high-throughput testing capability to analyze large data sets and validate community-based conclusions.

5. Conclusions

The amount of IgG produced following vaccination with the primary doses of the mRNA-1273 (Moderna) and BNT162b2 (Pfizer/BioNTech) vaccines correlates positively with the PRNT50 titers obtained when challenged with the SARS-CoV-2 Delta (B.1.617.2) variant, although this was not observed when challenged with the Omicron (BA.5) variant, meaning that immunity offered by antibodies produced by the first vaccine doses is effective at neutralizing the delta variant only. No significant correlation was observed with age and no significant difference was observed between the two vaccines administered. The conclusions of this work corroborate the importance of fine-tuning the vaccines to the current strains that dominate in the population and potentially tailoring to individual regions as well.

Author Contributions: S.H.C. performed neutralization assays, data analysis, and wrote the manuscript. S.R. developed the assay and performed neutralization assays and data analysis. R.M. performed neutralization assays. L.S. (Lauren Smith) assisted in manuscript preparation. L.S. (Lakshmanan Suresh) provided CLIA data and patient samples. A.J. provided overall management, experimental planning, and manuscript editing. All authors have read and agreed to the published version of the manuscript.

Funding: This work was funded by the UB Center for Advanced Technology in Big Data and Health Sciences.

Institutional Review Board Statement: Not applicable.

Informed Consent Statement: Not applicable.

Data Availability Statement: The data presented in this study are available on request from the corresponding author.

Acknowledgments: The authors would like to acknowledge Spyridon Stavrou, University at Buffalo, for intellectual input and protocol troubleshooting. The virus isolates were obtained through BEI Resources, NIAID, NIH: SARS-Related Coronavirus 2, Isolate hCoV-19/USA/COR-22-063113/2022 (Lineage BA.5; Omicron Variant), NR-58616, contributed by Richard J. Webby, and SARS-Related Coronavirus 2, Isolate hCoV-19/USA/MD-HP05647/2021 (Lineage B.1.617.2; Delta variant), NR-55672, contributed by Andrew S. Pekosz. This work was funded by the State of New York through the State University of New York at Buffalo's Center for Advanced Technologies in Big Data and Health Sciences. The opinions, results, findings and/or interpretations are the sole responsibility of the authors and do not represent the opinions, interpretations or policy of the State".

Conflicts of Interest: Lakshmanan Suresh is employed at KSL Diagnostics Inc., where CLIA testing used in this manuscript was performed.

References

1. Rahban, M.; Stanek, A.; Hooshmand, A.; Khamineh, Y.; Ahi, S.; Kazim, S.N.; Ahmad, F.; Muronetz, V.; Samy Abousenna, M.; Zolghadri, S.; et al. Infection of Human Cells by SARS-CoV-2 and Molecular Overview of Gastrointestinal, Neurological, and Hepatic Problems in COVID-19 Patients. *J. Clin. Med.* **2021**, *10*, 4028. [CrossRef] [PubMed]
2. Chung, M.K.; Zidar, D.A.; Bristow, M.R.; Cameron, S.J.; Chan, T.; Harding, C.V., III; Kwon, D.H.; Singh, T.; Tilton, J.C.; Tsai, E.J.; et al. COVID-19 and Cardiovascular Disease: From Bench to Bedside. *Circ. Res.* **2021**, *128*, 1214–1236. [CrossRef] [PubMed]
3. Castanares-Zapatero, D.; Chalon, P.; Kohn, L.; Dauvrin, M.; Detollenaere, J.; Noordhout, C.M.D.; Jong, C.P.-D.; Cleemput, I.; Heede, K.V.D. Pathophysiology and mechanism of long COVID: A comprehensive review. *Ann. Med.* **2022**, *54*, 1473–1487. [CrossRef] [PubMed]
4. Wang, M.Y.; Zhao, R.; Gao, L.J.; Gao, X.F.; Wang, D.P.; Cao, J.M. SARS-CoV-2: Structure, Biology, and Structure-Based Therapeutics Development. *Front. Cell. Infect. Microbiol.* **2020**, *10*, 587269. [CrossRef]
5. Baj, J.; Karakula-Juchnowicz, H.; Teresinski, G.; Buszewicz, G.; Ciesielka, M.; Sitarz, R.; Forma, A.; Karakula, K.; Flieger, W.; Portincasa, P.; et al. COVID-19: Specific and Non-Specific Clinical Manifestations and Symptoms: The Current State of Knowledge. *J. Clin. Med.* **2020**, *9*, 1753. [CrossRef]
6. Fortner, A.; Schumacher, D. First COVID-19 Vaccines Receiving the US FDA and EMA Emergency Use Authorization. *Discoveries* **2021**, *9*, e122. [CrossRef]
7. Bajpai, P.; Singh, V.; Chandele, A.; Kumar, S. Broadly Neutralizing Antibodies to SARS-CoV-2 Provide Novel Insights Into the Neutralization of Variants and Other Human Coronaviruses. *Front. Cell. Infect. Microbiol.* **2022**, *12*, 928279. [CrossRef]
8. Bewley, K.R.; Coombes, N.S.; Gagnon, L.; McInroy, L.; Baker, N.; Shaik, I.; St-Jean, J.R.; St-Amant, N.; Buttigieg, K.R.; Humphries, H.E.; et al. Quantification of SARS-CoV-2 neutralizing antibody by wild type plaque reduction neutralization, microneutralization and pseudotyped virus neutralization assays. *Nat. Protoc.* **2021**, *16*, 3114–3140. [CrossRef]
9. Long, Q.X.; Liu, B.Z.; Deng, H.J.; Wu, G.C.; Deng, K.; Chen, Y.K.; Liao, P.; Qiu, J.F.; Lin, Y.; Cai, X.F.; et al. Antibody responses to SARS-CoV-2 in patients with COVID-19. *Nat. Med.* **2020**, *26*, 845–848. [CrossRef]

10. GeurtsvanKessel, C.H.; Okba, N.M.A.; Igloi, Z.; Bogers, S.; Embregts, C.W.E.; Laksono, B.M.; Leijten, L.; Rokx, C.; Rijnders, B.; Rahamat-Langendoen, J.; et al. An evaluation of COVID-19 serological assays informs future diagnostics and exposure assessment. *Nat. Commun.* **2020**, *11*, 3436. [CrossRef]

11. Focosi, D.; Anderson, A.O.; Tang, J.W.; Tuccori, M. Convalescent Plasma Therapy for COVID-19: State of the Art. *Clin. Microbiol. Rev.* **2020**, *33*, e00072-20. [CrossRef]

12. Valcourt, E.J.; Manguiat, K.; Robinson, A.; Lin, Y.C.; Abe, K.T.; Mubareka, S.; Shigayeva, A.; Zhong, Z.; Girardin, R.C.; DuPuis, A.; et al. Evaluating Humoral Immunity against SARS-CoV-2: Validation of a Plaque-Reduction Neutralization Test and a Multilaboratory Comparison of Conventional and Surrogate Neutralization Assays. *Microbiol. Spectr.* **2021**, *9*, e0088621. [CrossRef] [PubMed]

13. Lee, K.J.; Choi, S.Y.; Lee, Y.M.; Kim, H.W. Neutralizing Antibody Response, Safety, and Efficacy of mRNA COVID-19 Vaccines in Pediatric Patients with Inflammatory Bowel Disease: A Prospective Multicenter Case-Control Study. *Vaccines* **2022**, *10*, 1265. [CrossRef] [PubMed]

14. Perez-Then, E.; Lucas, C.; Monteiro, V.S.; Miric, M.; Brache, V.; Cochon, L.; Vogels, C.B.F.; Malik, A.A.; Cruz, E.D.L.; Jorge, A.; et al. Neutralizing antibodies against the SARS-CoV-2 Delta and Omicron variants following heterologous CoronaVac plus BNT162b2 booster vaccination. *Nat. Med.* **2022**, *28*, 481–485. [CrossRef] [PubMed]

15. Shrestha, L.B.; Foster, C.; Rawlinson, W.; Tedla, N.; Bull, R.A. Evolution of the SARS-CoV-2 omicron variants BA.1 to BA.5: Implications for immune escape and transmission. *Rev. Med. Virol.* **2022**, *32*, e2381. [CrossRef]

16. Tseng, H.F.; Ackerson, B.K.; Bruxvoort, K.J.; Sy, L.S.; Tubert, J.E.; Lee, G.S.; Ku, J.H.; Florea, A.; Luo, Y.; Qiu, S.; et al. Effectiveness of mRNA-1273 vaccination against SARS-CoV-2 omicron subvariants BA.1, BA.2, BA.2.12.1, BA.4, and BA.5. *Nat. Commun.* **2023**, *14*, 189. [CrossRef] [PubMed]

17. Tartof, S.Y.; Slezak, J.M.; Puzniak, L.; Hong, V.; Frankland, T.B.; Ackerson, B.K.; Takhar, H.; Ogun, O.A.; Simmons, S.; Zamparo, J.M.; et al. BNT162b2 vaccine effectiveness against SARS-CoV-2 omicron BA.4 and BA.5. *Lancet Infect. Dis.* **2022**, *22*, 1663–1665. [CrossRef]

18. Pinto, D.; Sauer, M.M.; Czudnochowski, N.; Low, J.S.; Tortorici, M.A.; Housley, M.P.; Noack, J.; Walls, A.C.; Bowen, J.E.; Guarino, B.; et al. Broad betacoronavirus neutralization by a stem helix-specific human antibody. *Science* **2021**, *373*, 1109–1116. [CrossRef]

19. Zhou, T.; Wang, L.; Misasi, J.; Pegu, A.; Zhang, Y.; Harris, D.R.; Olia, A.S.; Talana, C.A.; Yang, E.S.; Chen, M.; et al. Structural basis for potent antibody neutralization of SARS-CoV-2 variants including B.1.1.529. *Science* **2022**, *376*, eabn8897. [CrossRef] [PubMed]

20. Hyams, C.; Challen, R.; Marlow, R.; Nguyen, J.; Begier, E.; Southern, J.; King, J.; Morley, A.; Kinney, J.; Clout, M.; et al. Severity of Omicron (B.1.1.529) and Delta (B.1.617.2) SARS-CoV-2 infection among hospitalised adults: A prospective cohort study in Bristol, United Kingdom. *Lancet Reg. Health Eur.* **2023**, *25*, 100556. [CrossRef] [PubMed]

21. Bates, T.A.; Leier, H.C.; Lyski, Z.L.; Goodman, J.R.; Curlin, M.E.; Messer, W.B.; Tafesse, F.G. Age-Dependent Neutralization of SARS-CoV-2 and P.1 Variant by Vaccine Immune Serum Samples. *JAMA* **2021**, *326*, 868–869. [CrossRef] [PubMed]

22. Hvidt, A.K.; Baerends, E.A.M.; Sogaard, O.S.; Staerke, N.B.; Raben, D.; Reekie, J.; Nielsen, H.; Johansen, I.S.; Wiese, L.; Benfield, T.L.; et al. Comparison of vaccine-induced antibody neutralization against SARS-CoV-2 variants of concern following primary and booster doses of COVID-19 vaccines. *Front. Med.* **2022**, *9*, 994160. [CrossRef] [PubMed]

Article

Hemagglutinin Antibodies in the Polish Population during the 2019/2020 Epidemic Season

Karol Szymański [1,*], Katarzyna Kondratiuk [1], Ewelina Hallmann [1,*], Anna Poznańska [2] and Lidia B. Brydak [1]

1   National Influenza Center, Department of Influenza Research, National Institute of Public Health NIH—National Research Institute, 00-791 Warsaw, Poland

2   Department of Population Health Monitoring and Analysis, National Institute of Public Health NIH—National Research Institute, 00-791 Warsaw, Poland

*   Correspondence: kszymanski@pzh.gov.pl (K.S.); ehallmann@pzh.gov.pl (E.H.)

Abstract: The aim of the study was to determine the level of antibodies against hemagglutinin of influenza viruses in the serum of subjects belonging to seven different age groups in the 2019/2020 epidemic season. The level of anti-hemagglutinin antibodies was tested using the hemagglutination inhibition (HAI) test. The tests included 700 sera from all over Poland. Their results confirmed the presence of antibodies against the following influenza virus antigens: A/Brisbane/02/2018 (H1N1)pdm09 (48% of samples), A/Kansas/14/2017/ (H3N2) (74% of samples), B/Colorado/06/ 2017 Victoria line (26% of samples), and B/Phuket/3073/2013 Yamagata line (63% of samples). The level of antibodies against hemagglutinin varied between the age groups. The highest average (geometric mean) antibody titer (68.0) and the highest response rate (62%) were found for the strain A/Kansas/14/2017/ (H3N2). During the epidemic season in Poland, only 4.4% of the population was vaccinated.

Keywords: influenza; vaccine; antibody; virus

check for **updates**

Citation: Szymański, K.; Kondratiuk, K.; Hallmann, E.; Poznańska, A.; Brydak, L.B. Hemagglutinin Antibodies in the Polish Population during the 2019/2020 Epidemic Season. *Viruses* **2023**, *15*, 760. https://doi.org/10.3390/v15030760

Academic Editors: Jason Yiu Wing Kam and Kenneth Lundstrom

Received: 30 January 2023
Revised: 10 March 2023
Accepted: 14 March 2023
Published: 16 March 2023

1. Introduction

Seasonal flu is a respiratory illness caused by a highly contagious virus [1]. It can spread from person to person in several ways: droplets together with respiratory secretions (coughing, blowing the nose, talking), direct contact, and airborne or indirect contact (contaminated surface) [2,3]. There are three types of human pathogenic influenza virus: A, B, and C, but only A and B are considered clinically relevant. Influenza A viruses consist of numerous subtypes that are divided based on the variety and combinations of glycoproteins found on the viral surface: hemagglutinin (HA), which allows the virion to anchor to the cell surface, and neuraminidase (NA), which allows the virus to be released from host cells [4]. As of now, there are 18 types of hemagglutinin and 11 types of neuraminidase. Types linked with humans are: H1, H2, H3, H5, H6, H7, H9, and H10 [5,6]. Influenza B viruses also contain those glycoproteins on their surface, but they are not divided into subtypes. It was not until the late 1980s that two lines of influenza B virus were isolated: Victoria and Yamagata [7,8]. Hemagglutinin and neuraminidase are antigens, which means they are recognized by the immune system and can trigger an immune response [9].

Due to the segmental nature of the genetic material, the influenza virus is highly mutagenic. Two types of genetic changes can be distinguished: antigenic shift and antigenic drift. Antigenic drift occurs as a result of a point mutation in genes, leading to an altered sequence of amino acids that changes the antigenic site. This is caused by seasonal influenza epidemics [10]. An epidemic is said to occur when the number of cases of a given disease within a specific area and at a specific time is clearly higher than in previous years. In the northern hemisphere, in a temperate climate, influenza epidemics usually occur in winter months [11]. In contrast, antigenic shift takes place when several viruses infect the same host cell and then exchange segments of genes encoding hemagglutinin and neuraminidase with each other. This gives rise to a new virus with new gene constellations, which may cause a pandemic [12]. Pandemics occur when a previously uncirculated influenza virus emerges, causing numerous illnesses due to the lack of immunity in the population and spreading over large areas [1].

Due to its structure, influenza A virus is most susceptible to antigenic variability and is the most common cause of pandemics [13], while influenza B virus causes seasonal epidemics. Influenza C infection is common, but is usually asymptomatic or causes a mild respiratory disease [1].

Due to seasonal influenza epidemics and the risk of a new pandemic, the virological and epidemiological monitoring of influenza virus is of fundamental importance. Currently, there are 149 National Influenza Centers (NICs) in 123 countries as part of the Global Influenza Surveillance and Response System (GISRS) [14].

Any change to the flu virus lowers the chance of a successful immune response. Therefore, it is important to immunize the body every season with a vaccine that contains virus strains selected based on virological data collected in a given hemisphere by the National Influenza Centers [15]. The composition of the flu vaccine must be updated regularly due to the variability of the virus, so it is important to take it every season. There are several methods of influenza virus diagnosis, e.g., virus isolation in chicken embryos; cell culture; antigen detection by IF, ELISA, RT-PCR, rRT-PCR; bedside tests; and serological methods; e.g., HAI [16]. The hemagglutination inhibition (HAI) assay is widely used to evaluate the antibody responses induced by the vaccine as well as for the antigenic characterization of influenza viruses. This is a conventional method used in various aspects of global influenza surveillance, diagnosis, antigen characterization, and vaccine evaluation [17].

Hemagglutination is a process whereby the surface glycoprotein—hemagglutinin—binds to the sialic acid sites on the surface of red blood cells (RBCs), thus creating a stable suspension of RBCs in solution. Anti-HA antibodies bind to hemagglutinin, blocking the possibility of hemagglutinin binding to RBCs, and as a result, hemagglutination is inhibited. By testing successive dilutions of a patient's serum, antibody levels can be measured [18]. Due to its simplicity, the HAI test has long been used to detect virus-neutralizing antibodies in serum. It is assumed that a titer ≥ 40 reduces the risk of influenza by 50% and is referred to as a seroprotective titer [19]. The titer of hemagglutination-inhibiting antibodies is currently the main immunological marker correlated with protection against influenza [20].

We are focusing on antihemagglutinin antibodies, because they are produced during the viral infection or three to six weeks after the vaccination and are responsible for blocking virus ability to adsorb to host cells. The viral hemagglutinin protein plays crucial role in the process of infecting its host. Hemagglutinin is responsible for membrane fusion, entry of virion to cell, and binding to host receptors. Approximately 80% of all viral envelope proteins are hemagglutinins.

This study was conducted to determine the average levels of antibodies and protective titers against influenza viruses in the Polish population. The results of the study provide an insight into the course of the season in terms of serology, which is also important for influenza surveillance in Poland because not every patient potentially exposed to influenza viruses had a PCR test. Without a swab sample from such patients, we are unable to determine whether they actually had contact with the influenza virus or just flu-like viruses with similar symptoms.

Determining the level of antibodies and determining whether the patient had a protective titer gives information about contact with the virus. Values below the protective titer indicate the lack of contact with the influenza virus, or if the person has been vaccinated, whether he is able to respond to the vaccine or not. When there are no antibodies after vaccination, the person is "non-respondent". The level of antibodies after the vaccination is decreasing during the epidemic season. It is possible to vaccinate a second time to achieve full protection for the duration of the season.

2. Materials and Methods

Sera were obtained from patients belonging to 7 age groups (0–4, 5–9, 10–14, 15–25, 26–44, 45–64, and 65 years or older) as those age groups are used in Poland in influenza surveillance and in reporting influenza cases to epidemiological departments. Samples were collected at voivodeship sanitary and epidemiological stations (VSES) in Poland, between 1 October 2019, and 31 March 2020. Anonymized samples were then sent to the Influenza Research Department, National Influenza Center. Until the test was performed, the samples were stored at $-30\ ^\circ$C.

In each age group, 100 serum samples were tested—700 samples in total. Samples were selected from each VSES to maintain representability for each region and checked for hemolysis in the sample. If noticed, sample was discarded. The hemagglutination inhibition (HAI) test was used to determine the antibody level. All viruses are corresponding to those used in flu vaccine for the 2019/2020 season.

Influenza viruses used in research:

- Subtype A/H1N1/: A/Brisbane/02/2018
- Subtype A/H3N2/: A/Kansas/14/2017
- Influenza type B, Victoria lineage: B/Colorado/06/2017
- Influenza type B, Yamagata lineage: B/Phuket/3073/2013

All viruses were obtained from World Influenza Centre at Francis Crick Institute, London, and then propagated in chicken embryos in NIC in Poland. Then titer of each virus was determined. Labelled vials containing viruses were stored at $-80\ ^\circ$C upon using in research.

Necessary viruses with high titer were selected to be used in this test. After thawing the viruses, their titer was checked once again on the day of performing the test. Then, a solution of titer 1:8 from each of the virus was prepared. For example, for virus titer 1:64, suspension of the virus was diluted 8 times. For 1:16—two times. After each dilution, the titer was checked, as it is important to use 1:8 titer only. After preparing all necessary solutions, they were stored at $4\ ^\circ$C upon adding them on the plates.

PBS and 0.1% calcium salt are prepared in–house. In this study, V-bottom, clear, microtitration plates were used.

Preparation of Receptor Destroying Enzyme (RDE) (Sigma-Aldrich, Jerusalem, Israel IL). Lyophilisate of RDE is resuspended in 5 mL of sterile water, and mixed firmly, then filled up to 100 mL with 0.1% of calcium salt of pH = 7.4. Then, 150 µL of prepared solution is pipetted into a sterile tube and stored at $-30\ ^\circ$C until required.

The study used chicken red blood cells. Blood cells delivered to the laboratory were suspended in Alsever's solution. Alsever's solution is prepared in-house. In order to obtain packed cells, 5 mL of cells were collected from Alsever's solution into a 50-mL centrifuge tube, topped up to 50 mL with PBS, then centrifuged for 10 min at 1200 RPM. After centrifugation, the supernatant was decanted, topped up again to 50 mL with PBS, and then washing and centrifugation at 1200 RPM for 5 min were repeated three times. In the next step, the centrifuged blood cells were transferred to a new 15-mL centrifuge tube and filled with PBS to 12 mL, then centrifuged for 10 min at 1200 RPM. The packed blood cells obtained in this way were used in further studies. In order to obtain the appropriate blood cell concentration for the HAI test, the following proportion was used: 1 mL PBS:5 µL packed cells.

Each of the sera was treated with RDE for 16 h at $37\ ^\circ$C prior to the hemagglutination inhibition test. For this purpose, 50 µL of serum from the patient was added to 150 µL of RDE, followed by incubation under appropriate conditions. After this step, to inactivate the enzyme, the mixture was incubated at $56\ ^\circ$C for 30 min.

In order to eliminate non-specific hemagglutination inhibitors, a mixture of blood cells and PBS was prepared in a ratio of 1 volume of packed blood cells to 20 volumes of serum after incubation with RDE. The serum-cell mixture was incubated for one hour at $4\ ^\circ$C. After incubation, the suspension was centrifuged at 1200 RPM for 10 min—the supernatant contained serum ready for further processing.

To row A, 50 µL of the patient's prepared serum was pipetted, and then a serial dilution was made in PBS. Then, the prepared solution of the virus, which has a titer of 1:8, was added to each well on the plate. After virus addition, the plate was incubated for 15 min at room temperature. After incubation, 50 µL of blood cell solution was added. Readings were taken after 30 min of incubation at room temperature.

The plate was lifted and tilted to allow RBCs to run down to the side of the well. Readings were performed according to pattern showed in the Figure in the Supplementary Materials.

The analysis of the results for each of the virus subtypes was carried out in 7 age groups. Numbers and percentages of patients with adequate antibodies and patients achieving a protective level (antibody titer $\geq$ 40) were determined. The average antibody level in the study group was calculated as a geometric mean of non-zero values (GMT) with 95% confidence intervals [CI]. The following statistical tests were employed in the analysis: chi-square test (for comparisons between age groups and epidemiological seasons in terms of antibody occurrence frequency and reaching protective level); Mann–Whitney test (comparison of antibody titer distributions between two epidemiological seasons); Kruskal–Wallis test (comparisons of antibody titer distributions between 7 age groups and 3 epidemiological seasons—post-hoc tests with Bonferroni correction for multiple comparisons were used to identify differing seasons). A significance level of 0.05 was assumed in all of the analyses.

Calculations were performed using the SPSS 12-PL statistical software.

3. Results

The serum of people belonging to all the age groups showed the presence of antibodies against all the four viruses analyzed. In total, 48% of the subjects had antibodies against the H1 subtype, H3—74%, B/Colorado—26%, and B/Phuket—63%. Figures 1 and 2 show the percentage of patients in individual age groups that had antibodies against a particular influenza virus.

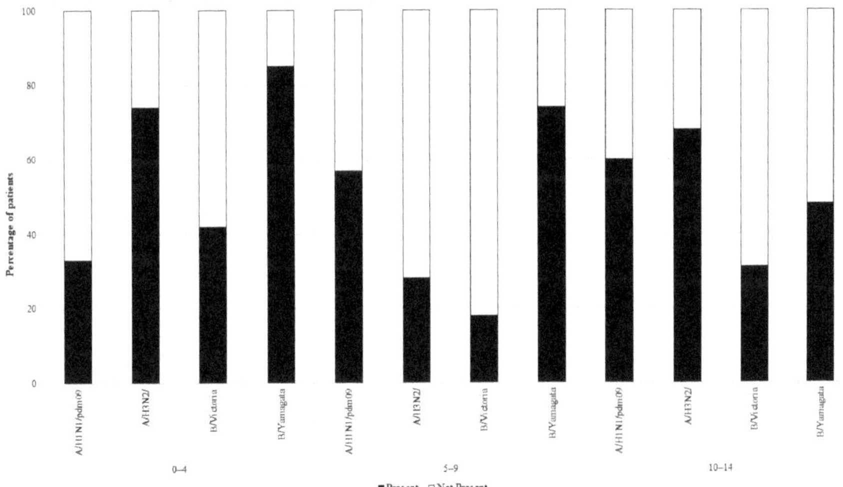

Figure 1. The presence of antibodies in the serum of patients aged 0–4 years, 5–9 years of age, and 10–14 years of age in the 2019/2020 epidemic season.

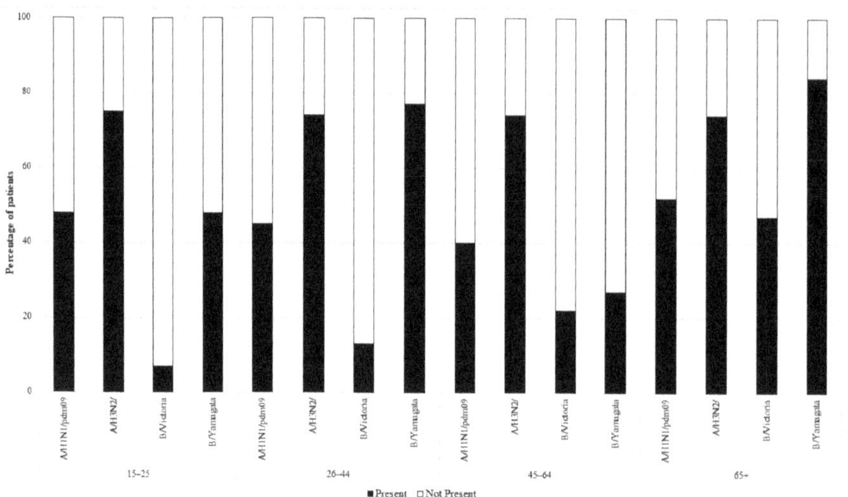

Figure 2. The presence of antibodies in the serum of patients aged 15–25, 26–44, 45–64, and 65+ years of age in the 2019/2020 epidemic season.

Using the chi-square test for analysis, statistically significant differences in the prevalence of antibodies (between seven age groups) were found for the following influenza virus types and subtypes:

- For subtype H1: $p = 0.001$, the proportion of subjects with antibodies ranged from 33% in the group of 0–4 years to 60% in the group of 10–14 years.
- For the Victoria line (B/Colorado): $p < 0.001$, the proportion of subjects with antibodies from 7% in the group of 15–25 years to 47% in the group of 65 years or more.
- For the Yamagata line (B/Phuket): $p < 0.001$, from 27% of subjects with antibodies in the 45–64 group to 85% in the 0–4 age group.

The analysis showed no statistically significant differences for the H3 subtype, the percentage of subjects with antibodies ranged from 68% in the 10–14 age group to 82% in the 5–9 group.

Statistically significant differences in the percentage of subjects with antibodies between ages up to 14 and over 14 were observed only for antibodies against B/Colorado and B/Phuket (analysis using the chi-square test):

- For B/Colorado $p = 0.015$, 22% of subjects over 14 years of age and 30% of children up to 14 years of age had antibodies.
- For B/Phuket $p = 0.007$, 59% of subjects over 14 years of age and 69% of children up to 14 years of age had antibodies.
- For the H1 subtype: 46% of subjects over 14 years of age and 50% of children up to 14 years of age had antibodies.
- For the H3 subtype: 74% of subjects over 14 years of age and 75% of children up to 14 years of age had antibodies.

Figure 3 shows geometric mean antibody titers (GMT) in the sera of patients who had antibodies. A statistical analysis using the Kruskal–Wallis test showed a statistically significant difference between the seven age groups for all the antibody types:

- For the H1 subtype: 335 people had antibodies, average level 54.4, statistical significance of differences—$p = 0.002$, the lowest level of antibodies in the group of 26–44 years—GMT = 40.0 [CI = 33.9–46.1] the highest in the group of 10–14 years—GMT = 88.8 [CI = 81.1–96.4].

- For the H3 subtype: 521 people had antibodies, average level 68.0, [CI = 65.9–70.1]. p = 0.004, the lowest level in the age groups 26–44 and 45–64 years of age—GMT = 55.5, [CI = CI: 49.9–61.2 and 49.8–61.2, respectively], the highest in the 5–9 age group—GMT = 89.3 [CI: 83.9–94.7].
- For the Victoria (B/Colorado) line: 180 people had antibodies, average level 25.3, [CI: 21.8–28.7], p < 0.001, the lowest level in the group of 26–44 years—GMT = 14.5 [CI: 4.5–24.5] and the highest in children up to 4 years of age—GMT = 41.3 [CI = 33.4–49.3].
- For the Yamagata line (B/Phuket): 443 people had antibodies, average level 40.9 [CI = 38.8–43.1], p < 0.001, the lowest level in the group of 46–64 years old—GMT = 23.3 [CI = 15.1–31.5], the highest in children up to 4 years of age—GMT = 51.9 [CI = 47.5–56.4].

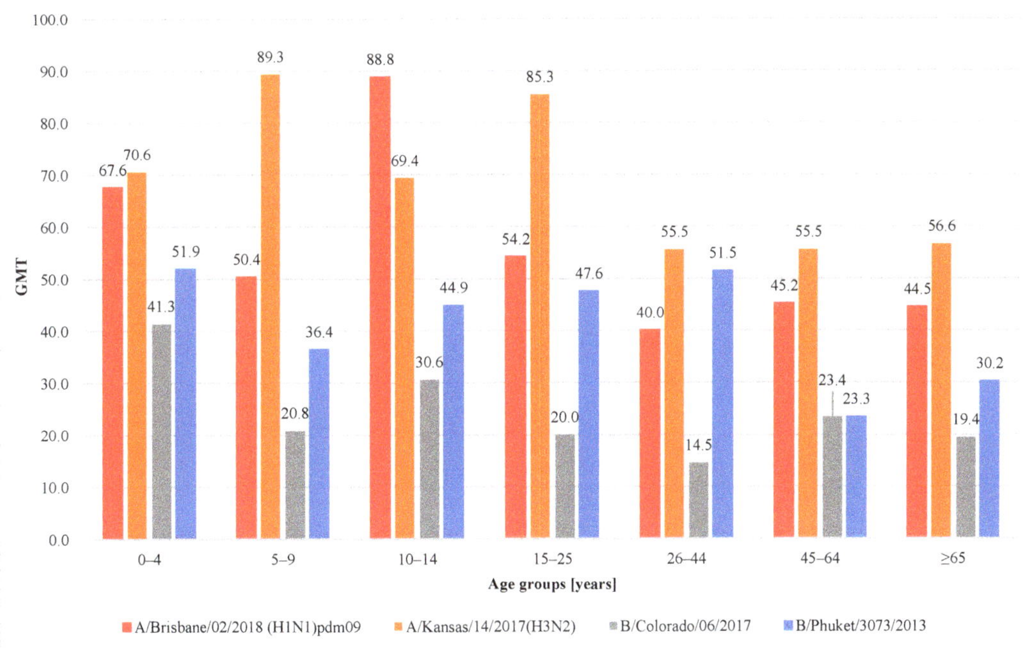

Figure 3. Geometric mean titers of anti-haemagglutinin antibodies (GMT) in the epidemic season 2019/2020 in age groups in Poland.

Figure 4 shows the response rates (percentages with protective anti-HA titers $\geq$ 40) for the four tested influenza viruses in seven different age groups. In each of the groups, subjects with a protective titer of anti-HA antibodies $\geq$ 40 were found. Statistical analysis (using the chi-square test) showed a statistically significant difference in the protection factors between the seven age groups for three subtypes. In the case of H3, the difference was on the border of statistical significance.

- For subtype H1: 239 people had antibodies with a titer $\geq$ 40, the protective factor was 34%; statistical significance of differences between age groups—p < 0.001 (rates from 20% in the 0–4 age group to 51% in the 10–14 age group).
- For subtype H3: 431 people had antibodies with a titer $\geq$ 40, protective factor was 62%; statistical significance of differences between age groups—p = 0.082 (rates from 55% in the 45–64 and $\geq$65 age groups to 72% in the 5–9 age group).
- For the Victoria line (B/Colorado): 79 people had antibodies with a titer $\geq$ 40, protective factor was 11%; statistical significance of differences between age groups—p = 0.001 (rates from 2% in the 15–25 group to 24% in the 0–4 age group).

- For the Yamagata line (B/Phuket): 297 people had antibodies with titers ≥ 40, protection factor was 42%; statistical significance of differences between age groups—$p = 0.001$ (rates from 7% in the group of 45–64 and ≥ 65 years to 68% in the group of 0–4 years).

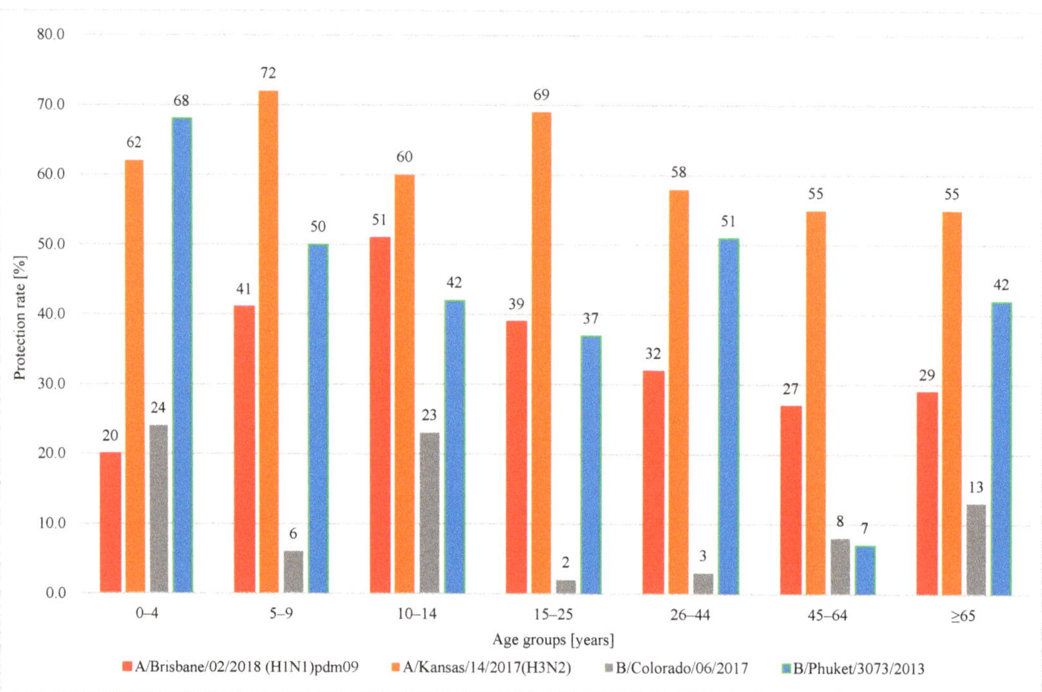

Figure 4. Percentage of cases with a protective titer of anti-haemagglutinin antibodies (%) in the 2019/2020 epidemic season in different age groups.

For influenza B virus of the Yamagata lineage, which remained unchanged in the vaccine composition in the epidemic season from 2017/2018 to 2019/2020, an analysis of differences in antibody levels between the three seasons was performed. The frequency of reaching the protective level (using the chi-square test) and the distribution of antibody titers were compared.

There were statistically significant differences between the three compared seasons in the percentage of subjects with antibodies at the level ≥ 40 ($p < 0.001$), with the highest value recorded in the 2018/19 season (67%) and the lowest in 2017/18 (34%) (Table 1).

Statistically significant differences occurred in six out of seven analyzed age groups ($p < 0.001$ in all cases), with the exception of the 26–44 group where the percentages amounting to 40%, 47%, and 51% did not show statistically significant differences.

The lowest percentages mostly concerned the 2017/18 season; lower values were recorded in the 2019/2020 season only in the two oldest age groups. The highest percentages usually concerned the 2018/2019 season; higher values were observed in the 2019/2020 season only for two age groups. The exceptions are as follows: the 0–4 age group and the 26–44 age group; in the latter case, however, the differences between the seasons were not statistically significant (Table 1).

Table 1. Comparison of three epidemic seasons for B/Phuket/3073/2013—Yamagata Lineage.

Age Group [Years]	Protection Rate			The Significance of the Differences (*p*-Value)
	2017/2018	2018/2019	2019/2020	
0–4	41%	55%	68%	<0.001
5–9	31%	60%	50%	<0.001
10–14	27%	93%	42%	<0.001
15–25	27%	68%	37%	<0.001
26–44	40%	47%	51%	Not Significant (NS)
45–64	25%	71%	7%	<0.001
>65	48%	75%	42%	<0.001
Total	34%	67%	42%	<0.001

Age Group [Years]	Antibodies Level (GMT) with 95% Confidence Intervals			The Significance of the Differences (*p*-Value)
	2017/2018	2018/2019	2019/2020	
0–4	54.4 [50.6–58.1]	64.9 [61.0–68.8]	69.4 [65.5–73.2]	0.011
5–9	63.2 [58.1–68.3]	52.8 [49.8–55.8]	49.9 [46.1–53.8]	NS
10–14	54.6 [49.2–60.1]	133.7 [130.2–137.2]	51.2 [46.9–55.6]	<0.001
15–25	47.6 [43.2–51.9]	62.2 [58.9–65.6]	62.7 [57.3–68.1]	0.015
26–44	57.2 [52.9–61.5]	58.3 [54.4–62.2]	89.2 [84.1–94.3]	<0.001
45–64	49.2 [44.2–54.1]	57.5 [54.5–60.5]	65.6 [49.6–81.6]	NS
>65	76.1 [71.6–80.7]	67.6 [64.1–71.1]	71.3 [65.2–77.4]	NS
Total	58.3 [56.5–60.1]	70.9 [69.4–72.3]	65.0 [63.0–67.0]	<0.001

Note: NS: Not Significant.

Mean antibody values (GMT) by season and age group as well as the statistical significance of differences between the three seasons are shown in Table 1. Statistically significant differences in terms of the level of antibodies against B/Phuket were observed in four age categories (0–4 10–14, 15–25, and 26–44 years old) and for all groups together. The analysis used the Kruskal–Wallis test supplemented with post-hoc tests to identify differing pairs: In the 0–4 age group, the seasons 2017/2018 and 2019/2020 differ statistically significantly, in which the average level of antibodies was 54.4 and 69.4, respectively ($p = 0.008$ after adjusting for multiple comparisons using the Bonferroni method).

- In the 10–14 age group, the level of antibodies in the 2018/19 season (average 133.7) significantly differed from the values in the other seasons (54.6 and 51.2), in both cases $p < 0.001$ (even after adjusting for multiple comparison by means of Bonferroni method).
- In the 15–25 age group, the level of antibodies in the 2017/2018 season (47.6) was significantly different from that observed in 2018/19 (62.2, $p = 0.020$, after taking into account the correction) and 2019/2020 (62.7, $p = 0.047$, after correction),
- In the 26–44 age group, the level of antibodies in the 2019/2020 season (89.2) was significantly different from that observed in 2017/18 (57.2) and 2018/19 (58.3)—in both cases $p < 0.001$, after the correction.

Taking an overall look at all the age groups, the level of antibodies in the 2017/2018 season (58.3) differed both from the one observed in the 2018/19 season (70.9, $p < 0.001$ after adjustment) and in the 2019/2020 season (65.0, $p = 0.011$, after correction).

An analysis of differences in the level of antibodies against influenza B virus of the Victoria line, which remained unchanged in the composition of the vaccine in the epidemic seasons from 2018/2019 to 2019/2020, was also carried out.

Statistically significant differences were recorded between the compared seasons in terms of the percentage of subjects with anti-B/Colorado antibodies $\geq$ 40 (chi-square test, $p < 0.001$)—the protection factor was 22% in the 2018/19 season compared to 11% in 2019/20 (Table 2). This effect consisted of significantly higher percentages in the 10–14, 15–25, and 26–44 age groups in the 2018/2019 season. In the oldest and youngest age groups, the differences were statistically insignificant (Table 2).

Table 2. Comparison of two past epidemic seasons for B/Colorado/06/2017—Victoria Lineage.

Age Group [Years]	Protection Rate		The Significance of the Differences (*p*-Value)
	2018/2019	2019/2020	
0–4	23%	24%	NS
5–9	11%	6%	NS
10–14	47%	23%	<0.001
15–25	10%	2%	0.019
26–44	37%	3%	<0.001
45–64	10%	8%	NS
>65	13%	13%	NS
Total	22%	11%	<0.001

Age Group [Years]	Antibodies Level (GMT) with 95% Confidence Intervals		The Significance of the Differences (*p*-Value)
	2018/2019	2019/2020	
0–4	48.1 [43.3–52.8]	77.7 [68.8–86.7]	0.008
5–9	41.7 [36.0–47.3]	40.0 [32.0–48.0]	NS
10–14	50.6 [47.0–54.1]	40.0 [35.9–44.1]	0.004
15–25	43.9 [37.4–50.3]	80.0 [43.1–116.9]	NS
26–44	64.6 [59.7–69.4]	40.0 [28.7–51.3]	NS
45–64	52.8 [45.6–60.0]	51.9 [42.0–61.8]	NS
>65	59.8 [51.4–68.2]	68.2 [56.0–80.4]	NS
Total	52.9 [50.7–55.0]	55.8 [51.7–60.0]	NS

Note: NS: Not Significant.

There were no statistically significant differences between the season 2018/2019 and 2019/2020 in the average level of antibodies against B/Colorado (Mann–Whitney test)—the average titer was 52.9 vs. 55.8, respectively (Table 2). Such differences occurred only in two age categories in children: the 0–4 age group, in which a significantly higher level of antibodies was recorded in the 2019/2020 season (77.7 vs. 48.1; $p = 0.008$), and the 10–14 age group with a significantly higher level in the 2018/2019 season (50.6 vs 40.0; $p = 0.004$).

4. Discussion

Antibodies against all four tested viruses were found in patients of all age groups. All the viruses analyzed were included in the influenza vaccine for the 2019/2020 season. However, due to the low vaccination rate in the population (only 4.4% of Poles received the flu vaccine in the 2019/2020 season) [21]. In comparison, the vaccination rate calculated from administered doses in the past seasons were characterized by an even lower rate: 2018/2019: 3.9%, 2017/2018: 3.6%. It should be assumed that the presence of antibodies in the serum is the result of contracting the disease.

A protective antibody titer is considered to be $\geq$40. High titers of antibodies may indicate a past infection. Immunoglobulin levels also increase after influenza vaccination. In the 2019/2020 epidemic season, the highest percentages of patients with a protective antibody titer were observed for virus A/Kansas/14/2017 (subtype A/H3N2/), and the lowest for B/Colorado/06/2017, although this strain has been in the vaccine since the season 2018/2019.

In the case of the influenza A virus, the A/H3N2/ subtype featured the highest percentage of subjects with a protective level of antibodies, despite the fact that this season the most molecularly confirmed strain was the A/H1N1/pdm09 subtype.

The main role of the influenza vaccine is to increase immunity against influenza virus infection. Vaccination is an effective method of preventing the disease and its complications and related mortality. The effectiveness of the influenza vaccine varies depending on the patient age, the effectiveness of the immune system response, as well as on the match between the circulating strains and the vaccine strains in a given year [6]. However, cross-immunity, i.e., alleviation of disease symptoms, can be observed even though the vaccine received was for a different variant of the virus [22], which does not occur in young children [23]. On the other hand, vaccination is more effective in children than in the elderly. Therefore, vaccinating children may be an important way to prevent the disease and post-disease complications in elderly people [24], whose immune response is weakened [20].

As research conducted in Poland in the period from December 2018 to April 2020 among the examined patients shows, influenza was found mainly in unvaccinated subjects [25]. Therefore, taking the flu vaccine every season is recommended not only for people belonging to the high-risk group. Seasonal influenza vaccination is also necessary because immunity declines over time after vaccination [26]. In addition, a new disease caused by the SARS-CoV-2 virus appeared in the 2019/2020 epidemic season. This virus has a similar route of transmission and causes a disease with similar symptoms [27,28]. That is why a correct diagnosis is important to ensure that patients are adequately treated. However, we did not notice any impact of wearing masks or other restrictions on antibody levels. It can be explained by the fact that restrictions in Poland were incorporated in April of 2020. Testing patients' sera is highly connected with the vaccination policy, as it informs us about immunity against influenza viruses in the population. Based on those findings, we can recommend actions to ministry of health of Poland.

Low protection rates confirm the low vaccination rate in Poland. The vaccination rate should be higher for the elderly and people at high risk; the vaccination rate recommended by the WHO should reach 75%. To achieve that, influenza vaccines should be mandatory for children up to 14 years of age, the elderly above 65 years of age, and all individuals in risk groups.

Supplementary Materials: The following supporting information can be downloaded at: https://www.mdpi.com/article/10.3390/v15030760/s1.

Author Contributions: Conceptualization, L.B.B. and K.S.; methodology K.S. and K.K.; validation, L.B.B. and A.P.; formal analysis K.S., E.H. and A.P.; investigation K.S.; resources K.S. and L.B.B.; data curation K.S. and A.P.; writing—original draft preparation K.S.; writing—review and editing, L.B.B.; visualization, K.S.; supervision, L.B.B.; project administration, L.B.B.; funding acquisition, L.B.B. All authors have read and agreed to the published version of the manuscript.

Funding: Funded by the National Institute of Public Health NIH—NRI, research grant No. BI-1/2020.

Institutional Review Board Statement: Ethical review and approval were waived for this study due to National Influenza Centre (NIC) cooperation with Voivodship Sanitary Epidemiological Stations (VSES) within Global Influenza Surveillance and Response System (GISRS). VSES are collecting samples from patients during epidemic season.

Informed Consent Statement: Samples were collected by VSES, and all documents regarding patients and their consent, belongs to VSES.

Data Availability Statement: Not applicable.

Acknowledgments: The authors would like to thank Lidia Malec, and Henryk Malec, from "Malec" Company for providing hatching eggs. The authors extend special thanks to the STASIN Poultry Plant: Krzysztof Borkowski, President of the Board; Paweł Okrzeja, Director; Grzegorz Borkowski, Director; and Members of the Board: Urszula Lipińska-Witkowska and Ewelina Paprota for providing turkey blood. We would like to thank the employees of the Provincial Sanitary and Epidemiological Stations for collecting, describing and transferring the sera to the NIPH NIH NRI.

Conflicts of Interest: The authors declare no conflict of interest.

References

1. Wang, Y.; Tang, C.Y.; Wan, X.F. Antigenic characterization of influenza and SARS-CoV-2 viruses. *Anal Bioanal Chem.* **2022**, *414*, 2841–2881. [CrossRef] [PubMed]
2. Paules, C.; Subbarao, K. Influenza. *Lancet* **2017**, *390*, 697–708. [CrossRef] [PubMed]
3. Brydak, L.B. Chapter VII: Influenza laboratory diagnosis. In *Influenza, Pandemic Flu, Myth or Real Threat?* Rhythm: Warsaw, Poland, 2008; pp. 125–140.
4. Kowalczyk, D.; Szymański, K.; Cieślak, K.; Brydak, L.B. Circulation of Antibodies Against Influenza Virus Hemagglutinins in the 2014/2015 Epidemic Season in Poland. *Adv. Exp. Med. Biol.* **2017**, *968*, 35–40. [CrossRef] [PubMed]
5. Labella, A.M.; Merel, S.E. Influenza. *Med. Clin. N. Am.* **2013**, *97*, 621–645. [CrossRef]
6. Webster, R.G.; Govorkova, E.A. Continuing challenges in influenza. *Ann. N. Y. Acad. Sci.* **2014**, *1323*, 115–139. [CrossRef]
7. Rota, P.A.; Wallis, T.R.; Harmon, M.W.; Rota, J.S.; Kendal, A.P.; Nerome, K. Cocirculation of two distinct evolutionary lineages of influenza type B virus since 1983. *Virology* **1990**, *175*, 59–68. [CrossRef]
8. Keilman, L.J. Seasonal Influenza (Flu). *Nurs. Clin. N. Am.* **2019**, *54*, 227–243. [CrossRef]
9. Kim, H.; Webster, R.G.; Webby, R.J. Influenza Virus: Dealing with a Drifting and Shifting Pathogen. *Viral Immunol.* **2018**, *31*, 174–183. [CrossRef]
10. Ghebrehewet, S.; MacPherson, P.; Ho, A. Influenza. *BMJ* **2016**, *355*, i6258. [CrossRef]
11. Bouvier, N.M.; Palese, P. The biology of influenza viruses. *Vaccine* **2008**, *26* (Suppl. 4), D49–D53. [CrossRef]
12. Gaitonde, D.Y.; Moore, F.C.; Morgan, M.K. Influenza: Diagnosis and Treatment. *Am. Fam. Physician* **2019**, *100*, 751–758. [PubMed]
13. WHO. 2022. Available online: https://www.who.int/initiatives/global-influenza-surveillance-and-response-system/national-influenza-centres (accessed on 24 June 2022).
14. GISAID. 2022. Available online: https://www.gisaid.org/references/human-influenza-vaccine-composition/ (accessed on 27 June 2022).
15. CDC. 2022. Available online: https://www.cdc.gov/flu/professionals/diagnosis/table-testing-methods.htm (accessed on 27 June 2022).
16. Wilson, G.; Ye, Z.; Xie, H.; Vahl, S.; Dawson, E.; Rowlen, K. Automated interpretation of influenza hemagglutination inhibition (HAI) assays: Is plate tilting necessary? *PLoS ONE* **2017**, *12*, e0179939. [CrossRef] [PubMed]
17. WHO. *WHO–Global Influenza Surveillance Network (2011) Manual for the Laboratory Diagnosis and Virological Surveillance of Influenza*; WHO Press: Geneva, Switzerland, 2011; pp. 1–153.
18. Krammer, F. The human antibody response to influenza A virus infection and vaccination. *Nat. Rev. Immunol.* **2019**, *19*, 383–397. [CrossRef] [PubMed]
19. Reber, A.; Katz, J. Immunological assessment of influenza vaccines and immune correlates of protection. *Expert Rev. Vaccines* **2013**, *12*, 519–536. [CrossRef] [PubMed]
20. Brydak, L.B. Szczepić się czy nie szczepić przeciwko grypie podczas pandemii SARS-CoV-2, Terapia. 2020. (In Polish)
21. Bednarska, K.; Nowak, M.A.; Kondratiuk, K.; Hallmann-Szelińska, E.; Brydak, L.B. Incidence of Circulating Antibodies Against Hemagglutinin of Influenza Viruses in the Epidemic Season 2013/2014 in Poland. *Adv. Exp. Med. Biol.* **2015**, *857*, 45–50. [CrossRef]
22. Kumar, V. Influenza in Children. *Indian J. Pediatr.* **2017**, *84*, 139–143. [CrossRef]
23. Smetana, J.; Chlibek, R.; Shaw, J.; Splino, M.; Prymula, R. Influenza vaccination in the elderly. *Hum. Vaccin Immunother.* **2018**, *14*, 540–549. [CrossRef]
24. Rzepka, A.; Mania, A. The clinical picture of influenza against other respiratory tract infections in a general practitioner practice. *Przegl. Epidemiol.* **2021**, *75*, 159–174. [CrossRef]
25. Buchy, P.; Badur, S. Who and when to vaccinate against influenza. *Int. J. Infect. Dis.* **2020**, *93*, 375–387. [CrossRef]

26. Solomon, D.A.; Sherman, A.C.; Kanjilal, S. Influenza in the COVID-19 Era. *JAMA* **2020**, *324*, 1342–1343. [CrossRef]
27. Chotpitayasunondh, T.; Fischer, T.K.; Heraud, J.M.; Hurt, A.C.; Monto, A.S.; Osterhaus, A.; Shu, Y.; Tam, J.S. Influenza and COVID-19, What does co-existence mean? *Influenza Other Respir. Viruses* **2021**, *15*, 407–412. [CrossRef] [PubMed]
28. WHO. 2022. Available online: https://www.who.int/europe/news-room/fact-sheets/item/influenza-vaccination-coverage-and-effectiveness (accessed on 1 March 2023).

Systematic Review

Convalescent Plasma Therapy for COVID-19: A Systematic Review and Meta-Analysis on Randomized Controlled Trials

Charalampos Filippatos [1], Ioannis Ntanasis-Stathopoulos [1], Kalliopi Sekeri [1], Anastasios Ntanasis-Stathopoulos [1], Maria Gavriatopoulou [1], Theodora Psaltopoulou [1], George Dounias [2], Theodoros N. Sergentanis [2,*] and Evangelos Terpos [1,*]

1   Department of Clinical Therapeutics, School of Medicine, National and Kapodistrian University of Athens, 11528 Athens, Greece
2   Department of Public Health Policy, School of Public Health, University of West Attica, 11521 Athens, Greece
*   Correspondence: tsergent@med.uoa.gr (T.N.S.); eterpos@med.uoa.gr (E.T.); Tel.: +30-213-216-2846 (E.T.)

Abstract: Background: While passive immunotherapy has been considered beneficial for patients with severe respiratory viral infections, the treatment of COVID-19 cases with convalescent plasma produced mixed results. Thus, there is a lack of certainty and consensus regarding its effectiveness. This meta-analysis aims to assess the role of convalescent plasma treatment on the clinical outcomes of COVID-19 patients enrolled in randomized controlled trials (RCTs). **Methods:** A systematic search was conducted in the PubMed database (end-of-search: 29 December 2022) for RCTs on convalescent plasma therapy compared to supportive care\standard of care. Pooled relative risk (RR) and 95% confidence intervals were calculated with random-effects models. Subgroup and meta-regression analyses were also performed, in order to address heterogeneity and examine any potential association between the factors that varied, and the outcomes reported. The present meta-analysis was performed following the Preferred Reporting Items for Systematic Reviews and Meta-Analyses (PRISMA) guidelines. **Results:** A total of 34 studies were included in the meta-analysis. Per overall analysis, convalescent plasma treatment was not associated with lower 28-day mortality [RR = 0.98, 95% CI (0.91, 1.06)] or improved 28-day secondary outcomes, such as hospital discharge [RR = 1.00, 95% CI (0.97, 1.03)], ICU-related or score-related outcomes, with effect estimates of RR = 1.00, 95% CI (0.98, 1.05) and RR = 1.06, 95% CI (0.95, 1.17), respectively. However, COVID-19 outpatients treated with convalescent plasma had a 26% less risk of requiring hospital care, when compared to those treated with the standard of care [RR = 0.74, 95% CI (0.56, 0.99)]. Regarding subgroup analyses, COVID-19 patients treated with convalescent plasma had an 8% lower risk of ICU-related disease progression when compared to those treated with the standard of care (with or without placebo or standard plasma infusions) [RR = 0.92, 95% CI (0.85, 0.99)] based on reported outcomes from RCTs carried out in Europe. Finally, convalescent plasma treatment was not associated with improved survival or clinical outcomes in the 14-day subgroup analyses. **Conclusions:** Outpatients with COVID-19 treated with convalescent plasma had a statistically significantly lower risk of requiring hospital care when compared to those treated with placebo or the standard of care. However, convalescent plasma treatment was not statistically associated with prolonged survival or improved clinical outcomes when compared to placebo or the standard of care, per overall analysis in hospitalized populations. This hints at potential benefits, when used early, to prevent progression to severe disease. Finally, convalescent plasma was significantly associated with better ICU-related outcomes in trials carried out in Europe. Well-designed prospective studies could clarify its potential benefit for specific subpopulations in the post-pandemic era.

Keywords: COVID-19; convalescent plasma; meta-analysis; randomized controlled trials; mortality; intensive care unit

check for **updates**

Citation: Filippatos, C.; Ntanasis-Stathopoulos, I.; Sekeri, K.; Ntanasis-Stathopoulos, A.; Gavriatopoulou, M.; Psaltopoulou, T.; Dounias, G.; Sergentanis, T.N.; Terpos, E. Convalescent Plasma Therapy for COVID-19: A Systematic Review and Meta-Analysis on Randomized Controlled Trials. *Viruses* **2023**, *15*, 765. https:// doi.org/10.3390/v15030765

Academic Editor: Jason Yiu Wing KAM

Received: 4 March 2023
Accepted: 7 March 2023
Published: 16 March 2023

1. Introduction

At the end of 2019, a surge of pneumonia cases in Wuhan, a city in the Hubei Province of China, led to the identification of a novel coronavirus as the cause. Its rapid spread resulted in an epidemic throughout China, followed by an increasing number of cases around the world. On 11 March 2020, the World Health Organization declared the novel coronavirus outbreak a pandemic. The disease associated with it was designated as COVID-19, which stands for coronavirus disease 2019, and the virus that caused it was designated as Severe Acute Respiratory Syndrome Coronavirus 2 (SARS-CoV-2) [1].

As of the end of 2022, COVID-19 disease management involves mostly supportive care, symptomatic treatment and prevention by vaccination. There have been only a few drugs or treatments proven to be effective specifically against the virus and the illness it causes, such as nirmatrelvir\ritonavir for high-risk patients.

Convalescent plasma, having been used to treat viral outbreaks of novel infectious diseases affecting the respiratory system in the past [2,3], was an early candidate [4,5]. The idea behind it is to transfuse blood plasma from a person who has recovered from a specific illness to someone who currently has the same illness in order to provide passive immunity and boost their fight against the pathogen, since such plasma contains antibodies to it [6,7].

With the potential for convalescent plasma to be beneficial, there was an urgency for clinical trials. The FDA provided emergency use authorization for its use and the WHO reinforced clinical trials to continue enrolment. Later updates included revisions on the matter, such as the focus and authorization being shifted to immunosuppressed patients or outpatients (FDA) or severe and high-risk patients in general (WHO) [8]. Moreover, the most up-to-date emergency authorization letter by the FDA states that convalescent plasma units used should be "high-titer", based on studies showing the superiority of high-titer convalescent plasma in terms of preventing severe COVID-19-related outcomes [9].

Because studies and reviews yielded conflicting results, there has been a persistent lack of certainty and consensus regarding its efficacy [10,11]. Therefore, we conducted a meta-analysis, focusing strictly on RCTs, to assess the effect of convalescent plasma treatment on the clinical outcomes of patients with COVID-19.

2. Materials and Methods

2.1. Search Strategy and Eligibility of Studies

The present meta-analysis was performed following the Preferred Reporting Items for Systematic Reviews and Meta-Analyses (PRISMA) guidelines [12]. The study protocol was discussed and agreed upon in advance by all authors.

A systematic search was conducted in the PubMed database, using the following algorithm: (COVID-19 OR SARS-CoV-2 OR "novel coronavirus") AND (convalescent OR convalescence) AND (plasma OR serum).

Eligible articles included randomized clinical trials on convalescent plasma treatment vs. supportive care or standard of care controls, with or without placebo. Case-control, cohort and cross-sectional studies, case series and case reports, reviews, in vitro and animal studies were not included in this meta-analysis. The selection of studies was conducted initially by two co-authors (CF and ANS) by independent work and any disagreements were resolved following consultation with a senior author (INS or TNS) and team consensus.

2.2. Data Abstraction and Effect Estimates

The data abstraction encompassed: general information (first author's name, publication year, PubMed and CT database ID), study characteristics (time period, follow-up period, geographic region, multicenter status, control type, participant numbers, percentage of males, age), intervention characteristics (time to intervention from symptom onset and total CP dose) and outcomes (mortality and clinical outcomes with reported effect estimates or fourfolds with plain data, adjustment details).

If one of the above was not found in the main article, the Supplementary Material was thoroughly screened. There was no shortage of required data for the purposes of the meta-analysis. Data were independently extracted, analyzed and recorded in separate data extraction sheets by two authors (CF and KS). The finalized data form was reached after consultation with a senior author (TNS) and team consensus.

Extracted effect estimates included relative risks alongside their 95% Cis (per outcome) or any other form that could be mathematically transformed or translated to relative risk. Mortality was extracted as a primary outcome for our work and hospitalization, and hospital discharge, ICU-related outcomes and score-related outcomes were secondary outcomes.

As far as score-related outcomes are concerned, all of them were based on or using variations of the 9-point WHO score for COVID-19. This is defined as: 0: no clinical or virological evidence of infection; 1: ambulatory, no activity limitation; 2: ambulatory, activity limitation; 3: hospitalized, no oxygen therapy; 4: hospitalized, oxygen mask or nasal prongs; 5: hospitalized, noninvasive mechanical ventilation (NIMV) or high-flow nasal cannula (HFNC); 6: hospitalized, intubation and invasive mechanical ventilation (IMV); 7: hospitalized, IMV + additional support such as pressors or extracardiac membranous oxygenation (ECMO); 8: death.

Finally, a titer subgroup analysis was carried out, between studies that fulfilled the latest EUA/FDA cut-offs for high-titer plasma units versus the rest. This is defined as a neutralizing antibody titer of $\geq$250 in the Broad Institute's neutralizing antibody assay or an S/C cutoff of $\geq$12 in the Ortho VITROS IgG assay.

In case the aforementioned information was not available, crude effect estimates and 95% CIs were calculated by means of fourfolds from plain data extracted from the articles.

2.3. Statistical Analyses

Statistical analyses included pooling of studies as well as post hoc meta-regression. Random-effects models were appropriately used to calculate the pooled effect estimates (relative risks). The convalescent plasma treatment arms were compared to the control arms. Between-study heterogeneity was assessed by Q-test and I^2 estimations. Subgroup analyses were performed based on adjustment, multicenter status, blinding status and the geographic region of each study.

The post hoc meta-regression analysis was performed for subgroups with a total of 10 or more data entries for the variables to be analyzed. The aim was to assess whether gender, age, time from symptom onset to intervention or total convalescent plasma dose modified the association between convalescent plasma transfusion and each reported outcome.

All statistical analyses were performed using STATA/SE version 13 (Stata Corp, College Station, TX, USA).

2.4. Assessment of Study Quality and Risk of Bias

All records included randomized clinical trials, either blinded or open label. Risk was assessed with the implementation of the RoB:2 algorithm by Cochrane to our analysis tools [11]. Specifically, two authors (KS and ANS) carried out the assessment procedure independently, and upon inspection of the results by a third author (CF), consensus was met.

Publication bias was evaluated in the analyses that included 10 or more study arms [13]. For this purpose, Egger's statistical test (statistical significance $p < 0.1$) [14,15] was implemented as well as the funnel plot inspection. The evaluation of publication bias was performed using STATA/SE version 13 (Stata Corp, College Station, TX, USA).

3. Results

3.1. Description of Eligible Studies

A total of 2374 records were identified from PubMed using the search algorithm (Section 2.1) and were assessed for eligibility. The flowchart (Figure 1) portrays the successive steps in the selection of eligible studies.

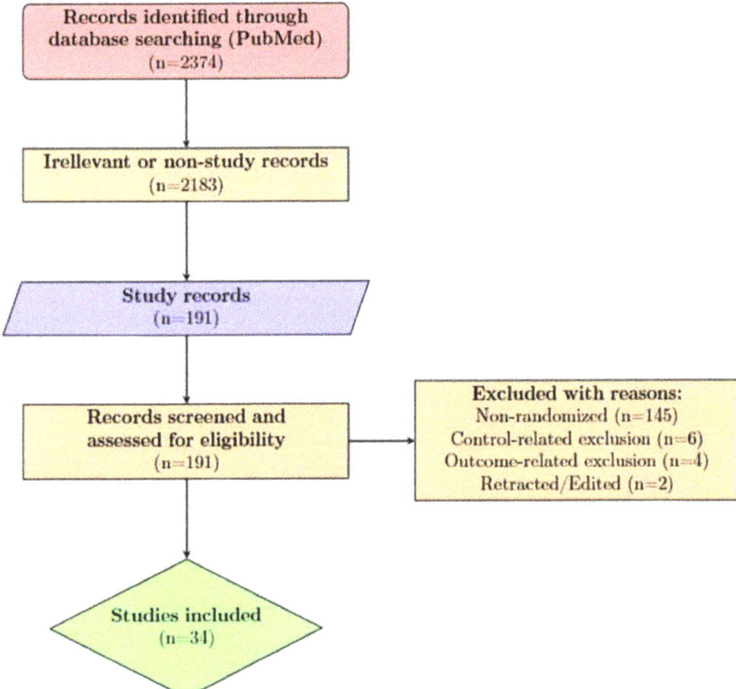

Figure 1. Study selection flowchart.

For the 28-day main cohort, 34 randomized controlled trials were included [16–49]. For the 14-day secondary cohort, 10 articles on randomized controlled trials provided the necessary data. All studies had convalescent plasma therapy arms vs. standard of care or supportive care arms, with some including standard plasma, non-convalescent plasma or fresh-frozen plasma to the control arms.

From the 28-day cohort studies, all of them reported mortality figures, except one (Alemany, 2022) [17]. Regarding secondary clinical outcomes, hospital discharge was reported on nine records, ICU-related outcomes were reported on twenty-one records, hospitalization was a reported outcome in six studies and score-related outcomes (WHO score for COVID-19) in six studies.

Tables 1 and 2 present the characteristics of the included studies regarding study design, patient and disease characteristics and interventions.

Table 1. General characteristics of the included studies.

Author and Year	Setting	Geographic Region	Multicenter	Blinded	Control
Abani (2021)	Hospitalized	More than one area	Yes	No	SoC
Agarwal (2020)	Hospitalized	India	Yes	No	SoC
Alemany (2022)	Outpatients	Europe	Yes	Yes	Placebo
AlQahtani (2021)	Hospitalized	Middle East	No	No	SoC
Avendaño-Solá (2021)	Hospitalized	Europe	Yes	No	SoC
Bajpai (2020)	Hospitalized	India	No	No	SoC
Bajpai (2022)	Hospitalized	India	Yes	No	SoC
Baldeón (2022)	Hospitalized	Latin America	Yes	Yes	Placebo
Bar (2021)	Hospitalized	USA	No	No	SoC
Bégin (2021)	Hospitalized	More than one area	Yes	No	SoC
Bennett-Guerrero (2021)	Hospitalized	USA	No	Yes	Placebo
Dekinger (2022)	Hospitalized	Europe	Yes	No	SoC

Table 1. *Cont.*

Author and Year	Setting	Geographic Region	Multicenter	Blinded	Control
Devos (2022)	Hospitalized	Europe	Yes	No	SoC
Estcourt (2021)	Hospitalized	More than one area	Yes	No	SoC
Gharbharan (2021)	Hospitalized	Europe	Yes	No	SoC
Gharbharan (2022)	Outpatients	Europe	Yes	Yes	Placebo
Holm (2021)	Hospitalized	Europe	No	No	SoC
Kirenga (2021)	Mixed	Africa	No	No	SoC
Korley (2021)	Outpatients	USA	Yes	Yes	SoC
Li (2020)	Hospitalized	East Asia	Yes	No	SoC
Libster (2021)	Outpatients	Latin America	Yes	Yes	Placebo
Manzini (2022)	Hospitalized	Europe	No	Yes	SoC
Menichetti (2021)	Hospitalized	Europe	Yes	No	SoC
O'Donnell (2021)	Hospitalized	More than one area	Yes	Yes	Placebo
Ortigoza (2022)	Hospitalized	USA	Yes	Yes	Placebo
Ray (2022)	Hospitalized	India	No	No	SoC
Rojas (2022)	Hospitalized	Latin America	Yes	Yes	SoC
Santis (2022)	Hospitalized	Latin America	Yes	No	SoC
Sekine (2022)	Hospitalized	Latin America	No	Yes	SoC
Self (2022)	Hospitalized	USA	Yes	Yes	Placebo
Simonovich (2021)	Hospitalized	Latin America	No	Yes	Placebo
Sullivan (2022)	Outpatients	USA	Yes	Yes	Placebo
Thorlacius-Ussing (2022)	Hospitalized	Europe	Yes	No	Placebo
van de Berg (2022)	Hospitalized	Africa	No	Yes	Placebo

SoC: standard of care; placebo control includes SoC.

Table 2. Intervention characteristics of the included studies.

Author and Year	CP (n)	Control (n)	Male %	Age ($\mu \pm \sigma$)	Time from Symptom Onset to Intervention ($\mu \pm \sigma$)	CP Dose (mL)
Abani (2021)	5795	5763	64%	63.50 ± 14.70	9.00 ± 4.45	550
Agarwal (2020)	235	229	76%	51.13 ± 19.53	8.35 ± 3.73	400
Alemany (2022)	188	188	54%	56.70 ± 7.44	4.40 ± 1.40	275
AlQahtani (2021)	20	20	80%	51.65 ± 19.45	10.00	400
Avendaño-Solá (2021)	179	171	65%	63.00 ± 15.30	5.65 ± 2.23	275
Bajpai (2020)	14	15	73%	48.20 ± 9.80	3.00	500
Bajpai (2022)	200	200	67%	55.52 ± 1.17	-	500
Baldeón (2022)	63	95	68%	74.34 ± 18.39	10.60 ± 4.90	-
Bar (2021)	40	39	46%	-	7.71 ± 4.53	-
Bégin (2021)	625	313	59%	67.50 ± 15.60	7.90 ± 3.70	500
Bennett-Guerrero (2021)	59	15	60%	65.70 ± 23.50	11.12 ± 9.12	480
Denkinger (2022)	68	66	68%	68.50 ± 11.30	7.00 ± 4.50	575
Devos (2022)	320	163	69%	62.00 ± 14.00	7.00 ± 4.46	450
Estcourt (2021)	1078	909	68%	60.77 ± 18.38	-	550
Gharbharan (2021)	43	43	72%	64.40 ± 13.45	10.35 ± 6.72	300
Gharbharan (2022)	207	209	78%	60.00 ± 7.44	5.00 ± 1.49	400
Holm (2021)	17	14	61%	69.95 ± 40.64	7.00 ± 3.23	675
Kirenga (2021)	69	67	71%	50.18 ± 17.61	6.30 ± 3.00	-
Korley (2021)	257	254	46%	51.90 ± 16.35	3.65 ± 2.24	250
Li (2020)	52	51	58%	70.00 ± 12.03	29.65 ± 14.29	-
Libster (2021)	80	80	38%	77.20 ± 8.60	1.65 ± 0.58	250
Manzini (2022)	60	60	72%	65.48 ± 11.96	8.35 ± 5.23	600
Menichetti (2021)	232	241	64%	64.00 ± 14.87	7.21 ± 2.98	400
O'Donnell (2021)	150	73	66%	60.30 ± 17.91	10 ± 4.49	-
Ortigoza (2022)	468	473	59%	62.65 ± 15.59	6.65 ± 3.71	250
Ray (2022)	40	40	71%	-	4.20 ± 2.21	400
Rojas (2022)	46	45	70%	51.76 ± 18.68	10.65 ± 2.96	500
Santis (2022)	36	71	72%	56.00 ± 16.16	9.00 ± 1.50	1800
Sekine (2022)	80	80	41%	58.74 ± 14.96	10.00 ± 3.00	300
Self (2022)	487	473	57%	59.65 ± 15.59	7.65 ± 3.72	300
Simonovich (2021)	228	105	67%	62.00 ± 14.89	7.65 ± 3.72	500
Sullivan (2022)	592	589	57%	43.35 ± 23.12	5.65 ± 2.23	250
Thorlacius-Ussing (2022)	98	46	72%	65.00 ± 14.98	10.65 ± 3.76	600
van de Berg (2022)	52	51	41%	77.20 ± 8.60	8.65 ± 3.76	250

Missing values were not reported either in the article or in the supplementary material. CP dose is the total convalescent plasma transfused.

3.2. Meta-Analysis

3.2.1. 28-Day Results

In total, 34 studies were included in the overall meta-analysis for the 28-day cohort [13–46]. The effect outcome for 28-day mortality was not statistically significant [RR = 1.00, 95% C.I. (0.95, 1.06)] (Figure 2). There were no statistically significant results in the adjustment, multicenter status, blinding status and geographic region subgroup analyses (Table 3).

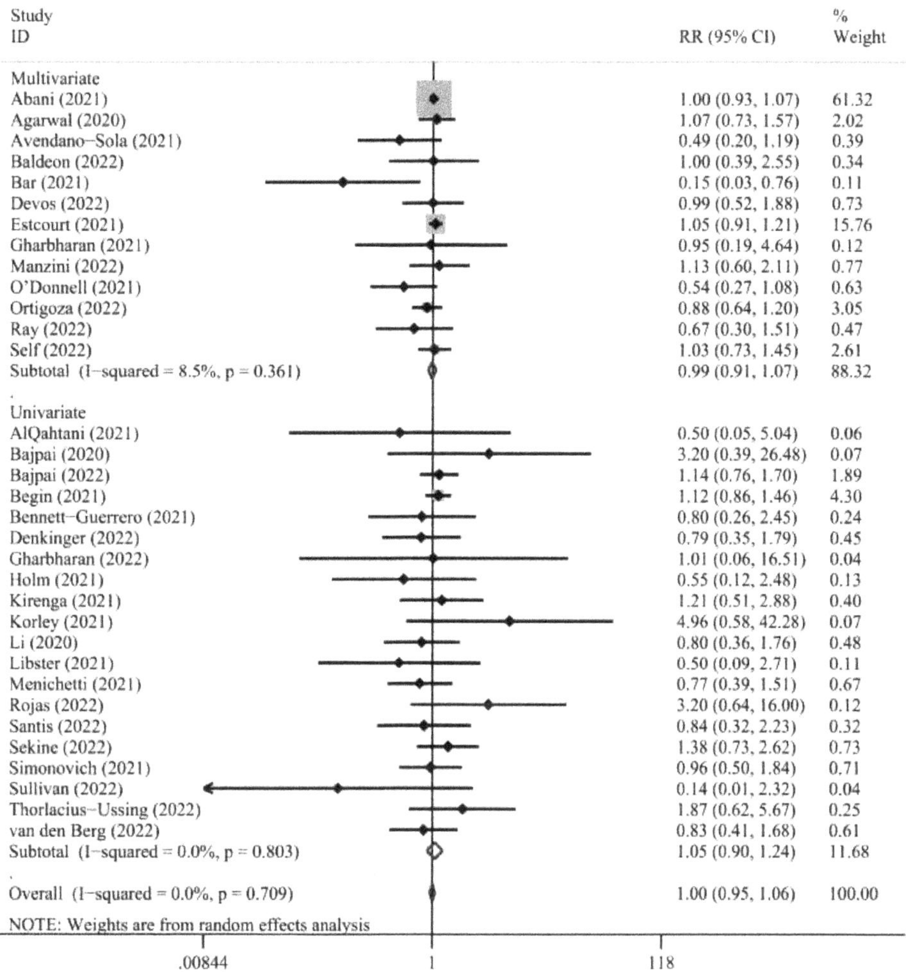

Figure 2. Forest plot describing the association between convalescent plasma treatment and 28-day mortality. Apart from the overall analysis, the subanalysis on adjustment type is presented.

A meta-analysis for the secondary clinical outcomes showed no statistically significant association between convalescent plasma therapy and hospital discharge [RR = 0.99, 95% C.I. (0.96, 1.03)] or score-related outcomes [RR = 1.06, 95% C.I. (0.97, 1.16)] (Tables 4 and 5). The ICU-related outcomes analysis yielded no statistically significant overall result [RR = 0.98, 95% C.I. (0.93, 1.02)] as well (Table 6).

Table 3. Results of the meta-analyses examining the association between convalescent plasma therapy and mortality (28-day); subgroup analyses by adjustment, multicenter status, blinding status and geographic region are presented.

	n	*RR*	*Heterogeneity* I^2, *p*
Overall analysis	33	0.98 (0.91, 1.06)	0.0%, 0.709
Subgroups by adjustment			
Multivariate	13	0.99 (0.91, 1.07)	8.5%, 0.361
Univariate	20	1.05 (0.95, 1.06)	0.0%, 0.803
Subgroups by multicenter status			
Multicenter	22	1.00 (0.95, 1.06)	0.0%, 0.703
Single-center	11	0.95 (0.74, 1.24)	0.0, 0.451
Subgroups by blinding status			
Blinded	15	0.97 (0.82, 1.15)	0.0%, 0.524
Open label	18	1.01 (0.95, 1.07)	0.0%, 0.667
Subgroups by geographic region			
Africa	2	0.96 (0.56, 1.67)	0.0%, 0.509
East Asia	1	0.80 (0.36, 1.76)	Not calculable
Europe	9	0.89 (0.67, 1.19)	0.0%, 0.778
India	4	1.06 (0.82, 1.38)	0.0%, 0.493
Latin America	6	1.10 (0.77, 1.57)	0.0%, 0.623
Middle East	1	0.50 (0.05, 5.04)	Not calculable
USA	6	0.85 (0.55, 1.31)	47.1%, 0.092
More than one area	4	1.02 (0.93, 1.12)	26.8%, 0.251

Highlighted rows denote statistically significant associations.

Table 4. Results of the meta-analyses examining the association between convalescent plasma therapy and hospital discharge (28-day); subgroup analyses by adjustment, multicenter status, blinding status and geographic region are presented.

	n	*RR*	*Heterogeneity* I^2, *p*
Overall analysis	9	0.99 (0.96, 1.03)	0.0%, 0.955
Subgroups by adjustment			
Multivariate	2	0.98 (0.94, 1.03)	0.0%, 0.455
Univariate	7	1.01 (0.96, 1.05)	0.0%, 0.949
Subgroups by multicenter status			
Multicenter	7	1.00 (0.96, 1.03)	0.0%, 0.859
Single-center	2	0.95 (0.74, 1.24)	0.0, 0.451
Subgroups by blinding status			
Blinded	15	0.97 (0.82, 1.15)	0.0%, 0.524
Open label	18	0.98 (0.87, 1.11)	0.0%, 1.000
Subgroups by geographic region			
Africa	1	0.98 (0.72, 1.34)	Not calculable
Europe	1	1.06 (0.87, 1.30)	Not calculable
India	1	1.03 (0.88, 1.21)	Not calculable
Latin America	2	1.00 (0.92, 1.09)	0.0%, 0.652
USA	2	0.99 (0.94, 1.06)	0.0%, 0.529
More than one area	2	1.00 (0.92 1.08)	21.5%, 0.259

Highlighted rows denote statistically significant associations.

Table 5. Results of the meta-analyses examining the association between convalescent plasma therapy and score-related outcomes (28-day); subgroup analyses by adjustment, multicenter status, blinding status and geographic region are presented.

	n	*RR*	*Heterogeneity I^2, p*
Overall analysis	7	1.06 (0.97, 1.16)	17.2%, 0.299
Subgroups by adjustment			
Multivariate	2	1.25 (0.87, 1.78)	61.0%, 0.110
Univariate	5	1.01 (0.94, 1.09)	0.0%, 0.601
Subgroups by multicenter status			
Multicenter	4	1.15 (1.02, 1.29)	0.0%, 0.451
Single-center	3	0.99 (0.91, 1.07)	0.0%, 0.854
Subgroups by blinding status			
Blinded	4	1.12 (0.98, 1.27)	0.0%, 0.394
Open label	3	1.03 (0.92, 1.17)	21.5%, 0.280
Subgroups by geographic region			
Africa	2	0.98 (0.90, 1.07)	0.0%, 0.749
East Asia	1	1.20 (0.82, 1.75)	Not calculable
Europe	1	1.16 (0.91, 1.47)	Not calculable
Latin America	1	1.60 (1.03, 2.49)	Not calculable
USA	1	1.33 (0.37, 4.77)	Not calculable
More than one area	1	1.09 (0.93, 1.27)	Not calculable

Highlighted rows denote statistically significant associations.

Table 6. Results of the meta-analyses examining the association between convalescent plasma therapy and ICU-related outcomes (28-day); subgroup analyses by adjustment, multicenter status, blinding status and geographic region are presented.

	n	*RR*	*Heterogeneity I^2, p*
Overall analysis	20	0.98 (0.93, 1.02)	0.0%, 0.542
Subgroups by adjustment			
Multivariate	3	1.00 (0.86, 1.15)	70.7%, 0.033
Univariate	17	0.98 (0.93, 1.02)	0.0%, 0.542
Subgroups by multicenter status			
Multicenter	14	0.98 (0.93, 1.04)	8.0%, 0.365
Single-center	6	0.84 (0.57, 1.24)	0.0%, 0.703
Subgroups by blinding status			
Blinded	10	0.96 (0.71, 1.31)	8.0%, 0.807
Open label	10	0.99 (0.92, 1.06)	27.4%, 0.192
Subgroups by geographic region			
Africa	1	0.33 (0.04, 2.88)	Not calculable
Europe	7	0.92 (0.85, 0.99)	0.0%, 0.897
India	1	0.88 (0.63, 1.23)	Not calculable
Latin America	3	0.77 (0.46, 1.29)	0.0%, 0.529
Middle East	1	0.67 (0.22, 2.03)	Not calculable
USA	3	0.84 (0.43, 1.64)	0.0%, 0.460
More than one area	4	1.05 (0.96, 1.15)	32.8%, 0.215
Subgroups by ICU-related outcome			
ICU admission	4	0.97 (0.74, 1.26)	0.0%, 0.501
IMV or ECMO or death	1	1.10 (0.98, 1.24)	Not calculable
Intubation or death	1	1.16 (0.94, 1.43)	Not calculable
IMV	1	0.33 (0.04, 2.88)	Not calculable
Invasive ventilatory support	1	0.88 (0.42, 1.83)	Not calculable
MV	1	0.50 (0.09, 2.74)	Not calculable
MV or ICU admission	1	0.75 (0.17, 3.31)	Not calculable
MV or death	1	1.10 (0.62, 1.97)	Not calculable
MV or ECMO	1	0.49 (0.15, 1.58)	Not calculable
NIV or high flow O2 or IMV or ECMO or death	1	0.91 (0.84, 0.99)	Not calculable
PaO2/FiO2 of <150 mm Hg or death	1	0.91 (0.67, 1.23)	Not calculable
Ventilation treatment	6	0.98 (0.90, 1.06)	0.0%, 0.784

Highlighted rows denote statistically significant associations.

Furthermore, a subgroup analysis was conducted according to the levels of anti-SARS-CoV-2 antibodies in the CP. Studies were grouped as "high-titer" or "non-high-titer" as per the latest EUA/FDA guideline cut-offs. The subgroup analysis for the titer level did not reveal any statistically significant associations (Supplementary Figures S5, S10, S14, S20 and S25).

However, when analyzing by geographic region, studies carried out in Europe [RR = 0.92, 95% C.I. (0.85, 0.99)] showed a statistically significant association between convalescent plasma therapy and ICU-related outcomes (Table 6 and Figure 3).

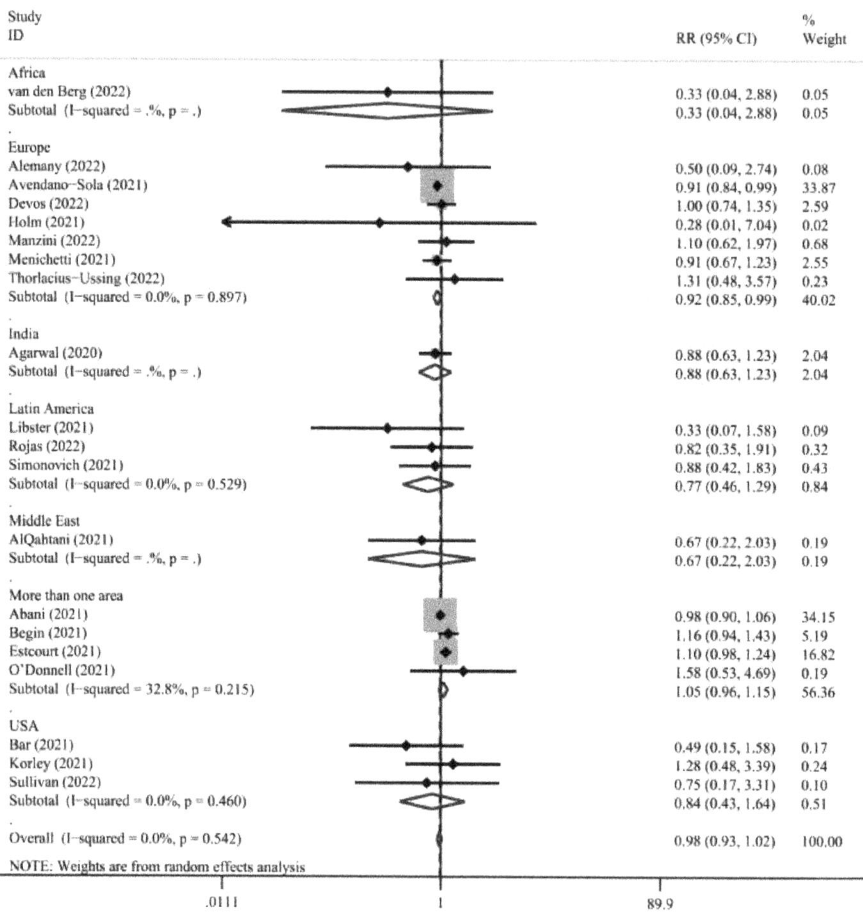

Figure 3. Forest plot describing the association between convalescent plasma treatment and 28-day ICU-related outcomes. Apart from the overall analysis, the subanalysis per geographic region is presented.

The subanalysis on hospitalization outcomes [RR = 0.74, 95% C.I. (0.56, 0.99)] was also statistically significant, showing that outpatients treated with convalescent plasma had a 26% less risk of needing hospital care than those treated with the standard of care (Table 7, Figure 4).

Table 7. Subanalysis on hospitalization (28-day).

	n	*RR*	*Heterogeneity* I^2, *p*
Overall analysis	6	0.74 (0.56, 0.99)	49.8%, 0.076
Subgroups by adjustment			
Multivariate	0	-	-
Univariate	6	0.74 (0.56, 0.99)	49.8%, 0.076
Subgroups by multicenter status			
Multicenter	5	0.72 (0.52, 1.00)	59.6%, 0.042
Single-center	1	0.91 (0.38, 2.17)	Not calculable
Subgroups by blinding status			
Blinded	5	0.72 (0.52, 1.00)	59.6%, 0.042
Open label	1	0.91 (0.38, 2.17)	Not calculable
Subgroups by geographic region			
Africa	1	0.91 (0.38, 2.17)	Not calculable
Europe	2	0.87 (0.51, 1.48)	51.2%, 0.152
Latin America	1	0.52 (0.29, 0.94)	Not calculable
USA	2	0.67 (0.35, 1.27)	75.1%, 0.045

Table 7 Results of the meta-analyses examining the association between convalescent plasma therapy and hospitalization outcomes (28-day); subgroup analyses by adjustment, multicenter status, blinding status and geographic region are presented. Highlighted rows denote statistically significant associations.

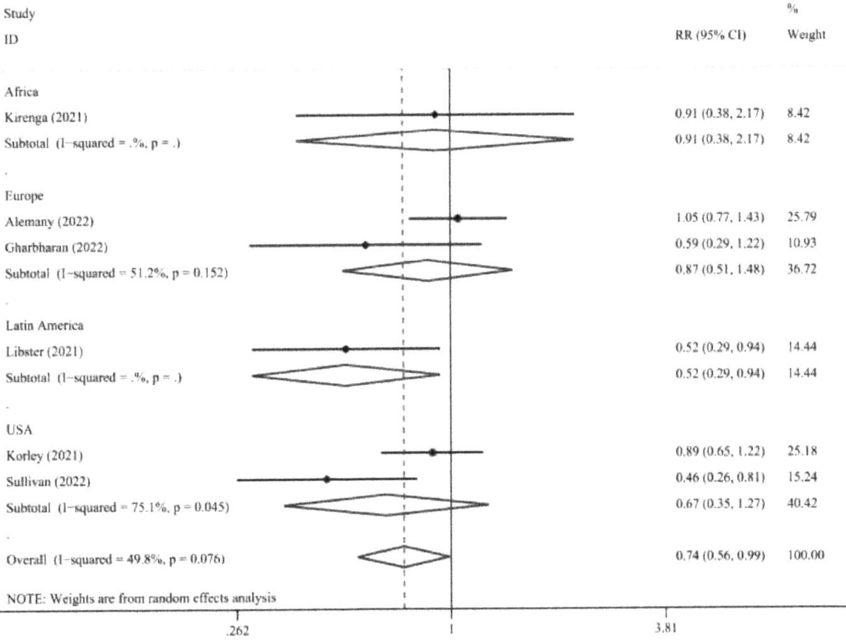

Figure 4. Forest plot describing the association between convalescent plasma treatment and 28-day hospitalization outcomes. Apart from the overall analysis, the subanalysis on geographic region is presented.

3.2.2. 14-Day Results

In total, 10 studies were included in the overall meta-analysis for the 14-day cohort. The effect outcome for 14-day mortality was not statistically significant [RR = 0.98, 95% C.I. (0.91, 1.06)] (Figure 5 and Table 8). There were no statistically significant results in the adjustment, multicenter status, blinding status or geographic region subgroups (Table 8).

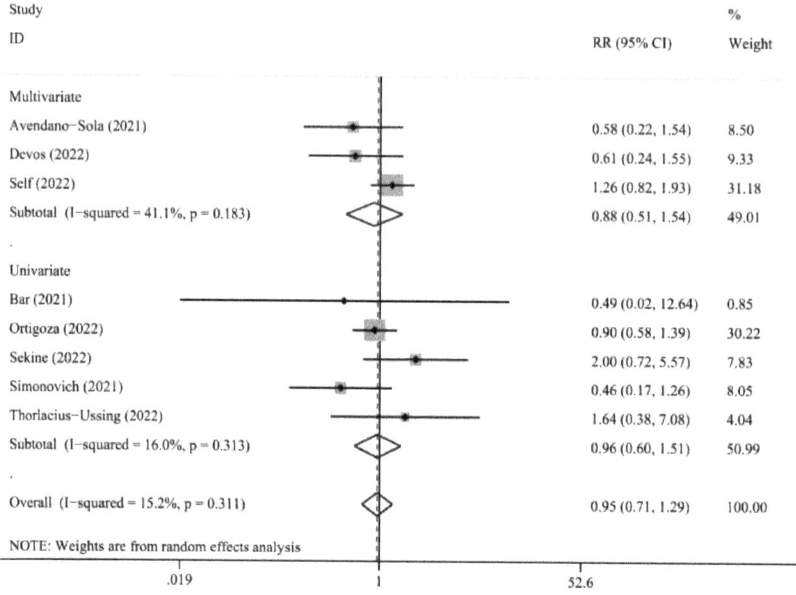

Figure 5. Forest plot describing the association between convalescent plasma treatment and 14-day mortality. Apart from the overall analysis, the subanalysis on adjustment type is presented.

Table 8. Results of the meta-analyses examining the association between convalescent plasma therapy and overall mortality (14-day); subgroup analyses by adjustment, multicenter status, blinding status and geographic region are presented.

	n	RR	Heterogeneity I^2, p
Overall analysis	8	0.95 (0.71, 1.29)	15.2%, 0.311
Subgroups by adjustment			
Multivariate	3	0.88 (0.51, 1.54)	41.1%, 0.183
Univariate	5	0.96 (0.60, 1.51)	6.0%, 0.313
Subgroups by multicenter status			
Multicenter	5	0.98 (0.75, 1.30)	1.4%, 0.398
Single-center	3	0.89 (0.28, 2.83)	51.9%, 0.125
Subgroups by blinding status			
Blinded	4	1.03 (0.68, 1.57)	43.1%, 0.153
Open label	4	0.70 (0.38, 1.28)	0.0%, 0.666
Subgroups by geographic region			
Europe	3	0.71 (0.31, 1.31)	0.0%, 0.467
Latin America	2	0.96 (0.23, 4.04)	75.1%, 0.045
USA	3	1.06 (0.78, 1.44)	0.0%, 0.500

Highlighted rows denote statistically significant associations.

A meta-analysis for the secondary clinical outcomes showed no statistically significant association between convalescent plasma therapy and hospital discharge [RR = 0.96, 95% C.I. (0.89, 1.03)] (Table 9). A subgroup analysis for the titer level was not statistically significant as well (Supplementary Figures S30 and S35).

Table 9. Results of the meta-analyses examining the association between convalescent plasma therapy and hospital discharge (14-day); subgroup analyses by adjustment, multicenter status, blinding status and geographic region are presented.

	n	*RR*	*Heterogeneity* I^2, p
Overall analysis	4	0.96 (0.89, 1.03)	0.0%, 0.995
Subgroups by adjustment			
Multivariate	-	-	-
Univariate	4	0.96 (0.89, 1.03)	0.0%, 0.995
Subgroups by multicenter status			
Multicenter	2	0.96 (0.88, 1.04)	0.0%, 0.795
Single-center	2	0.96 (0.81, 1.14)	0.0%, 0.964
Subgroups by blinding status			
Blinded	3	0.96 (0.88, 1.04)	0.0%, 0.999
Open label	1	0.93 (0.75, 1.16)	Not calculable
Subgroups by geographic region			
Europe	1	0.93 (0.75, 1.16)	Not calculable
Latin America	2	0.96 (0.81, 1.14)	0.0%, 0.964
USA	1	0.96 (0.87, 1.05)	Not calculable

Highlighted rows denote statistically significant associations.

3.3. Meta-Regression Analysis

The post hoc meta-regression aimed to assess whether gender, age, time from symptom onset to intervention or total cp dose modified the association between convalescent plasma treatment and each reported outcome. This analysis yielded no statistically significant associations (Tables 10 and 11). It was carried out only for the 28-day analysis cohort and specifically only for the overall mortality, hospital discharge and ICU-related outcomes, as other categories had less than 10 study arms.

Table 10. Meta-regression on mortality (28-day). Results of meta-regression analysis examining the role of potential modifiers in the association between convalescent plasma treatment and 28-day mortality.

Variables	Increment	*n*	Exponentiated Coefficient	*p*
Male%	10% increase	33	1.06 (0.93, 1.21)	0.368
Mean age	10 y increase	31	0.92 (0.76, 1.12)	0.405
Time from symptom onset to intervention	1 day more	31	1.00 (0.97, 1.04)	0.945
Total CP dose	100 mL more	27	1.01 (0.96, 1.07)	0.691

Table 11. Results of meta-regression analysis examining the role of potential modifiers in the association between convalescent plasma treatment and 28-day ICU-related outcomes.

Variables	Increment	*n*	Exponentiated Coefficient	*p*
Male%	10% increase	20	1.02 (0.85, 1.23)	0.789
Mean age	10 y increase	19	1.07 (0.86, 1.35)	0.514
Time from symptom onset to intervention	1 day more	19	1.03 (0.99, 1.06)	0.157
Total CP dose	100 mL more	18	1.05 (1.00, 1.11)	0.064

3.4. Quality Assessment and Risk of Bias

All included studies were randomized control trials, blinded or open label. For the evaluation of quality and risk of bias of each one, the RoB:2 tool by Cochrane was used [13]. Table 12 presents the risk of bias assessment for the included studies.

Table 12. Risk of bias assessment based on the RoB:2 algorithm.

	Randomization Process	Deviations from Intended Interventions	Missing Outcome Data	Measurement of the Outcome	Selection of the Reported Result	Overall
Abani (2021)	Low risk	Some concerns	Low risk	Low risk	Low risk	Some concerns
Agarwal (2020)	Low risk	Low risk	Low risk	Low risk	Low risk	Low risk
Alemany (2022)	Low risk	Low risk	Low risk	Low risk	Low risk	Low risk
AlQahtani (2021)	Low risk	Low risk	Low risk	Low risk	Low risk	Low risk
Avendaño-Solá (2021)	Low risk	Low risk	Low risk	Low risk	Low risk	Low risk
Bajpai (2020)	Some concerns	Low risk	Low risk	Low risk	Low risk	Some concerns
Bajpai (2022)	Low risk	Low risk	Low risk	Low risk	Low risk	Low risk
Baldeón (2022)	Low risk	Low risk	Low risk	Low risk	Low risk	Low risk
Bar (2021)	Some concerns	Low risk	Low risk	Low risk	Low risk	Some concerns
Bégin (2021)	Low risk	Low risk	Low risk	Low risk	Low risk	Low risk
Bennett-Guerrero (2021)	Low risk	Low risk	Low risk	Low risk	Low risk	Low risk
Denkinger (2022)	Some concerns	Low risk	Low risk	Low risk	Low risk	Some concerns
Devos (2022)	Low risk	Low risk	Low risk	Low risk	Low risk	Low risk
Estcourt (2021)	Low risk	Low risk	Low risk	Low risk	Low risk	Low risk
Gharbharan (2021)	Low risk	Some concerns	Low risk	Low risk	Low risk	Some concerns
Gharbharan (2022)	Low risk	Low risk	Low risk	Low risk	Low risk	Low risk
Holm (2021)	Some concerns	Some concerns	Low risk	Low risk	Low risk	Some concerns
Kirenga (2021)	Low risk	Low risk	Low risk	Low risk	Low risk	Low risk
Korley (2021)	Low risk	Low risk	Low risk	Low risk	Low risk	Low risk
Li (2020)	Some concerns	Low risk	Low risk	Low risk	Low risk	Some concerns
Libster (2021)	Low risk	Low risk	Low risk	Low risk	Low risk	Low risk
Manzini (2022)	Some concerns	Low risk	Low risk	Low risk	Low risk	Some concerns
Menichetti (2021)	Low risk	Low risk	Low risk	Low risk	Low risk	Low risk
O'Donnell (2021)	Low risk	Low risk	Low risk	Low risk	Some concerns	Some concerns
Ortigoza (2022)	Low risk	Low risk	Low risk	Low risk	Low risk	Low risk
Ray (2022)	High risk	Some concerns	Low risk	Low risk	Low risk	High risk
Rojas (2022)	Low risk	Low risk	Low risk	Low risk	Low risk	Low risk
Santis (2022)	Low risk	Some concerns	Some concerns	Low risk	Low risk	Some concerns
Sekine (2022)	Low risk	Low risk	Low risk	Low risk	Low risk	Low risk
Self (2022)	Low risk	Low risk	Low risk	Low risk	Low risk	Low risk
Simonovich (2021)	Low risk	Low risk	Low risk	Low risk	Low risk	Low risk
Sullivan (2021)	Low risk	Low risk	Low risk	Low risk	Low risk	Low risk
Thorlacius-Ussing (2022)	Low risk	Low risk	Low risk	Low risk	Low risk	Low risk
van de Berg (2022)	Low risk	Low risk	Low risk	Low risk	Low risk	Low risk

In total, 23/34 studies (67.7%) were assessed as having a low risk of bias, 10/34 studies (29.4%) raised some concerns and only one was deemed to have a high risk of bias [40]. More specifically:

- Six studies (17.7%) raised some concerns on their randomization process, mostly due to lack of information on allocation concealment;
- Five studies (14.7%) raised some concerns on whether there were deviations from the intended interventions;
- Only one study raised concerns on potential selection of the reported result;
- One study had a high risk of bias due to vital randomization process concerns.

3.5. Publication Bias

A publication bias assessment was performed on outcomes reported in 10 or more studies with the use of Egger's test [14,15]. These were the 28-day mortality and 28-day ICU-related outcomes.

For the 28-day mortality analysis, for a total of 33 studies, the *p*-value for the bias coefficient generated by Egger's regression test for small-study effects was $p = 0.247$ (Supplemental Figure S35). For the 28-day ICU-related outcomes, for a total of 20 studies, the aforementioned *p*-value was $p = 0.337$ (Supplemental Figure S36). In both cases, this means that there were no small-study effects, and thus no publication bias.

4. Discussion

The present meta-analysis, comprising data from 34 individual randomized controlled trials, found no statistically significant association between convalescent plasma treatment and 28-day or 14-day mortality, hospital discharge, hospitalization, ICU-related or score-related outcomes. When analyzing by subgroups, though, the European cohort for the ICU-related outcomes yielded a statistically significant result [RR = 0.92, 95% CI (0.85, 0.99)], showing that convalescent plasma treatment was beneficial in protecting patients from ICU-related disease progression. Specifically, patients treated with convalescent plasma had an 8% less risk of presenting an ICU-related outcome (such as the need for ventilation treatment, intubation, ECMO), when compared to those treated with the standard of care or supportive care (with or without placebo/standard plasma infusion). While this result is interesting and significant, it can largely be attributed to the contribution of the weight of the Avendaño-Solá (2021) study [17]. Moreover, convalescent plasma was found to be beneficial in protecting outpatients from hospitalization. After analyzing hospitalization outcomes, a statistically significant result [RR = 0.74, 95% C.I. (0.56, 0.99)] was yielded, meaning that outpatients treated with convalescent plasma had a 26% lower risk of needing to be hospitalized than those treated with the standard of care.

Carrying out a subgroup analysis for titer levels (high-titer vs. non-high titer) was challenging, as each study used different antibody measurements and cut-off levels for high-titer labeling. Moreover, achieving in-study heterogeneity among the titers of the plasma units administered was also significant. These led to a statistically nonsignificant and mostly inconclusive result. There was a scarcity of outcome data regarding secondary clinical outcomes, such as hospital discharge (9/34 studies), hospitalization (6/34 studies) and score-related outcomes (6/34 studies). The plasma titer between studies varied and so did COVID-19 disease severity at randomization and study size. Serostatus at the time of treatment was not possible to assess and analyze, as only a percentage of studies provided robust and uniform data for it. Furthermore, records were extracted solely from the PubMed database.

In addition, the RECOVERY trial (Abani, 2021) has raised some concerns during our risk of bias assessment and is worth mentioning, as its weight skewed the results. This is due to the fact that it failed to completely adhere to its design, as 9% of the patients did not receive the allocated intervention (plasma infusion). While this raises questions about the robustness of the results, the aforementioned population percentage was excluded from the comparison analysis between the convalescent plasma group and the control group.

Despite the aforementioned notable limitations, the present work possesses a plethora of important strengths. Overall heterogeneity was low and not significant both in the 28-day ($I^2 = 0.0\%$, $p = 0.709$) and 14-day ($I^2 = 15.2\%$, $p = 0.311$) cohorts. In the statistically significant ICU-related European subgroup, heterogeneity was also low and not significant ($I^2 = 0.0\%$, $p = 0.897$). Overall heterogeneity was 49.8% for the hospitalization outcomes subanalysis, but it was marginally not statistically significant ($p = 0.076$). While region, sex, age, time from symptom onset to intervention and total convalescent plasma dose can be considerable sources of heterogeneity, subgroup analyses and meta-regression showed no statistically significant association between them and treatment effectiveness. The extensive abstraction and analysis of separate and discrete clinical outcomes and thorough risk of bias assessment are also parts of this study's strengths. Contrary to other meta-analyses [50–52], our work focuses strictly on randomized controlled trials, thus lying in the highest part of the hierarchy of evidence pyramid.

Moreover, screening was extensive and detailed, pairing information from each trial article and its official registry page. This led to avoiding errors such as misclassifying [50–52] the article record by Rasheed et al. [53] as an RCT, when it was a control-matched cohort study. Furthermore, thorough auditing led to excluding two trials, which were retracted/edited as far as their patient allocation method was concerned.

When comparing our work to others, the results for overall mortality (a commonly reported primary outcome) are similar. Axfors et al. conducted a systematic review and meta-analysis on 33 published and unpublished trial papers and showed a non-statistically significant association as well [54]. Other published meta-analyses were comprised of considerably fewer studies, such as the study by Piscova et al. [50] with five trials and m, the analysis by Snow et al. [51] with seventeen trials and the study by Janiaud et al. including ten trials [55]. The meta-analysis by Kloypan et al. [52] showed a statistically significant association between convalescent plasma therapy and overall mortality but it was subject to notable limitations. The primary outcome of all-cause mortality at any given time point included nonrandomized trials and observational studies, whereas the Rasheed trial was misclassified. Another difference lies in our secondary outcomes analysis, where the aforementioned systematic reviews and meta-analyses failed to yield statistically significant results. This can be attributed to the big pool of studies (and thus variety and data available), outcome assessment and categorization and extensive subgroup analyses.

Finally, subgroup analyses on immunocompromised patients were not feasible due to the scarcity of available data from randomized studies in the field. A recently published randomized controlled trial by Dekinger et al. [49] evaluated the role of convalescent plasma in a subgroup of 56 patients with hematological or solid cancer and severe COVID-19. The administration of convalescent plasma significantly improved survival and reduced the time to clinical improvement. Patients with cancer under active treatment present attenuated humoral responses to COVID-19 vaccination, and thus they are at high risk for severe SARS-CoV-2 infection [56–59]. Other trials on vulnerable populations for severe COVID-19-related outcomes showed signs of benefits with [35] or without statistically significant results [30,33]. A recent systematic review and meta-analysis including trials, cohort studies, case series and case reports found that convalescent plasma therapy was associated with a mortality benefit in patients who were immunocompromised and were diagnosed with COVID-19 [60,61].

5. Conclusions

Convalescent plasma treatment was not associated with a statistically significant reduced risk of overall 28-day or 14-mortality or any other clinical outcome. It was associated, though, with a statistically significant beneficial effect on 28-day ICU-related outcomes in the European study cohort and 28-day hospitalization. The aforementioned evidence hints against the use of convalescent plasma for the treatment of COVID-19 in the general population, but it highlights potential clinical benefits when studying subpopulations (e.g., European ICU cohorts, outpatients). As such, further study on specific subpopulations and outcomes could establish consensus on determining the clinical benefits of convalescent plasma therapy.

Supplementary Materials: The following supporting information can be downloaded at: https://www.mdpi.com/article/10.3390/v15030765/s1. Figure S1. 28-day mortality, by adjustment; Figure S2. 28-day mortality, by multicenter status; Figure S3. 28-day mortality, by blinding status; Figure S4. 28-day mortality, by geographic region; Figure S5. 28-day mortality, by titer; Figure S6. 28-day hospitalization, by adjustment; Figure S7. 28-day hospitalization, by multicenter status; Figure S8. 28-day hospitalization, by blinding status; Figure S9. 28-day hospitalization, by geographic region; Figure S10. 28-day hospitalization, by titer; Figure S11. 28-day hospital discharge, by adjustment; Figure S12. 28-day hospital discharge, by multicenter status; Figure S13. 28-day hospital discharge, by blinding status; Figure S14. 28-day hospital discharge, by geographic region; Figure S15. 28-day hospital discharge, by titer; Figure S16. 28-day ICU-related outcomes, by adjustment; Figure S17. 28-day ICU-related outcomes, by multicenter status; Figure S18. 28-day ICU-related outcomes, by adjustment; Figure S19. 28-day ICU-related outcomes, by geographic region; Figure S20. 28-day ICU-related outcomes, by ICU status; Figure S21. 28-day ICU-related outcomes, by titer; Figure S22. 28-day score-related outcomes, by adjustment; Figure S23. 28-day score-related outcomes, by multicenter status; Figure S24. 28-day score-related outcomes, by blinding status; Figure S25. 28-day score-related outcomes, by geographic region; Figure S26. 28-day score-related outcomes, by titer; Figure S27. 14-day mortality, by adjustment; Figure S28. 14-day mortality, by multicenter status; Figure S29. 14-day mortality, by blinding status; Figure S30. 14-day mortality, by geographic region; Figure S31. 14-day mortality, by titer; Figure S32. 14-day hospital discharge, by adjustment; Figure S33. 14-day hospital discharge, by multicenter status; Figure S34. 14-day hospital discharge, by blinding status; Figure S35. 14-day hospital discharge, by multicenter status; Figure S36. 14-day hospital discharge, by titer; Figure S37. Funnel plot, portraying publication bias for 28-day mortality; Figure S38. Funnel plot, portraying publication bias for 28-day ICU-related outcomes

Author Contributions: Conceptualization, T.P. and E.T.; methodology, T.N.S. and I.N.-S.; software, C.F. and T.N.S.; validation, T.N.S., G.D., M.G. and T.P.; formal analysis, C.F. and T.N.S.; investigation, C.F., I.N.-S., K.S. and A.N.-S.; data curation, C.F., K.S. and A.N.-S.; writing—original draft preparation, C.F. and I.N.-S.; writing—review and editing, K.S., A.N.-S., M.G., T.P., G.D., T.N.S. and E.T.; visualization, T.N.S.; supervision, T.N.S. and E.T. All authors have read and agreed to the published version of the manuscript.

Funding: This research received no external funding.

Institutional Review Board Statement: Not applicable.

Informed Consent Statement: Not applicable.

Data Availability Statement: Data available upon reasonable request from the corresponding author.

Conflicts of Interest: The authors declare no relevant conflict of interest.

References

1. Kenneth McIntosh. COVID-19: Clinical Features. Available online: https://www.uptodate.com/contents/covid-19-clinical-features (accessed on 30 December 2022).
2. Mair-Jenkins, J.; Saavedra-Campos, M.; Baillie, J.K.; Cleary, P.; Khaw, F.M.; Lim, W.S.; Makki, S.; Rooney, K.D.; Nguyen-Van-Tam, J.S.; Beck, C.R.; et al. The Effectiveness of Convalescent Plasma and Hyperimmune Immunoglobulin for the Treatment of Severe Acute Respiratory Infections of Viral Etiology: A Systematic Review and Exploratory Meta-analysis. *J. Infect. Dis.* **2015**, *211*, 80–90. [CrossRef] [PubMed]
3. van Griensven, J.; De Weiggheleire, A.; Delamou, A.; Smith, P.G.; Edwards, T.; Vandekerckhove, P.; Bah, E.I.; Colebunders, R.; Herve, I.; Lazaygues, C.; et al. The Use of Ebola Convalescent Plasma to Treat Ebola Virus Disease in Resource-Constrained Settings: A Perspective from the Field. *Clin. Infect. Dis.* **2016**, *62*, 69–74. [CrossRef]
4. Barone, P.; DeSimone, R.A. Convalescent plasma to treat coronavirus disease 2019 (COVID-19): Considerations for clinical trial design. *Transfusion* **2020**, *60*, 1123–1127. [CrossRef]
5. Rojas, M.; Rodríguez, Y.; Monsalve, D.M.; Acosta-Ampudia, Y.; Camacho, B.; Gallo, J.E.; Rojas-Villarraga, A.; Ramírez-Santana, C.; Díaz-Coronado, J.C.; Manrique, R.; et al. Convalescent plasma in COVID-19: Possible mechanisms of action. *Autoimmun. Rev.* **2020**, *19*, 102554. [CrossRef] [PubMed]
6. Psaltopoulou, T.; Sergentanis, T.N.; Pappa, V.; Politou, M.; Terpos, E.; Tsiodras, S.; Pavlakis, G.N.; Dimopoulos, M.A. The Emerging Role of Convalescent Plasma in the Treatment of COVID-19. *HemaSphere* **2020**, *4*, e409. [CrossRef]
7. Rosati, M.; Terpos, E.; Ntanasis-Stathopoulos, I.; Agarwal, M.; Bear, J.; Burns, R.; Hu, X.; Korompoki, E.; Donohue, D.; Venzon, D.J.; et al. Sequential Analysis of Binding and Neutralizing Antibody in COVID-19 Convalescent Patients at 14 Months After SARS-CoV-2 Infection. *Front. Immunol.* **2021**, *12*, 793953. [CrossRef] [PubMed]
8. WHO. *7th Update of WHO's Living Guidelines on COVID-19 Therapeutics*; WHO: Geneva, Switzerland, 2022.
9. Pappa, V.; Bouchla, A.; Terpos, E.; Thomopoulos, T.P.; Rosati, M.; Stellas, D.; Antoniadou, A.; Mentis, A.; Papageorgiou, S.G.; Politou, M.; et al. A Phase II Study on the Use of Convalescent Plasma for the Treatment of Severe COVID-19—A Propensity Score-Matched Control Analysis. *Microorganisms* **2021**, *9*, 806. [CrossRef]
10. Estcourt, L.; Callum, J. Convalescent Plasma for COVID-19—Making Sense of the Inconsistencies. *N. Engl. J. Med.* **2022**, *386*, 1753–1754. [CrossRef]
11. Sullivan, D.J.; Gebo, K.A.; Hanley, D.F. Correspondence: Convalescent Plasma for COVID-19—Making Sense of the Inconsistencies. *N. Engl. J. Med.* **2022**, *387*, 955–956. [CrossRef]
12. Higgins, J.P.; Thomas, J.; Chandler, J.; Cumpston, M.; Li, T.; Page, M.J.; Welch, V.A. *Cochrane Handbook for Systematic Reviews of Interventions Version 6.3*; Cochrane: Oxford, UK, 2022.
13. Sterne, J.A.; Savović, J.; Page, M.J.; Elbers, R.G.; Blencowe, N.S.; Boutron, I.; Cates, C.J.; Cheng, H.Y.; Corbett, M.S.; Eldridge, S.M.; et al. RoB 2: A revised tool for assessing risk of bias in randomised trials. *BMJ* **2019**, *366*, l4898. [CrossRef]
14. Egger, M.; Smith, G.D.; Schneider, M.; Minder, C. Bias in meta-analysis detected by a simple, graphical test. *BMJ* **1997**, *315*, 629–634. [CrossRef] [PubMed]

15. Sterne, J.A.C.; Egger, M. Funnel plots for detecting bias in meta-analysis: Guidelines on choice of axis. *J. Clin. Epidemiol.* **2001**, *54*, 1046–1055. [CrossRef]
16. RECOVERY Collaborative Group. Convalescent plasma in patients admitted to hospital with COVID-19 (RECOVERY): A randomised controlled, open-label, platform trial. *Lancet* **2021**, *397*, 2049–2059. [CrossRef] [PubMed]
17. Agarwal, A.; Mukherjee, A.; Kumar, G.; Chatterjee, P.; Bhatnagar, T.; Malhotra, P. Convalescent plasma in the management of moderate COVID-19 in adults in India: Open label phase II multicentre randomized controlled trial (PLACID Trial). *BMJ* **2020**, *371*, m3939. [CrossRef] [PubMed]
18. Alemany, A.; Millat-Martinez, P.; Corbacho-Monné, M.; Malchair, P.; Ouchi, D.; Ruiz-Comellas, A.; Ramírez-Morros, A.; Codina, J.R.; Simon, R.A.; Videla, S.; et al. High-titre methylene blue-treated convalescent plasma as an early treatment for outpatients with COVID-19: A randomised, placebo-controlled trial. *Lancet* **2022**, *10*, 278–288. [CrossRef]
19. AlQahtani, M.; Abdulrahman, A.; Almadani, A.; Alali, S.Y.; Al Zamrooni, A.M.; Hejab, A.H.; Conroy, R.M.; Wasif, P.; Otoom, S.; Atkin, S.L.; et al. Randomized controlled trial of convalescent plasma therapy against standard therapy in patients with severe COVID-19 disease. *Sci. Rep.* **2021**, *11*, 9927. [CrossRef] [PubMed]
20. Avendaño-Solá, C.; Ramos-Martínez, A.; Muñez-Rubio, E.; Ruiz-Antorán, B.; de Molina, R.M.; Torres, F.; Fernández-Cruz, A.; Calderón-Parra, J.; Payares-Herrera, C.; de Santiago, A.D.; et al. A multicenter randomized open-label clinical trial for convalescent plasma in patients hospitalized with COVID-19 pneumonia. *J. Clin. Investig.* **2021**, *131*, e152740. [CrossRef] [PubMed]
21. Bajpai, M.; Maheshwari, A.; Kumar, S.; Chhabra, K.; Kale, P.; Narayanan, A.; Gupta, A.; Gupta, E.; Trehanpati, N.; Agarwal, R.; et al. Comparison of safety and efficacy of convalescent plasma with fresh frozen plasma in severe COVID-19 patients. *SciELO* **2020**. [CrossRef] [PubMed]
22. Bajpai, M.; Maheshwari, A.; Dogra, V.; Kumar, S.; Gupta, E.; Kale, P.; Saluja, V.; Thomas, S.S.; Trehanpati, N.; Bihari, C.; et al. Efficacy of convalescent plasma therapy in the patient with COVID-19: A randomised control trial (COPLA-II trial). *BMJ Open* **2022**, *12*, e055189. [CrossRef]
23. Baldeón, M.E.; Maldonado, A.; Ochoa-Andrade, M.; Largo, C.; Pesantez, M.; Herdoiza, M.; Granja, G.; Bonifaz, M.; Espejo, H.; Mora, F.; et al. Effect of convalescent plasma as complementary treatment in patients with moderate COVID-19 infection. *Transfus. Med.* **2022**, *32*, 153–161. [CrossRef]
24. Bar, K.J.; Shaw, P.A.; Choi, G.H.; Aqui, N.; Fesnak, A.; Yang, J.B.; Soto-Calderon, H.; Grajales, L.; Starr, J.; Andronov, M.; et al. A randomized controlled study of convalescent plasma for individuals hospitalized with COVID-19 pneumonia. *J. Clin. Investig.* **2021**, *131*, e155114. [CrossRef]
25. Bégin, P.; Callum, J.; Jamula, E.; Cook, R.; Heddle, N.M.; Tinmouth, A.; Zeller, M.P.; Beaudoin-Bussières, G.; Amorim, L.; Bazin, R.; et al. Convalescent plasma for hospitalized patients with COVID-19: An open-label, randomized controlled trial. *Nat. Med.* **2021**, *27*, 2012–2024. [CrossRef]
26. Bennett-Guerrero, E.; Romeiser, J.L.; Talbot, L.R.; Ahmed, T.; Mamone, L.J.; Singh, S.M.; Hearing, J.C.; Salman, H.; Holiprosad, D.D.; Freedenberg, A.T.; et al. Severe Acute Respiratory Syndrome Coronavirus 2 Convalescent Plasma Versus Standard Plasma in Coronavirus Disease 2019 Infected Hospitalized Patients in New York: A Double-Blind Randomized Trial. *Crit. Care Med.* **2021**, *49*, 1015–1025. [CrossRef]
27. Devos, T.; Van Thillo, Q.; Compernolle, V.; Najdovski, T.; Romano, M.; Dauby, N.; Jadot, L.; Leys, M.; Maillart, E.; Loof, S.; et al. Early high antibody titre convalescent plasma for hospitalised COVID-19 patients: DAWn-plasma. *Eur. Respir. J.* **2022**, *59*, 2101724. [CrossRef]
28. Writing Committee for the REMAP-CAP Investigators. Effect of Convalescent Plasma on Organ Support-Free Days in Critically Ill Patients with COVID-19: A Randomized Clinical Trial. *JAMA* **2021**, *326*, 1690–1702. [CrossRef]
29. Gharbharan, A.; Jordans, C.C.E.; GeurtsvanKessel, C.; Hollander, J.G.D.; Karim, F.; Mollema, F.P.N.; Schukken, J.E.S.; Dofferhoff, A.; Ludwig, I.; Koster, A.; et al. Effects of potent neutralizing antibodies from convalescent plasma in patients hospitalized for severe SARS-CoV-2 infection. *Nat. Commun.* **2021**, *12*, 3189. [CrossRef]
30. Gharbharan, A.; Jordans, C.; Zwaginga, L.; Papageorgiou, G.; van Geloven, N.; van Wijngaarden, P.; Hollander, J.D.; Karim, F.; van Leeuwen-Segarceanu, E.; Soetekouw, R.; et al. Outpatient convalescent plasma therapy for high-risk patients with early COVID-19: A randomized placebo-controlled trial. *Clin. Microbiol. Infect.* **2023**, *29*, 208–214. [CrossRef] [PubMed]
31. Holm, K.; Lundgren, M.N.; Kjeldsen-Kragh, J.; Ljungquist, O.; Böttiger, B.; Wikén, C.; Öberg, J.; Fernström, N.; Rosendal, E.; Överby, A.K.; et al. Convalescence plasma treatment of COVID-19: Results from a prematurely terminated randomized controlled open-label study in Southern Sweden. *BMC Res. Notes* **2021**, *14*, 440. [CrossRef] [PubMed]
32. Kirenga, B.; Byakika-Kibwika, P.; Muttamba, W.; Kayongo, A.; Loryndah, N.O.; Mugenyi, L.; Kiwanuka, N.; Lusiba, J.; Atukunda, A.; Mugume, R.; et al. Efficacy of convalescent plasma for treatment of COVID-19 in Uganda. *BMJ Open Respir. Res.* **2021**, *8*, e001017. [CrossRef] [PubMed]
33. Korley, F.K.; Durkalski-Mauldin, V.; Yeatts, S.D.; Schulman, K.; Davenport, R.D.; Dumont, L.J.; El Kassar, N.; Foster, L.D.; Hah, J.M.; Jaiswal, S.; et al. Early Convalescent Plasma for High-Risk Outpatients with COVID-19. *N. Engl. J. Med.* **2021**, *385*, 1951–1960. [CrossRef]
34. Li, L.; Zhang, W.; Hu, Y.; Tong, X.; Zheng, S.; Yang, J.; Kong, Y.; Ren, L.; Wei, Q.; Mei, H. Effect of Convalescent Plasma Therapy on Time to Clinical Improvement in Patients with Severe and Life-threatening COVID-19: A Randomized Clinical Trial. *JAMA* **2020**, *324*, 460–470. [CrossRef] [PubMed]

35. Libster, R.; Marc, G.P.; Wappner, D.; Coviello, S.; Bianchi, A.; Braem, V.; Esteban, I.; Caballero, M.T.; Wood, C.; Berrueta, M.; et al. Early High-Titer Plasma Therapy to Prevent Severe COVID-19 in Older Adults. *N. Engl. J. Med.* **2021**, *384*, 610–618. [CrossRef]

36. Manzini, P.M.; Ciccone, G.; De Rosa, F.G.; Cavallo, R.; Ghisetti, V.; D'Antico, S.; Galassi, C.; Saccona, F.; Castiglione, A.; Birocco, N.; et al. Convalescent or standard plasma versus standard of care in the treatment of COVID-19 patients with respiratory impairment: Short and long-term effects. A three-arm randomized controlled clinical trial. *BMC Infect. Dis.* **2022**, *22*, 879. [CrossRef]

37. Menichetti, F.; Popoli, P.; Puopolo, M.; Alegiani, S.S.; Tiseo, G.; Bartoloni, A.; De Socio, G.V.; Luchi, S.; Blanc, P.; Puoti, M.; et al. Effect of High-Titer Convalescent Plasma on Progression to Severe Respiratory Failure or Death in Hospitalized Patients with COVID-19 Pneumonia: A Randomized Clinical Trial. *JAMA Netw. Open* **2021**, *4*, e2136246. [CrossRef]

38. O'Donnell, M.R.; Grinsztejn, B.; Cummings, M.J.; Justman, J.E.; Lamb, M.R.; Eckhardt, C.M.; Philip, N.M.; Cheung, Y.K.; Gupta, V.; João, E.; et al. A randomized double-blind controlled trial of convalescent plasma in adults with severe COVID-19. *J. Clin. Investig.* **2021**, *131*, e150646. [CrossRef] [PubMed]

39. Ortigoza, M.B.; Yoon, H.; Goldfeld, K.S.; Troxel, A.B.; Daily, J.P.; Wu, Y.; Li, Y.; Wu, D.; Cobb, G.F.; Baptiste, G.; et al. Efficacy and Safety of COVID-19 Convalescent Plasma in Hospitalized Patients: A Randomized Clinical Trial. *JAMA Intern. Med.* **2022**, *182*, 115–126. [CrossRef] [PubMed]

40. Ray, Y.; Paul, S.R.; Bandopadhyay, P.; D'Rozario, R.; Sarif, J.; Raychaudhuri, D.; Bhowmik, D.; Lahiri, A.; Vasudevan, J.S.; Maurya, R.; et al. A phase 2 single center open label randomised control trial for convalescent plasma therapy in patients with severe COVID-19. *Nat. Commun.* **2022**, *13*, 383. [CrossRef]

41. Rojas, M.; Rodríguez, Y.; Hernández, J.C.; Díaz-Coronado, J.C.; Vergara, J.A.D.; Vélez, V.P.; Mancilla, J.P.; Araujo, I.; Yepes, J.T.; Ricaurte, O.B.; et al. Safety and efficacy of convalescent plasma for severe COVID-19: A randomized, single blinded, parallel, controlled clinical study. *BMC Infect. Dis.* **2022**, *22*, 575. [CrossRef]

42. De Santis, G.C.; Oliveira, L.C.; Garibaldi, P.M.; Almado, C.E.; Croda, J.; Arcanjo, G.G.; Oliveira, A.; Tonacio, A.C.; Langhi, D.M.; Bordin, J.O.; et al. High-Dose Convalescent Plasma for Treatment of Severe COVID-19. *Emerg. Infect. Dis.* **2022**, *28*, 548–555. [CrossRef]

43. Sekine, L.; Arns, B.; Fabro, B.R.; Cipolatt, M.M.; Machado, R.R.G.; Durigon, E.L.; Parolo, E.; Pellegrini, J.A.S.; Viana, M.V.; Schwarz, P.; et al. Convalescent plasma for COVID-19 in hospitalised patients: An open-label, randomized clinical trial. *Eur. Respir. J.* **2022**, *59*, 2101471. [CrossRef]

44. Self, W.H.; Wheeler, A.P.; Stewart, T.G.; Schrager, H.; Mallada, J.; Thomas, C.B.; Cataldo, V.D.; O'Neal, H.R., Jr.; Shapiro, N.I.; Higgins, C.; et al. Neutralizing COVID-19 Convalescent Plasma in Adults Hospitalized with COVID-19 A Blinded, Randomized, Placebo-Controlled Trial. *Chest Infect. Orig. Res.* **2022**, *162*, 982–994.

45. Simonovich, V.A.; Pratx, L.D.B.; Scibona, P.; Beruto, M.V.; Vallone, M.G.; Vázquez, C.; Savoy, N.; Giunta, D.H.; Pérez, L.G.; Sánchez, M.D.L.; et al. A Randomized Trial of Convalescent Plasma in COVID-19 Severe Pneumonia. *N. Engl. J. Med.* **2021**, *384*, 619–629. [CrossRef]

46. Sullivan, D.J.; Gebo, K.A.; Shoham, S.; Bloch, E.M.; Lau, B.; Shenoy, A.G.; Mosnaim, G.S.; Gniadek, T.J.; Fukuta, Y.; Patel, B.; et al. Early Outpatient Treatment for COVID-19 with Convalescent Plasma. *N. Engl. J. Med.* **2022**, *386*, 1700–1711. [CrossRef]

47. Thorlacius-Ussing, L.; Brooks, P.T.; Nielsen, H.; Jensen, B.A.; Wiese, L.; Sækmose, S.G.; Johnsen, S.; Gybel-Brask, M.; Johansen, I.S.; Bruun, M.T.; et al. A randomized placebo-controlled trial of convalescent plasma for adults hospitalized with COVID-19 pneumonia. *Sci. Rep.* **2022**, *12*, 16385. [CrossRef] [PubMed]

48. van den Berg, K.; Glatt, T.N.; Vermeulen, M.; Little, F.; Swanevelder, R.; Barrett, C.; Court, R.; Bremer, M.; Nyoni, C.; Swarts, A.; et al. Convalescent plasma in the treatment of moderate to severe COVID-19 pneumonia: A randomized controlled trial (PROTECT-Patient Trial). *Sci. Rep.* **2022**, *12*, 2552. [CrossRef]

49. Denkinger, C.M.; Janssen, M.; Schäkel, U.; Gall, J.; Leo, A.; Stelmach, P.; Weber, S.F.; Krisam, J.; Baumann, L.; Stermann, J.; et al. Anti-SARS-CoV-2 antibody-containing plasma improves outcome in patients with hematologic or solid cancer and severe COVID-19: A randomized clinical trial. *Nat. Cancer* **2022**. [CrossRef] [PubMed]

50. Piscoya, A.; Ng-Sueng, L.F.; del Riego, A.P.; Cerna-Viacava, R.; Pasupuleti, V.; Thota, P.; Roman, Y.M.; Hernandez, A.V. Efficacy and harms of convalescent plasma for treatment of hospitalized COVID-19 patients: A systematic review and meta-analysis. *Arch. Med. Sci.* **2021**, *17*, 1251–1261. [CrossRef] [PubMed]

51. Snow, T.A.C.; Saleem, N.; Ambler, G.; Nastouli, E.; McCoy, L.E.; Singer, M.; Arulkumaran, N. Convalescent plasma for COVID-19: A meta-analysis, trial sequential analysis, and meta-regression. *Br. J. Anaesth.* **2021**, *127*, 834–844. [CrossRef]

52. Kloypan, C.; Saesong, M.; Sangsuemoon, J.; Chantharit, P.; Mongkhon, P. CONVALESCENT plasma for COVID-19: A meta-analysis of clinical trials and real-world evidence. *Eur. J. Clin. Investig.* **2021**, *51*, e13663. [CrossRef]

53. Rasheed, A.M.; Fatak, D.F.; Hashim, H.A.; Maulood, M.F.; Kabah, K.K.; Almusawi, Y.A.; Abdulamir, A.S. The therapeutic potential of convalescent plasma therapy on treating critically-ill COVID-19 patients residing in respiratory care units in hospitals in Baghdad, Iraq. *Infez. Med.* **2020**, *28*, 357–366.

54. Axfors, C.; Janiaud, P.; Schmitt, A.M.; Hooft, J.V.; Smith, E.R.; Haber, N.A.; Abayomi, A.; Abduljalil, M.; Abdulrahman, A.; Acosta-Ampudia, Y.; et al. Association between convalescent plasma treatment and mortality in COVID-19: A collaborative systematic review and meta-analysis of randomized clinical trials. *BMC Infect. Dis.* **2021**, *21*, 1170. [CrossRef] [PubMed]

55. Janiaud, P.; Axfors, C.; Schmitt, A.M.; Gloy, V.; Ebrahimi, F.; Hepprich, M.; Smith, E.R.; Haber, N.A.; Khanna, N.; Moher, D.; et al. Association of Convalescent Plasma Treatment with Clinical Outcomes in Patients With COVID-19: A Systematic Review and Meta-analysis. *JAMA* **2021**, *325*, 1185–1195. [CrossRef] [PubMed]

56. Rosati, M.; Terpos, E.; Bear, J.; Burns, R.; Devasundaram, S.; Ntanasis-Stathopoulos, I.; Gavriatopoulou, M.; Kastritis, E.; Dimopoulos, M.-A.; Pavlakis, G.N.; et al. Low Spike Antibody Levels and Impaired BA.4/5 Neutralization in Patients with Multiple Myeloma or Waldenstrom's Macroglobulinemia after BNT162b2 Booster Vaccination. *Cancers* **2022**, *14*, 5816. [CrossRef] [PubMed]

57. Terpos, E.; Fotiou, D.; Karalis, V.; Ntanasis-Stathopoulos, I.; Sklirou, A.D.; Gavriatopoulou, M.; Malandrakis, P.; Iconomidou, V.A.; Kastritis, E.; Trougakos, I.P.; et al. SARS-CoV-2 humoral responses following booster BNT162b2 vaccination in patients with B-cell malignancies. *Am. J. Hematol.* **2022**, *97*, 1300–1308. [CrossRef] [PubMed]

58. Ntanasis-Stathopoulos, I.; Karalis, V.; Gavriatopoulou, M.; Malandrakis, P.; Sklirou, A.D.; Eleutherakis-Papaiakovou, E.; Migkou, M.; Roussou, M.; Fotiou, D.; Alexopoulos, H.; et al. Second Booster BNT162b2 Restores SARS-CoV-2 Humoral Response in Patients With Multiple Myeloma, Excluding Those Under Anti-BCMA Therapy. *HemaSphere* **2022**, *6*, e764. [CrossRef]

59. Levine, A.C.; Fukuta, Y.; Huaman, M.; Ou, J.; Meisenberg, B.; Patel, B.; Paxton, J.; Hanley, D.F.; Rijnders, B.; Gharbharan, A.; et al. COVID-19 Convalescent Plasma Outpatient Therapy to Prevent Outpatient Hospitalization: A Meta-analysis of Individual Participant Data From Five Randomized Trials. *medRxiv* **2022**. [CrossRef]

60. Senefeld, J.W.; Franchini, M.; Mengoli, C.; Cruciani, M.; Zani, M.; Gorman, E.K.; Focosi, D.; Casadevall, A.; Joyner, M.J. COVID-19 Convalescent Plasma for the Treatment of Immunocompromised Patients: A Systematic Review and Meta-analysis. *JAMA Netw. Open* **2023**, *6*, e2250647. [CrossRef]

61. Thompson, M.A.; Henderson, J.P.; Shah, P.K.; Rubinstein, S.M.; Joyner, M.J.; Choueiri, T.K.; Flora, D.B.; Griffiths, E.A.; Gulati, A.P.; Hwang, C.; et al. Association of Convalescent Plasma Therapy With Survival in Patients With Hematologic Cancers and COVID-19. *JAMA Oncol.* **2021**, *7*, 1167–1175. [CrossRef]

viruses

MDPI

Article

T Cell Responses Correlate with Self-Reported Disease Severity and Neutralizing Antibody Responses Predict Protection against SARS-CoV-2 Breakthrough Infection

Zhen Zhao [1,†], Attila Kumanovics [2,†], Tanzy Love [3], Stacy E. F. Melanson [4,5], Qing H. Meng [6], Alan H. B. Wu [7], Joesph Wiencek [8], Fred S. Apple [9,10], Caitlin R. Ondracek [11], David D. Koch [12], Robert H. Christenson [13,*] and Yan Victoria Zhang [14,*]

[1] Department of Laboratory Medicine and Pathology, Weill Cornell Medicine, New York, NY 10065, USA
[2] Department of Laboratory Medicine and Pathology, Mayo Clinic, Rochester, MN 55905, USA
[3] Department of Biostatistics and Computational Biology, University of Rochester, Rochester, NY 14642, USA
[4] Department of Pathology, Brigham and Women's Hospital, Boston, MA 02115, USA
[5] Harvard Medical School, Boston, MA 02115, USA
[6] Department of Laboratory Medicine, The University of Texas MD Anderson Cancer Center, Houston, TX 77030, USA
[7] Department of Laboratory Medicine, University of California, San Francisco, CA 94143, USA
[8] Department of Pathology, Microbiology and Immunology, Vanderbilt School of Medicine, Nashville, TN 37240, USA
[9] Department of Laboratory Medicine and Pathology, Hennepin Healthcare/Hennepin County Medical Center, Minneapolis, MN 55415, USA
[10] Hennepin Healthcare Research Institute, Minneapolis, MN 55404, USA
[11] American Association for Clinical Chemistry, Washington, DC 22203, USA
[12] Department of Pathology and Laboratory Medicine, Emory University, Atlanta, GA 30303, USA
[13] Department of Pathology, University of Maryland School of Medicine, 685 W. Baltimore Street, Baltimore, MD 21201, USA
[14] Department of Pathology and Laboratory Medicine, University of Rochester Medical Center, 601 Elmwood Avenue, Box 626, Rochester, NY 14642, USA
[*] Correspondence: rchristenson@umm.edu (R.H.C.); victoria_zhang@urmc.rochester.edu (Y.V.Z.); Tel.: +585-276-4192 (Y.V.Z.)
[†] These authors contributed equally to this work.

Citation: Zhao, Z.; Kumanovics, A.; Love, T.; Melanson, S.E.F.; Meng, Q.H.; Wu, A.H.B.; Wiencek, J.; Apple, F.S.; Ondracek, C.R.; Koch, D.D.; et al. T Cell Responses Correlate with Self-Reported Disease Severity and Neutralizing Antibody Responses Predict Protection against SARS-CoV-2 Breakthrough Infection. *Viruses* 2023, *15*, 709. https://doi.org/10.3390/v15030709

Academic Editor: Jason Yiu Wing KAM

Received: 29 January 2023
Revised: 6 March 2023
Accepted: 7 March 2023
Published: 9 March 2023

Abstract: Objectives: The objective of this prospective study was to investigate the role of adaptive immunity in response to SARS-CoV-2 vaccines. Design and Methods: A cohort of 677 vaccinated individuals participated in a comprehensive survey of their vaccination status and associated side effects, and donated blood to evaluate their adaptive immune responses by neutralizing antibody (NAb) and T cell responses. The cohort then completed a follow-up survey to investigate the occurrence of breakthrough infections. Results: NAb levels were the highest in participants vaccinated with Moderna, followed by Pfizer and Johnson & Johnson. NAb levels decreased with time after vaccination with Pfizer and Johnson & Johnson. T cell responses showed no significant difference among the different vaccines and remained stable up to 10 months after the study period for all vaccine types. In multivariate analyses, NAb responses (<95 U/mL) predicted breakthrough infection, whereas previous infection, the type of vaccine, and T cell responses did not. T cell responses to viral epitopes (<0.120 IU/mL) showed a significant association with the self-reported severity of COVID-19 disease. Conclusion: This study provides evidence that NAb responses to SARS-CoV-2 vaccination correlate with protection against infection, whereas the T cell memory responses may contribute to protection against severe disease but not against infection.

Keywords: COVID-19; SARS-CoV-2; vaccine; T cell response; neutralizing antibody; breakthrough infection

1. Introduction

The COVID-19 pandemic has affected every facet of life across the globe for more than two years. As vaccines become more readily available and vaccination rates improve, both public and scientific communities are eager to learn more about the effectiveness of the various vaccines as objectively measured through humoral and cellular immune responses.

Long-term immunological memory to SARS-CoV-2 vaccines is crucial for the development of population-level immunity. Previous reports on the rapid waning of SARS-CoV-2 antibody levels and the loss of neutralizing capacity against the Delta, Omicron, and other Variants of Concern (VoC) [1,2], as well as the occurrence of minimally to moderately affected vaccine efficacy, have led to questions regarding the efficacy of humoral immunity post natural infection and vaccination [3]. In contrast, functional T cell responses in both frequency and intensity remained robust after one year post natural infection, and at least six months post vaccination [4,5]. SARS-CoV-specific T cells were still detectable 17 years after infection [6]. T cell-mediated immunity is thought to play a critical role in the immune response to SARS-CoV-2 infection, in preventing severe disease after natural infection. However, it is difficult to decipher the level of humoral and cellular immune responses required to protect against infection or severe disease [7]. Despite the reports of largely preserved vaccine-induced T cell response against VoCs [1,2] and minimally to moderately affected vaccine efficacy against VoCs in protecting individuals from severe disease, hospitalization, and death [8–10], the impact of T cell responses against various SARS-CoV-2 outcomes, such as breakthrough infection and severe disease, in vaccinated populations is currently unknown [7].

While many studies have investigated the durability of immune responses to SARS-CoV-2 vaccines, to our knowledge, studies that include large, diverse populations with various vaccines administered for evaluations of both humoral and cellular immune correlates of protection have not been conducted. The objective of this prospective study was to investigate whether neutralizing antibody (NAb) and T cell responses induced by different, widely used SARS-CoV-2 vaccines are durable and protective in a large, diverse cohort, as well as to illustrate the utility of NAb and T cell responses in protecting against SARS-CoV-2 infection and severe disease.

2. Methods

2.1. Study Design

This study was organized by the American Association for Clinical Chemistry (AACC) and included an online survey available between 9 September and 20 October 2021, as well as an onsite blood sample collection from 27 to 30 September 2021, during the AACC annual scientific meeting. AACC members were informed via email and/or social media about the study enrollment. The survey was designed to gather information from laboratory professionals about COVID-19 vaccination and its side effects, and these findings were recently published [11]. A follow-up survey (Supplementary Table S1) was sent to participants four months after the blood collection that took place from 9 to 23 February 2022, in order to gather information on breakthrough infection [12]. The study was approved by the University of Maryland Institutional Review Board and informed consent was obtained for participants.

2.2. Inclusion and Exclusion Criteria

AACC members and conference attendees were invited to participate [11]. The only exclusion criteria were age (<18 years) and pregnancy. Due to the heterogeneity and relatively small sample size, participants who received other or unknown vaccine types and those that had both a previous infection and a booster were not included in the analysis. There were 677 subjects included in the analysis dataset (Figure 1).

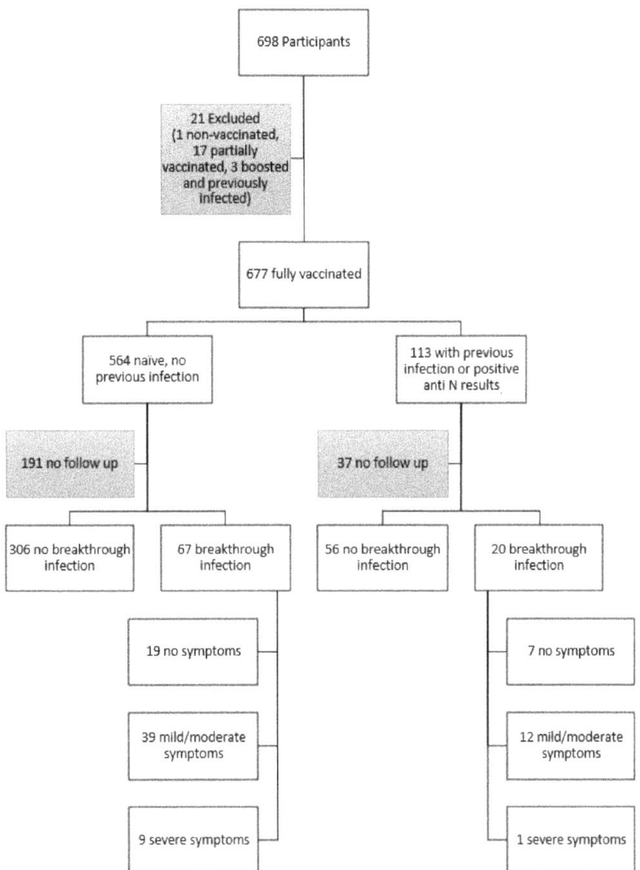

Figure 1. Consort chart for the subjects who had blood collected in September 2021. Breakthrough infections are defined as a follow-up positive COVID test reported in February 2022. Breakthrough infections and their severities are separated into those who had a previous COVID infection, and those who were infection-naïve at the time of blood draw.

2.3. Specimen Collection and Storage

Venous blood was collected into a serum separator tube (BD 368774), fully clotted, and then centrifuged (1500× g for 15 min at room temperature) within two hours of collection. Serum samples were aliquoted, stored at 4 °C at the sample collection site, and then transported to the Centers for Disease Control and Prevention (CDC) within four hours for aliquoting on TECAN (Mannedorf, Switzerland). The samples were stored in 350 µL aliquots at −80 °C until analysis (Eppendorf CryoStorage Vials, 0.5 mL, Mfr. No. 0030079400). Whole blood was collected into a lithium–heparin tube and stored at 4 °C for up to eight hours prior to overnight shipment to the testing clinical laboratory

(Quest Diagnostics, Secaucus, NJ, USA) for the T cell immunity analysis, according to the manufacturer's guidelines [13,14]. The testing was performed at time points consistent with the samples' collection time points across the three-day sample collection period.

2.4. Laboratory Analysis

2.4.1. Anti-SARS-CoV-2 Nucleocapsid (Anti-N) Antibody Assay

The Platelia SARS-CoV-2 Total Ab Enzyme-Linked Immunosorbent Assay (ELISA) assay (Bio-Rad Laboratories, Hercules, CA, USA) is a one-step antigen (Ag)-capture format ELISA, used for the qualitative detection of total anti-N antibodies (IgM/IgG/IgA) in human serum or plasma specimens. This assay received the FDA's Emergency Use Authorization (EUA) [15]. The assay was performed according to the manufacturer's instructions at a clinical laboratory at the University of Maryland.

2.4.2. SARS-CoV-2 NAb Assay

Semi-quantitative detection of SARS-CoV-2 total NAb was performed using the GenScript cPass SARS-CoV-2 Neutralization Antibody Detection Kit (GenScript, Piscataway, NJ, USA), which is a functional ELISA kit that received the FDA's EUA [16]. The assay was performed according to the manufacturer's instructions at Quest Diagnostics (Chantilly, VA, USA). The cutoff value for the cPass SARS-CoV-2 NAb Detection Kit is 47 Units/mL (30% signal inhibition). The test was calibrated for the semi-quantitative detection of anti-SARS-CoV-2 NAb by using the SARS-CoV-2 NAb Calibrator (GenScript, Piscataway, NJ, USA).

2.4.3. QuantiFERON SARS-CoV-2 Assay

The QuantiFERON 2 plate ELISA kit (Qiagen, Hilden, Germany) was used on a Dynex Agility (Dynex Technologies Inc., Chantilly, VA, USA) at Quest Diagnostics (Chantilly, VA, USA). Since this assay was allowed for research use only at the time of analysis, the laboratory performed an internal validation (Supplemental documents). The Limit of detection (LOD) and Limit of quantitation were established as 0.038 IU/mL and 0.061 IU/mL interferon-gamma (INF-γ), respectively. The LOD was applied as the cutoff value for a positive response.

The sample tubes were incubated at 36–38 °C for 16–24 h and centrifuged for 15 min at 2000–3000× g. The steps included the transfer of samples to the QuantiFERON SARS-CoV-2 Starter Set Blood Collection Tubes (QFN SARS-CoV-2 BCTs, consisting of SARS-CoV-2 Ag1 and SARS-CoV-2 Ag2 tubes), the QuantiFERON SARS-CoV-2 Extended Set BCTs (QFN SARS-CoV-2 Extended Set BCT, consisting of SARS-CoV-2 Ag3 tubes), and the QuantiFERON Control Set BCTs (Qiagen, Hilden, Germany), according to the manufacturer's guidelines [13] and as described previously [17]. Following ELISA, background INF-γ levels were subtracted to obtain quantitative results (INF-γ in IU/mL) for analysis. Further details are supplied in the Supplemental Materials.

2.5. Data Analysis

Basic demographic information was compared using descriptive statistics. Previous positive COVID-19 cases were defined based on self-reported positive PCR/Ag tests and/or a positive Platelia SARS-CoV-2 Total Ab ELISA assay (Anti-N antibody assay). Comparisons were made using the Kruskal–Wallis rank sum ANOVA (for continuous data) or Chi-squared tests (for categorical data). With 677 subjects, there was 80% power to compare differences of at least a 0.15 effect size (small–moderate) between vaccine types, boosters, and previous infections.

Linear models were fit, expressing the change in log biomarker values over months since the last dose. Estimates of the change in log biomarker values from the reference (two to four months since last dose) and their significance are reported. Univariate models include only time, since the vaccine dose and multivariate models also include sex, age, and the number of self-reported preexisting conditions as covariates, along with the other multivariate models below. Excluding those boosted or with previous infections, there were 533 remaining subjects, which provides 80% power to detect changes in biomarker values over time that explain at least 3.5% of the variability (small effect size).

Breakthrough infection and its severity was self-reported in the follow-up survey. Comparisons of biomarker levels were made between subjects with different breakthrough infection severities using a one-way ANOVA. The mean biomarker levels for each severity group are reported with univariate and multivariate *p*-value testing for differences; multivariate models include age, sex, time from vaccination to blood being drawn, previous infection status before blood being drawn, and time after blood being drawn to breakthrough infection. The threshold for statistical significance was $\alpha = 0.05$. Logistic regression was fit to predict the appearance of breakthrough infections and, separately, the severity of symptoms from vaccine type or biomarker assays. The prediction thresholds for the biomarker assays were either based on the LOD of the assay or were data-driven. For the data-driven approach, the thresholds were chosen by fitting a receiver operating characteristic (ROC) curve to the assay, to determine the optimal discriminator between those who had breakthrough infections in the follow-up period and between different symptom severity groups. We report the adjusted odds ratios between the two groups, in terms of the chance of each outcome and its significance, for both univariate and multivariate models; multivariate models include age, sex, time from vaccine to blood being drawn, previous infection status before blood being drawn, and time after blood being drawn to breakthrough infection. The 449 subjects who provided survey responses regarding breakthrough infections yielded 80% power to detect changes of at least 1.8 times the odds ratio (small effect size). To compare symptom severity, there were 87 positive cases, which provides 80% power to detect differences in severity of symptoms of at least six times the odds (large effect size).

3. Results

3.1. Participant Demographics and Characterization

Of the 698 people who participated in the on-site study, 677 were fully vaccinated and had completed the survey questionnaire at the time of sample collection. Among the 677 individuals, 564 indicated no previous infection (Pfizer, $n = 314$; Moderna, $n = 181$; Johnson & Johnson (J&J), $n = 38$; and booster, $n = 31$) and 113 reported a previous infection (Table 1). The 31 subjects that had received a booster at the time of sample collection collectively received 28 Pfizer–BioNTech (25 homologous and 3 heterologous), 2 Moderna (homologous), and 1 J&J (heterologous) booster doses. The previous infection group included 46 subjects that were positive for anti-N antibodies, and 52 subjects that were negative for anti-N antibodies; anti-N testing was not performed on the remaining 15 participants (Supplementary Table S2). The group with self-reported previous infection or positive anti-N response was not separated by vaccine type. There was no observable significant difference based on sex, age, race, or ethnicity among the five different groups (Table 1).

Table 1. Participant characterization and demographics.

Demographic [#]		Blood Draw (*n* = 698)	No Known Previous Infection and Negative Anti-N				Self-Reported Previous Infection or Positive Anti-N (*n* = 113)	Positive Anti-N * (*n* = 46)	*p* Value
			Pfizer (*n* = 314)	Moderna (*n* = 181)	Johnson & Johnson (*n* = 38)	Booster (*n* = 31)			
Sex	Male, N (%)	328 (46.9)	141 (44.9)	89 (49.2)	17 (44.7)	12 (38.7)	60 (53.1)	28 (60.9)	0.48
Age	Median (IQR)	50 (40–59)	50 (39–58)	50 (40–60)	49 (38–58)	50 (40–66)	49 (40–58)	49 (40–58)	0.66
Race	Caucasian, N (%)	528 (75.6)	239 (76.1)	137 (75.7)	30 (78.9)	25 (80.6)	89 (78.8)	37 (80.4)	
	African American /Black, N (%)	41 (5.8)	20 (6.4)	10 (5.5)	3 (7.9)	0 (0.0)	6 (5.3)	4 (8.7)	
	Asian, N (%)	67 (9.5)	30 (9.6)	18 (9.9)	3 (7.9)	2 (6.5)	9 (8.0)	4 (8.7)	0.86
	Native Hawaiian /Pacific Islander, N (%)	6 (0.8)	3 (1.0)	3 (1.7)	0 (0.0)	0 (0.0)	0 (0.0)	0 (0.0)	
	Unknown/Other /Prefer Not to Say, N (%)	56 (8.0)	22 (7.0)	13 (7.2)	2 (5.2)	4 (12.9)	9 (8.0)	1 (2.2)	
Ethnicity	Hispanic and/or Latino, N (%)	92 (13.1)	33 (10.5)	22 (12.2)	6 (15.8)	5 (16.1)	22 (19.5)	5 (10.9)	
	Non-Hispanic /Non-Latino, N (%)	575 (82.3)	268 (85.4)	151 (83.4)	30 (78.9)	25 (80.6)	87 (77.0)	38 (82.6)	0.56
	Prefer not to reply, N (%)	31 (4.4)	13 (4.1)	8 (4.4)	2 (5.3)	1 (3.2)	4 (3.5)	3 (6.5)	

* specimens from 653 participants were measured for anti-N antibodies. [#] All blood draws are presented in the first column and 677 non-excluded subjects are separated in the other columns. The *p*-value compares five groups: the three vaccine-only groups, the boosted group, and the vaccinated and previously infected group.

3.2. Follow-Up Survey

Sixty-six percent (449/677) of the fully vaccinated individuals (two-doses of a primary-series COVID-19 vaccine, or one dose of a single-dose, primary-series COVID-19 vaccine approved or authorized for use in the United States) participated in the follow-up February 2022 survey, of which, 66.1% (373/564) of participants in the infection-naïve group and 67.3% (76/113) of participants with previous infections completed the follow-up survey (Figure 1). In the follow-up survey, participants were asked if they tested positive after the blood draw; 18.0% (67/373) in the naïve group and 26.3% (20/76) in the previously infected group reported having breakthrough infections (positive Ag or PCR test). There was no significant difference in the rate of breakthrough infections between those who were infection-naïve vaccinated or previously infected and those who were vaccinated at the blood draw ($p = 0.129$). Among the breakthrough infections, 28.4%, 58.2%, and 13.4% (19, 39, and 9 out of 67) in the infection-naïve vaccinated group and 35.0%, 60.0% and 5.0% (7, 12, and 1 out of 20) in the previously infected and vaccinated group reported having infections with symptoms that caused no, mild/moderate, and severe limitations, respectively (Figure 1). The severity of the symptoms was based on definitions from the CDC (Vaccine Adverse Event Reporting System (VAERS), VAERS | Vaccine Safety | CDC) [18]. There was no significant difference in the severity of breakthrough infections between those who were infection-naïve and vaccinated or those who were previously infected and vaccinated at the blood draw ($p = 0.552$).

3.3. Post-Vaccination NAb and T Cell Responses at the Time of Sample Collection

NAb and T cell response were evaluated in the infection-naïve participants. Due to the small number of participants who had received a vaccine booster at that time (Table 1), they were not included in the analyses. In participants who were infection-naïve at the time of blood draw, 531 (99.6%) and 527 (98.7%) samples were measured for NAb and T cell responses, respectively. There was a poor correlation ($r = 0.19$–0.22) between T cell and NAb responses. Two to four months since the last dose of vaccination (PLDV) was used as a reference time window, due to the small number of individuals in the zero to two month window. At two to four months PLDV, levels of the measured assays were not significantly different in between individuals who received Pfizer and Moderna ($p = 0.557$, 0.342, 0.255, 0.199, 0.159 for the five different measures). However, those who received the J&J vaccine had significantly lower NAb levels ($p < 0.001$) (Supplementary Table S4). Compared to two to four months PLDV, log NAb levels were significantly decreased over time in participants after receiving Pfizer, and were moderately decreased after receiving J&J, but not after Moderna (Figure 2 and Supplementary Table S3). In contrast, the T cell responses to Ag1, Ag2, Ag3, and the sum of Ag 1–3 were not significantly different up to 10 months PLDV between Pfizer, Moderna, and J&J vaccines, except for sum of Ag 1–3 and Ag3, which were significantly higher at zero to two months PLDV compared to two to four months PLDV for Pfizer and/or J&J ($p < 0.05$). There was no significant difference for either the sum of Ag 1–3 or Ag3 after two to four months PLDV during the study period (Figure 2 and Supplementary Table S3). Similar results were also observed in multivariate models adjusted for sex, age, and the number of self-reported pre-existing conditions (Supplementary Table S3). In the models used to assess NAb responses, age is associated with a significant decrease in response ($p < 0.001$), while age is not a significant predictor of the T cell responses to Ag 1–3 ($p = 0.255$) (Supplementary Figure S1).

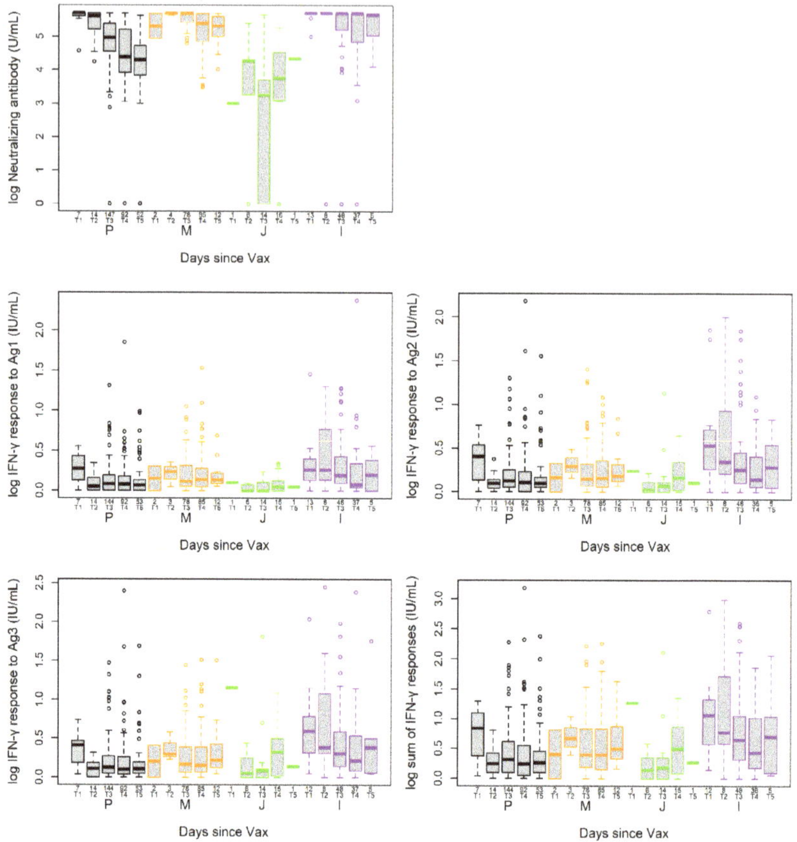

Figure 2. Distributions for T cell and NAb responses in participants receiving different SARS-CoV-2 vaccines. Each box shows the IQR with the median as the interior bar; the outer bars stretch out to the minimum and maximum, with the exception that points outside the fence are shown as individual circles. P: Pfizer; M: Moderna; J: Johnson & Johnson; I: previous infection or positive anti-N; T1: 0–2 months; T2: 2–4 months; T3: 4–6 months; T4: 6–8 months; T5: 8–10.5 months. Numbers indicated the time period indicators (e.g., T1–T5) are the number of participants for that group.

3.4. NAb and T Cell Responses and Breakthrough Infection

Among the participants (*n* = 449) who answered the follow up survey questions, 19.4% (87/449) reported breakthrough infections with positive Ag or PCR tests. All of these individuals were vaccinated, but most (430/449) had only completed the primary vaccination series without booster shots at the time of blood draw.

T cell and Nab responses were evaluated between participants with or without breakthrough infection via univariate and multivariate analysis. No significant differences were detected between the two groups for these two assays (Table 2). Furthermore, no significant differences in these assays were observed comparing participants with breakthrough infections having different severity of symptoms (no, mild/moderate, severe) (Table 2, Supplementary Figure S2).

Table 2. Median (IQR) values for the measured assays for participants with or without breakthrough infections including all vaccination and previous infection statuses.

449 of the 677 Completed the Follow-Up Questionnaire	No Breakthrough Infection (N = 362)	Breakthrough Infection (N = 87)	Among Those with Breakthrough			Uni * p Value	Multi p Value
			No Limitation (n = 26)	Mild-Moderate Limitations (n = 51)	Severe Limitations (n = 10)		
NAb (%)	88 (69–95)	83 (57–96)	87.5 (70.5–97)	82 (55.5–93.5)	72 (53–85.8)	0.475	0.405
NAb (UrnL)	194.9 (86.3–300)	151.4 (58.7–300)	185.7 (74.8–300)	148.4 (50.4–285.8)	91.6 (46.9–198.9)	0.285	0.197
Ag1, median (IQR)	0.11 (0.04–0.30)	0.11 (0.03–0.30)	0.09 (0.02–0.28)	0.14 (0.04–0.30)	0.11 (0.01–0.41)	0.547	0.572
Ag2, median (IQR)	0.16 (0.06–0.40)	0.16 (0.05–0.45)	0.15 (0.05–0.36)	0.20 (0.05–0.47)	0.08 (0.01–0.19)	0.629	0.621
Ag3, median (IQR)	0.17 (0.06–0.50)	0.18 (0.06–0.48)	0.17 (0.05–0.51)	0.19 (0.07–0.48)	0.17 (0.01–0.29)	0.873	0.874
Sum, median (IQR)	0.46 (0.16–1.24)	0.51 (0.12–1.10)	0.47 (0.11–1.05)	0.54 (0.15–1.11)	0.37 (0.04–0.99)	0.766	0.762

NAb = Neutralizing Antibody, Ag1–3 = T cell response to Ags 1–3 (IU/mL), Sag = sum of T cell responses to Ags1–3 (IU/mL). * The univariate p-value compares assay levels among different degrees of limitation during infection. The multivariate p-value compares the same three groups, while adjusting for age, sex, time from vaccine to blood draw, previous infection status before blood draw, time after blood draw to breakthrough infection.

Logistic regression was fit to predict the breakthrough infections and severity of symptoms. In the univariate model, those with NAb < 95 U/mL had 78% ($p = 0.020$) higher odds of having breakthrough infections at the time of the follow-up survey. Participants with T cell responses to Ag2 > 0.120 IU/mL had 13% ($p = 0.045$) lower odds of having severe limitations if they contracted a COVID-19 breakthrough infection at the time of the follow-up survey (Table 3). There was no significant difference in the prediction of breakthrough infections based on T cell responses. Likewise, there were no significant differences in the prediction of breakthrough infection severity based on NAb, T cell responses to Ag1, Ag3 or sum Ag1–3. The above findings were also confirmed using the multivariate models (Table 3). Furthermore, in the multivariate model, participants with T cell responses to Ag2 > 0.120 IU/mL had 31% higher odds of having mild-moderate limitations ($p = 0.022$).

Table 3. Risk of breakthrough infections using thresholds chosen based on ROC analysis.

			Univariate			
COVID-19	n	NAb (>95 U/mL)	sum Ag (>0.385 IU/mL)	Ag1 (>0.435 IU/mL)	Ag2 (>0.120 IU/mL)	Ag3 (>0.055 IU/mL)
No breakthrough	362	**1.78**	0.88	1.15	0.89	1.22
No limitations	26	1.07	0.95	0.84	0.93	0.90
Mild-moderate limitations	51	1.03	1.15	1.04	1.23	1.22
Severe limitations	10	0.91	0.92	1.14	**0.87**	0.91
Hospitalizations	0	-	-	-	-	-
			Multivariate			
COVID-19	n	NAb (>95 U/mL)	sum Ag (>0.385 IU/mL)	Ag1 (>0.435 IU/mL)	Ag2 (>0.120 IU/mL)	Ag3 (>0.055 IU/mL)
No breakthrough	362	**1.95**	0.91	1.21	0.92	1.27
No limitations	26	1.04	0.91	0.85	0.90	0.88
Mild-moderate limitations	51	1.08	1.23	1.03	**1.31**	1.28
Severe limitations	10	0.90	0.89	1.14	**0.85**	0.89
Hospitalizations	0	-	-	-	-	-

Multivariate models also adjust for age, sex, time since last vaccine of blood draw, previous infection status at blood draw, and time between blood draw and infection. Values are (adjusted) odds ratios for the event in the second column. Significant changes from the reference time are in bold with $p < 0.05$.

In both the univariate and multivariate models, no statistically significant differences were observed between the odds of breakthrough infection or infection severity when the LOD of each assay was used as the threshold for the prediction model (Supplementary Table S5). No statistically significant differences were observed between the odds of breakthrough infection or infection severity when comparing different vaccines (Supplementary Table S5). The risk of breakthrough infections, based on the status of self-reported previous infections and/or positive anti-N results, was also analyzed. In univariate models, participants with negative anti-N results had 57% ($p = 0.031$) lower odds of having a breakthrough infection. However, self-reported previous infection alone or combined with anti-N were not significant predictors of breakthrough and the multivariate models did not confirm the significant finding about anti-N.

4. Discussion

The AACC 2021 Annual Scientific Meeting & Clinical Lab Expo in September provided a useful opportunity to examine post-vaccine SARS-CoV-2 immune responses in a large cohort of volunteers that were diverse in areas such as their age, gender, race, ethnicity, vaccine types administered, and geography [11]. In this prospective study, we demonstrated that T cell responses to SARS-CoV-2 were uniformly durable up to 10 months during the study period, whereas the durability of NAb responses varied according to vaccine type. NAb responses to SARS-CoV-2 vaccination correlated with protection, and T cell memory responses may contribute to the protection against severe disease, but not against infection.

Most of the previous studies and vaccine design/development strategies focused on maximizing the SARS-CoV-2 vaccine-induced humoral responses, particularly NAb [19]. The evidence for a protective role of NAb in SARS-CoV-2 infection includes preclinical animal models, passive immunization, first with convalescent patient sera and then with therapeutic monoclonal antibody preparations [20,21], and, finally, population studies on patients who have recovered from infection, or vaccination studies [19,22]. The lack of standardization of NAb assays, however, makes the comparison of these studies difficult and there is no consensus on the protective neutralization level against COVID-19 [19]. In addition, the decline in NAb levels raises concerns about the long-term effectiveness of SARS-CoV-2 vaccination.

T cell-mediated immunity may be just as important in the long-lasting immune response to SARS-CoV-2 infection and vaccination as humoral responses. Virus-specific T cells were shown to be correlates of protective immunity against respiratory viruses such as the cause of the 2009 pandemic, H1N1 influenza [23]. T cell-mediated immunity also plays a role in the immune response to SARS-CoV-2 infection in preventing severe disease after natural infection [7]. In patients with humoral immune deficiency due to either genetic disease or undergoing B cell-depleting therapies, the value of T cell-mediated immunity is also clearly demonstrated [24,25]. T cell responses in resolving primary SARS-CoV-2 infection are also shown in experimental animal (mouse and macaque) infection models [20]. However, the contribution of T cells is difficult to analyze for both practical and conceptual reasons. First, there is no single way to measure T cell responses [7]. Second, on the practical side, most T cell assays require live cells, which often necessitates the isolation and storage of cells. Peripheral blood mononuclear cell isolation and cryopreservation introduces additional complexity and variability to the results. Because of these challenges, most large studies have omitted T cell testing, or it has been undertaken on only a small subset of samples. We utilized a recently available whole-blood interferon (IFN)-γ release assay for SARS-CoV-2, and evaluated a large cohort of participants.

Our results confirmed previous findings [8,23,26–28] that the level of NAb gradually declines after vaccination (especially after Pfizer), whereas T cell responses were sustained up to 10–12 months post-vaccination for the three vaccine types. We also confirmed that NAb responses correlated with protection against infection but not disease severity. We also demonstrated that the T cell immunity responses did not predict breakthrough infection, but T cell responses to Ag2 indicated significant protection against self-reported severe COVID-19 disease. Protection against severe disease can only be detected using stimulants in QFN SARS-CoV-2 Ag2, but not Ag1 and 3. The Ag1 tube contains CD4+ T cell epitopes derived from the S1 subunit of the Spike protein, whereas the Ag2 tube contains both CD4+ and CD8+ T cell epitopes from the S1 and S2 subunits of the Spike protein. The Ag3 tube consists of CD4+ and CD8+ T cell epitopes from S1 and S2, plus immunodominant CD8+ T cell epitopes derived from whole genome. Although the exact composition of antigenic peptides is not available from the manufacturer, the main difference between Ag1 and 2 is the inclusion of CD8 epitopes, suggesting that detecting CD8 responses is the critical distinction. Developing a NAb response requires coordinated T and B cell responses. Helper CD4+ T cell responses are required for most NAb responses, therefore, NAb responses may partially indicate proper CD4 responses ("CD4 T helper" responses), but the cytotoxic CD8+ responses must be directly tested. Our results suggest that measuring these CD8 responses provides additional predictive information regarding disease severity. This finding is supported by experimental studies in macaques [29].

Translating these population-level risks to individual patients will remain challenging, however, as the outcome of SARS-CoV 2 infection depends on numerous individual host and pathogen variables. Viral variants are clearly critical, as was demonstrated by the emergence of Delta, Omicron, and other forms of SARS-CoV-2 that are different from the original strain. Although the T cell assay in the present study utilized the specific Ags derived from the full genome of the ancestral SARS-CoV-2 virus, but not of the VoCs, unlike serological response (and assays), the T cell responses are inherently resistant to viral mutations [3]. Accordingly, while mutations that escape antibody binding have commonly been documented for respiratory viruses such as influenza, and now for SARS-CoV-2, complete immune escape at the level of T cell responses has not been reported, to our knowledge, for any human acute respiratory infection. An additional feature of T cell biology is that whereas circulating preformed antibodies can directly and immediately bind to the virus for neutralization, T cell activation requires the processing of endogenously synthesized viral proteins, presented by the HLA molecules. This suggests that T cells may not prevent infection but may play a pivotal role in reducing viral load and lowering pathogenicity by eliminating the infected cells. Previous studies have shown that the majority of T cell responses are preserved against VoCs, suggesting that memory T cells may play a critical role against infection in light of the considerable escape of Omicron from NAb. It is congruent with our findings that T cell responses were associated with, and could predict the risk of, severe breakthrough infection, but not the prevention of the breakthrough infection. Studying whether SARS-CoV-2-specific T cell responses are more relevant to protection against or the prevention of severe diseases will provide important insights for creating better correlative studies investigating vaccine effectiveness, as well as will help define future vaccine strategies [7].

Our study has several limitations. First, the study only involved a single time point of sample collection, and the time of blood collection was not coordinated with the time of vaccination. However, the large sample size gave enough statistical power to perform longitudinal and multivariate analyses. Second, studying disease severity in contrast to infection rate requires a larger sample size [7] due to the low proportion of severe diseases. None of the study participants who reported a breakthrough infection required hospitalization. Third, the SARS-CoV-2 variant information of the breakthrough infection was not determined. However, since the breakthrough infections in our participants occurred during the time period of 9–23 February 2022, we speculate that the majority of those infections were caused by the Delta and Omicron variants [30]. Fourth, the assays used in this study targeted the original SARS-CoV-2 but not VoCs. Targeted assays may produce more specific results, garnering better associations with the clinical outcomes. In this study, the assays used tested stimulation-based responses rather than in vivo responses. Fifth, the definitions of breakthrough infection and limitations were self-reported in the follow-up survey, which could be subjective. Sixth, NAb responses were measured using a surrogate neutralization assay, and only measured the NAb binding to RBD; therefore, the breadth of responses was not measured. Finally, the anti-N serological responses may decline as well after natural infection, potentially decreasing the overlap between self-reported results and serological testing.

Despite its limitations, our study proposes laboratory correlates for breakthrough infection and self-reported disease severity after SARS-CoV-2 vaccination. We showed that NAb responses can serve as predictors of breakthrough infection, whereas the T cell assay may aid in the prediction of breakthrough disease severity after SARS-CoV-2 vaccination.

Supplementary Materials: The following supporting information can be downloaded at: https://www.mdpi.com/article/10.3390/v15030709/s1, Table S1: Participants who reported previous infection and/or anti-N positive; Table S2: Change in log-Biomarker over time for each vaccine type for individuals who are infection naïve and not boosted; Table S3: Change in log-Biomarker at 2–4 months after complete vaccination between vaccine types for individuals who are infection naïve and not boosted; Table S4: Risk of breakthrough infections by vaccine type or with positive (measured above LOD) test reported; Table S5: Risk of breakthrough infections based on the status of self-reported previous infection (A) and/or positive anti-N results (B); Figure S1: Marginal effect of age on NAb responses and sum of T cell responses to Antigen 1-3; Figure S2: T Cell responses or levels of NAb levels in those with breakthrough infections versus months after blood draw when breakthrough infected.

Author Contributions: Z.Z. and A.K. contributed equally to this paper and are joint first authors. R.H.C. and Y.V.Z. contributed equally to this paper and are joint corresponding authors. T.L. was responsible for all the statistical analysis and all other members participated in the design of the study. C.R.O. managed the data and helped with the references. Z.Z. and A.K. wrote the first draft of the manuscript. All authors provided comments for the draft. The corresponding author attests that all listed authors meet authorship criteria and that no others meeting the criteria have been omitted. All authors have read and agreed to the published version of the manuscript.

Funding: This study was supported by the American Association for Clinical Chemistry as the COVID-19 Immunity Study and its corporate sponsors.

Institutional Review Board Statement: The study was conducted in accordance with the Declaration of Helsinki, and approved by the Institutional Review Board (or Ethics Committee) of University of Maryland Baltimore (protocol code HM-HP-00097761-4 and date of approval of 6/16/2022).

Informed Consent Statement: Informed consent was obtained from all subjects involved in the study.

Data Availability Statement: Data is available upon request.

Acknowledgments: We would like to acknowledge Loretta Doan, Hubert Vesper, Janetta Bryksin, Karen Schulz, all other AACC staff and volunteer members for their contributions to the design and execution of the COVID-19 Immunity Study. We would also like to acknowledge the support for a statistician from the Department of Pathology and Laboratory Medicine at the University of Rochester Medical Center.

Conflicts of Interest: F.S.A. declares Consultant with HyTest Ltd., AWE Medical Group; Associate Editor for Clinical Chemistry; Advisory Boards for Werfen, Siemens Healthineers, Qorvo, Abbott Vascular; Honorarium for Speaking at Industry Conferences: Siemens Healthineers, Beckman Coulter; PI on Industry Funded Grants (non-salaried) on cardiac biomarkers through Hennepin Healthcare Research Institute: Abbott Diagnostics, Abbott POC, BD, Beckman Coulter, Ortho-Clinical Diagnostics, Roche Diagnostics, Siemens Healthcare, ET Healthcare, Qorvo. J.W. has two pending patents with Provisional Patent Application Serial No. 63/015,215 and PCT/US2021/029187. R.H.C. is a scientific advisor for and receives honoraria from the following Siemens Healthineers, Beckman Coulter, Roche Diagnostics, Becton Dickinson, Quidel, and Sphingotech. Z.Z. has sponsored research supported by Novartis, Waters, Siemens, Polymedco, Waters, Roche and ET Healthcare and has received consulting/speaker fee from Siemens, Roche and ET Healthcare. None of these associations were involved in this manuscript. No conflict of interests or personal relationship declared from other authors that are related to the report in this paper.

Abbreviations

American Association for Clinical Chemistry (AACC), Antigen (Ag), Emergency Use Authorization (EUA), Enzyme-Linked Immunosorbent Assay (ELISA), Human leukocyte antigen (HLA), Interferon (IFN), Limit of Detection (LOD), Neutralization Antibody (NAb), Post the last dose of vaccination (PLDV), QuantiFERON (QFN), Receiver operating characteristic curve (ROC), Severe acute respiratory syndrome coronavirus 2 (SARS-CoV-2), Vaccine Adverse Event Reporting System (VAERS), Variants of Concern (VoC).

References

1. Keeton, R.; Tincho, M.B.; Ngomti, A.; Baguma, R.; Benede, N.; Suzuki, A.; Khan, K.; Cele, S.; Bernstein, M.; Karim, F.; et al. T cell responses to SARS-CoV-2 spike cross-recognize Omicron. *Nature* **2022**, *603*, 488–492. [CrossRef]
2. Gao, Y.; Cai, C.; Grifoni, A.; Müller, T.R.; Niessl, J.; Olofsson, A.; Humbert, M.; Hansson, L.; Österborg, A.; Bergman, P.; et al. Ancestral SARS-CoV-2-specific T cells cross-recognize the Omicron variant. *Nat. Med.* **2022**, *28*, 472–476. [CrossRef]
3. Vardhana, S.; Baldo, L.; Morice, W.G.; Wherry, E.J. Understanding T cell responses to COVID-19 is essential for informing public health strategies. *Sci. Immunol.* **2022**, *7*, eabo1303. [CrossRef]
4. Feng, C.; Shi, J.; Fan, Q.; Wang, Y.; Huang, H.; Chen, F.; Tang, G.; Li, Y.; Li, P.; Li, J.; et al. Protective humoral and cellular immune responses to SARS-CoV-2 persist up to 1 year after recovery. *Nat. Commun.* **2021**, *12*, 4984. [CrossRef] [PubMed]

5.   Hou, H.; Zhang, Y.; Tang, G.; Luo, Y.; Liu, W.; Cheng, C.; Jiang, Y.; Xiong, Z.; Wu, S.; Sun, Z.; et al. Immunologic memory to SARS-CoV-2 in convalescent COVID-19 patients at 1 year postinfection. *J. Allergy Clin. Immunol.* **2021**, *148*, 1481–1492. [CrossRef] [PubMed]

6.   Le Bert, N.; Tan, A.T.; Kunasegaran, K.; Tham, C.Y.L.; Hafezi, M.; Chia, A.; Chng, M.H.Y.; Lin, M.; Tan, N.; Linster, M.; et al. SARS-CoV-2-specific T cell immunity in cases of COVID-19 and SARS, and uninfected controls. *Nature* **2020**, *584*, 457–462. [CrossRef]

7.   Kent, S.J.; Khoury, D.S.; Reynaldi, A.; Juno, J.A.; Wheatley, A.K.; Stadler, E.; Wherry, E.J.; Triccas, J.; Sasson, S.C.; Cromer, D.; et al. Disentangling the relative importance of T cell responses in COVID-19: Leading actors or supporting cast? *Nat. Rev. Immunol.* **2022**, *22*, 387–397. [CrossRef]

8.   Tartof, S.Y.; Slezak, J.M.; Puzniak, L.; Hong, V.; Frankland, T.B.; Ackerson, B.K.; Takhar, H.S.; Ogun, O.A.; Simmons, S.R.; Zamparo, J.M.; et al. Effectiveness of a third dose of BNT162b2 mRNA COVID-19 vaccine in a large US health system: A retrospective cohort study. *Lancet Reg. Health-Am.* **2022**, *9*, 100198. [CrossRef]

9.   Pfizer and BioNTech Provide Update on Omicron Variant | Pfizer. Available online: https://www.pfizer.com/news/press-release/press-release-detail/pfizer-and-biontech-provide-update-omicron-variant (accessed on 18 October 2022).

10. Danza, P.; Koo, T.H.; Haddix, M.; Fisher, R.; Traub, E.; Oyong, K.; Balter, S. SARS-CoV-2 Infection and Hospitalization Among Adults Aged ≥18 Years, by Vaccination Status, Before and During SARS-CoV-2 B.1.1.529 (Omicron) Variant Predominance—Los Angeles County, California, 7 November 2021–8 January 2022. *MMWR. Morb. Mortal. Wkly. Rep.* **2022**, *71*, 177–181. [CrossRef]

11. Melanson, S.E.F.; Zhao, Z.; Kumanovics, A.; Love, T.; Meng, Q.H.; Wu, A.H.B.; Apple, F.; Ondracek, C.R.; Schulz, K.M.; Wiencek, J.R.; et al. Tolerance for three commonly administered COVID-19 vaccines by healthcare professionals. *Front. Public Health* **2022**, *10*, 975781. [CrossRef]

12. Vaccine Adverse Event Reporting System (VAERS). Available online: https://vaers.hhs.gov/index.html (accessed on 18 October 2022).

13. QuantiFERON ELISA Instructions for Use (Handbook)—QIAGEN. Available online: https://www.qiagen.com/us/resources/resourcedetail?id=0eb3ad62-171e-4106-af9a-fd28d763ff66&lang=en (accessed on 18 October 2022).

14. QuantiFERON SARS-CoV-2 Extended Set Blood Collection Tubes Instructions for Use (Handbook)—QIAGEN. Available online: https://www.qiagen.com/us/resources/resourcedetail?id=60729925-a7b2-4cf3-8d28-52e4106af16b&lang=en (accessed on 18 October 2022).

15. Platelia SARS-CoV-2 Total Ab-Instructions for Use. Available online: https://commerce.bio-rad.com/webroot/web/pdf/inserts/CDG/en/16008267_2020_04_EN.pdf (accessed on 18 October 2022).

16. COVID-19 Detection | cPassTM Kit Technology. Available online: https://www.genscript.com/covid-19-detection-cpass.html (accessed on 18 October 2022).

17. Jaganathan, S.; Stieber, F.; Rao, S.N.; Nikolayevskyy, V.; Manissero, D.; Allen, N.; Boyle, J.; Howard, J. Preliminary Evaluation of QuantiFERON SARS-CoV-2 and QIAreach Anti-SARS-CoV-2 Total Test in Recently Vaccinated Individuals. *Infect. Dis. Ther.* **2021**, *10*, 2765–2776. [CrossRef]

18. VAERS | Vaccine Safety | CDC. Available online: https://www.cdc.gov/vaccinesafety/ensuringsafety/monitoring/vaers/index.html (accessed on 18 October 2022).

19. Khoury, D.S.; Cromer, D.; Reynaldi, A.; Schlub, T.E.; Wheatley, A.K.; Juno, J.A.; Subbarao, K.; Kent, S.J.; Triccas, J.A.; Davenport, M.P. Neutralizing antibody levels are highly predictive of immune protection from symptomatic SARS-CoV-2 infection. *Nat. Med.* **2021**, *27*, 1205–1211. [CrossRef] [PubMed]

20. McMahan, K.; Yu, J.; Mercado, N.B.; Loos, C.; Tostanoski, L.H.; Chandrashekar, A.; Liu, J.; Peter, L.; Atyeo, C.; Zhu, A.; et al. Correlates of protection against SARS-CoV-2 in rhesus macaques. *Nature* **2020**, *590*, 630–634. [CrossRef]

21. Pantaleo, G.; Correia, B.; Fenwick, C.; Joo, V.S.; Perez, L. Antibodies to combat viral infections: Development strategies and progress. *Nat. Rev. Drug Discov.* **2022**, *21*, 676–696. [CrossRef] [PubMed]

22. Evans, J.P.; Zeng, C.; Carlin, C.; Lozanski, G.; Saif, L.J.; Oltz, E.M.; Gumina, R.J.; Liu, S.-L. Neutralizing antibody responses elicited by SARS-CoV-2 mRNA vaccination wane over time and are boosted by breakthrough infection. *Sci. Transl. Med.* **2022**, *14*, eabn8057. [CrossRef] [PubMed]

23. Sridhar, S.; Begom, S.; Bermingham, A.; Hoschler, K.; Adamson, W.; Carman, W.; Bean, T.; Barclay, W.; Deeks, J.J.; Lalvani, A. Cellular immune correlates of protection against symptomatic pandemic influenza. *Nat. Med.* **2013**, *19*, 1305–1312. [CrossRef]

24. Apostolidis, S.A.; Kakara, M.; Painter, M.M.; Goel, R.R.; Mathew, D.; Lenzi, K.; Rezk, A.; Patterson, K.R.; Espinoza, D.A.; Kadri, J.C.; et al. Cellular and humoral immune responses following SARS-CoV-2 mRNA vaccination in patients with multiple sclerosis on anti-CD20 therapy. *Nat. Med.* **2021**, *27*, 1990–2001. [CrossRef]

25. Bange, E.M.; Han, N.A.; Wileyto, P.; Kim, J.Y.; Gouma, S.; Robinson, J.; Greenplate, A.R.; Hwee, M.A.; Porterfield, F.; Owoyemi, O.; et al. CD8+ T cells contribute to survival in patients with COVID-19 and hematologic cancer. *Nat. Med.* **2021**, *27*, 1280–1289. [CrossRef]

26. Baum, A.; Ajithdoss, D.; Copin, R.; Zhou, A.; Lanza, K.; Negron, N.; Ni, M.; Wei, Y.; Mohammadi, K.; Musser, B.; et al. REGN-COV2 antibodies prevent and treat SARS-CoV-2 infection in rhesus macaques and hamsters. *Science* **2020**, *370*, 1110–1115. [CrossRef]

27. Sormani, M.P.; Inglese, M.; Schiavetti, I.; Carmisciano, L.; Laroni, A.; Lapucci, C.; Da Rin, G.; Serrati, C.; Gandoglia, I.; Tassinari, T.; et al. Effect of SARS-CoV-2 mRNA vaccination in MS patients treated with disease modifying therapies. *Ebiomedicine* **2021**, *72*, 103581. [CrossRef]
28. Tartof, S.Y.; Slezak, J.M.; Fischer, H.; Hong, V.; Ackerson, B.K.; Ranasinghe, O.N.; Frankland, T.B.; Ogun, O.A.; Zamparo, J.M.; Gray, S.; et al. Effectiveness of mRNA BNT162b2 COVID-19 vaccine up to 6 months in a large integrated health system in the USA: A retrospective cohort study. *Lancet* **2021**, *398*, 1407–1416. [CrossRef] [PubMed]
29. Liu, J.; Yu, J.; McMahan, K.; Jacob-Dolan, C.; He, X.; Giffin, V.; Wu, C.; Sciacca, M.; Powers, O.; Nampanya, F.; et al. CD8 T cells contribute to vaccine protection against SARS-CoV-2 in macaques. *Sci. Immunol.* **2022**, *7*, eabq7647. [CrossRef] [PubMed]
30. COVID-19 Variant Data | Department of Health. Available online: https://coronavirus.health.ny.gov/covid-19-variant-data (accessed on 18 October 2022).

Communication

Use of Oral Antivirals Ritonavir-Nirmatrelvir and Molnupiravir in Patients with Multiple Myeloma Is Associated with Low Rates of Severe COVID-19: A Single-Center, Prospective Study

Vassiliki Spiliopoulou, Ioannis Ntanasis-Stathopoulos , Panagiotis Malandrakis , Maria Gavriatopoulou, Foteini Theodorakakou, Despina Fotiou , Magdalini Migkou , Maria Roussou, Evangelos Eleutherakis-Papaiakovou, Efstathios Kastritis , Meletios A. Dimopoulos and Evangelos Terpos *

Department of Clinical Therapeutics, School of Medicine, National and Kapodistrian University of Athens, 11528 Athens, Greece
* Correspondence: eterpos@med.uoa.gr or eterpos@hotmail.com; Tel.: +30-2132162586

Abstract: In patients with multiple myeloma (MM), SARS-CoV-2 infection has been associated with a severe clinical course and high mortality rates due to the concomitant disease- and treatment-related immunosuppression. Specific antiviral treatment involves viral replication control with monoclonal antibodies and antivirals, including molnupiravir and the ritonavir-boosted nirmatrelvir. This prospective study investigated the effect of these two agents on SARS-CoV-2 infection severity and mortality in patients with MM. Patients received either ritonavir-nirmatrelvir or molnupiravir. Baseline demographic and clinical characteristics, as well as levels of neutralizing antibodies (NAbs), were compared. A total of 139 patients was treated with ritonavir-nirmatrelvir while the remaining 30 patients were treated with molnupiravir. In total, 149 patients (88.2%) had a mild infection, 15 (8.9%) had a moderate infection, and five (3%) had severe COVID-19. No differences in the severity of COVID-19-related outcomes were observed between the two antivirals. Patients with severe disease had lower neutralizing antibody levels before the COVID-19 infection compared to patients with mild disease ($p = 0.04$). Regarding treatment, it was observed that patients receiving belantamab mafodotin had a higher risk of severe COVID-19 ($p < 0.001$) in the univariate analysis. In conclusion, ritonavir-nirmatrelvir and molnupiravirmay prevent severe disease in MM patients with SARS-CoV-2 infection. This prospective study indicated the comparable effects of the two treatment options, providing an insight for further research in preventing severe COVID-19 in patients with hematologic malignancies.

Keywords: multiple myeloma; molnupiravir; ritonavir-nirmatrelvir; SARS-CoV-2; belantamab mafodotin

Citation: Spiliopoulou, V.; Ntanasis-Stathopoulos, I.; Malandrakis, P.; Gavriatopoulou, M.; Theodorakakou, F.; Fotiou, D.; Migkou, M.; Roussou, M.; Eleutherakis-Papaiakovou, E.; Kastritis, E.; et al. Use of Oral Antivirals Ritonavir-Nirmatrelvir and Molnupiravir in Patients with Multiple Myeloma Is Associated with Low Rates of Severe COVID-19: A Single-Center, Prospective Study. *Viruses* 2023, *15*, 704. https://doi.org/10.3390/v15030704

Academic Editor: Jason Yiu Wing KAM

Received: 29 December 2022
Revised: 23 February 2023
Accepted: 7 March 2023
Published: 8 March 2023

1. Introduction

Severe SARS-CoV-2 infection (Coronavirus disease 2019, COVID-19) is characterized by an initial viral phase, often followed by a severe inflammatory phase. In patients with MM, SARS-CoV-2 infection has been associated with a severe clinical course and high mortality rates, due to the concomitant disease- and treatment-related immunosuppression [1]. Furthermore, MM patients respond poorly to vaccination despite adequate immunization, especially those on treatment with anti-CD38 or anti-BCMA therapies [2–5]. Thus, MM patients are at a higher risk for breakthrough infection compared with noncancer patients or patients with solid tumor, and supportive measures along with prophylactic use of monoclonal antibodies against SARS-CoV-2 are needed [6–8]. At the time of SARS-CoV-2 infection, the use of antiviral therapy may improve patient outcomes [9]. Specific antiviral treatment involves viral replication control with monoclonal antibodies and antivirals, including molnupiravir and the ritonavir-boosted nirmatrelvir [10]. Both molnupiravir and ritonavir/nirmatrelvir have shown to reduce the risk for severe COVID-19, including hospitalization and/or death, compared with placebo in unvaccinated individuals [11,12].

More recent data including vaccinated patients with mild COVID-19 support the early administration of antivirals for preventing severe outcomes [13]. Although available evidence supports the use of antivirals in patients with SARS-CoV-2 infection to prevent severe disease, relevant data on MM patients are scarce. This prospective study investigated the effect of these two agents on SARS-CoV-2 infection severity and mortality in patients with MM.

2. Materials and Methods

Consecutive patients with MM and SARS-CoV-2 infection were prospectively enrolled in the study from February 2022 to October 2022. During this period, the Omicron SARS-CoV-2 variant, including BA.2, BA.4, BA.5, and BQ.1, was dominant in the country. All patients had microbiologically confirmed SARS-CoV-2 with polymerase chain reaction (PCR). The patients received either ritonavir-nirmatrelvir or molnupiravir according to the national guidelines and the availability of each drug. Treatment with antivirals was initiated within the first five days from COVID-19 symptoms onset in all patients with no need of supplemental oxygen. All patients were at high risk for severe COVID-19 due to the underlying MM. There were no special characteristics determining which drug was given to each patient. Baseline demographic and clinical characteristics, as well as levels of neutralizing antibodies (NAbs), were collected and compared. The effect of different treatments on SARS-CoV-2 infection severity and mortality was examined. The study was approved by the institutional review board.

Neutralizing antibodies (NAbs) against SARS-CoV-2 were determined using an FDA approved methodology (enzyme-linked immunosorbent assay, cPass SARS-CoV-2 NAbs Detection Kit; GenScript, Piscataway, NJ, USA). Anti-SARS-CoV-2 neutralizing antibodies are of particular importance because they inhibit the binding of the receptor-binding domain (RBD) of the surface spike (S) protein to the human angiotensin-converting enzyme 2 (ACE2) receptor [14].

Statistical analysis: Frequencies and percentages were used to describe categorical data, while means and standard deviations were used to describe scale measurements. The Fisher's exact test was applied to examine the differences in the outcome based on the drug [15]. After applying the Shapiro–Wilk criterion for normality assessment, analysis of variance was applied to examine differences in NAbs count, followed by multiple comparisons under the Tukey's HSD criterion. A binary logistic-regression model was applied to define ORs of the variables examined, instead of a multinomial logistic-regression model, as the treatment with belantamab was not recorded in patients with a moderate outcome. The significance level was set at 0.05 in all cases and the analysis was carried out with the SPSS v 26.0 software.

3. Results

A total of 169 patients infected with SARS-CoV-2 were included. Of those, 74 were female, the average age was 64.4 years, and the mean body mass index (BMI) was 26.91 kg/m^2. Table 1 shows the baseline characteristics of included patients before SARS-CoV-2 infection. The performance status (PS) is described as the status of symptoms and functions with respect to ambulatory status and need for care. The majority of the patients (78.7%) were in PS 0, which indicates normal activity, and the other 21.3% were in PS 1, which indicates some symptoms but that the patient is still ambulatory. Regarding the medical history, 14 patients (8.3%) were diagnosed with diabetes mellitus (DM), 71 (42%) had hypertension, six (3.6%) had coronary artery disease (CAD), and 16 (9.5%) had chronic obstructive pulmonary disease (COPD).

All patients, except for one, were vaccinated, mostly with the BNT162b2Pfizer/BioNTech vaccine (96.4%). A total of 153 patients had already received three doses, 7 had received four doses, and 8 had received two vaccine shots before SARS-CoV-2 diagnosis. The median time of the last vaccine dose to COVID-19 diagnosis was 6 months (range 1–11 months). No other pharmacological intervention was adopted as prophylaxis against SARS-CoV-2.

NAbs levels were not significantly different according to MM stage or line of treatment. Only patients under anti-BCMA therapy had significantly lower NAbs titers compared to the others ($p < 0.05$).

Table 1. Patients' characteristics before SARS-CoV-2 infection.

		Mean	SD
Nabs before SARS-CoV-2 infection (%)		65.77	29.19
		N (N = 169)	%
Performance Status (PS)	0	133	78.7%
	1	36	21.3%
Age	30–50 years	19	11.2%
	50–70 years	96	56.8%
	>70 years	54	32%
Diagnosis	Multiple Myeloma (MM)	160	94.7%
	Smoldering Multiple Myeloma (SMM)	9	5.3%
Line of therapy in MM (N = 160)	1st	100	62.5%
	2nd	36	22.5%
	3rd	14	8.8%
	4th	7	4.4%
	5th	3	1.8%

Regarding treatment for symptomatic MM, several drug combinations were administered to patients depending on their medical condition and history of the disease. These included regimens based on proteasome inhibitors, immunomodulatory drugs, anti-CD38 monoclonal antibodies, anti-BCMA treatments (belantamab mafodotin), and other treatments including selinexor and cyclophosphamide (Table 2). For most of the patients (n = 100) this was their 1st line of treatment, for 36 it was their 2nd, for 14 their 3rd, for 7 their 4th, and for 3 patients their 5th line of treatment. A total of 138 patients (81.7%) was receiving dexamethasone as part of their treatment for MM.

Table 2. Patients' treatment combinations concerning multiple myeloma: Proteasome inhibitor (PI), immunomodulatory drug (IMiD), SINE (selective inhibitor of nuclear transport).

Regimen	N
PI-based regimen	23 (14.4%)
IMiD-based regimen	43 (26.9%)
PI- and IMiD-based regimen	35 (21.9%)
Anti-CD38-based regimen	38 (23.7%)
Anti-BCMA-based regimen	17 (10.6%)
SINE compound based regimen	3 (1.9%)
Other	1 (0.6%)

As far as the administration of SARS-CoV-2 antivirals is concerned, 139 patients (82.2%) were treated with ritonavir-nirmatrelvir, while the remaining 30 patients (17.8%) were treated with molnupiravir. The duration of antiviral treatment was equal to five days in all but three cases. These three patients were hospitalized before the completion of the antiviral regimen and treatment was interrupted. Antivirals were well tolerated and no major adverse events were noted. Diarrhea grade 1 was reported in 10 patients (6%). All 10 patients were receiving ritonavir-nirmatrelvir. Regarding COVID-19 outcomes as defined by the WHO, 149 (88.2%) patients had mild infection with several signs and symptoms, such as fever, cough, and headache, but not shortness of breath or dyspnea. Fifteen (8.9%) patients experienced moderate infection with evidence of lower respiratory disease during clinical assessment or imaging. Five (3%) patients developed severe COVID-19 and required oxygen support, with complications such as respiratory failure, acute respiratory distress syndrome, sepsis and septic shock, thromboembolism, and/or multi-organ failure.

Table 3 shows that severe cases were treated differently, as expected, compared to mild or moderate cases. An exception was observed regarding the use of corticosteroids for COVID-19, since it appears that they were given almost equally to patients with moderate and severe outcomes.

Table 3. Patients' responses to the treatment for SARS-CoV-2 infection (hospitalized or non-hospitalized patients).

	Outcome					
	Mild (N = 149)		**Moderate (N = 15)**		**Severe (N = 5)**	
	N	**%**	**N**	**%**	**N**	**%**
Hospitalization	0	0%	1	6.7%	5	100%
Intubation	0	0%	0	0%	2	40%
Tocilizumab	0	0%	0	0%	5	100%
Corticosteroids	1	0.7%	13	86.7%	5	100%

Table 4 shows that no differences in the severity of COVID-19 outcomes were observed between the two antivirals ($p = 0.236$). COVID-19 infection was resolved in all patients, except for three fatal cases. No differences were observed in the baseline characteristics among patients who received molnupiravir and ritonavir-nirmatrelvir.

Table 4. Antiviral drug type and outcome of the SARS-CoV-2 infection (M: molnupiravir, R-N: ritonavir-nirmatrelvir).

		Outcome					
		Mild		**Moderate**		**Severe**	
		N	**%**	**N**	**%**	**N**	**%**
Antiviral Drug	M (N = 30)	27	90%	1	3%	2	7%
	R-N (N = 139)	122	88%	14	10%	3	2%

Table 5 shows the distribution of gender and myeloma stage according to the outcome of COVID-19 infection. Gender and myeloma stage according to the revised international staging system for myeloma (R-ISS) did not seem to differentiate patient outcomes.

As shown in Table 6, no significant differences were observed in the infection outcomes according to BMI, age, or past medical history defined as the presence of DM, COPD, CAD, or hypertension.

An initial approach to differences in patient outcomes showed that regarding PS, a statistically significant difference of mild versus severe cases ($p = 0.041$) was observed, while the difference between mild and moderate cases was borderline non-significant ($p = 0.052$). A difference between severe and mild cases was observed concerning NAbs response levels before SARS-CoV-2 infection (Table 7). As shown in Figure 1, patients with severe disease appeared to have lower NAbs levels before SARS-CoV-2 infection (median time from last NAbs measurement to infection date: 16 days; range: 8–24 days) compared to patients with mild disease (median: 18 days; range 4–27 days, $p = 0.04$). The mean and standard deviations for NAbs levels were 67.49% ± 28.85%, 59% ± 24.69% and 35.4% ± 37.57% for the groups of mild, moderate, and severe outcomes, respectively. Only four patients (2.4%) had lower than 10% neutralizing antibody levels before COVID-19 diagnosis, and two of them experienced a severe infection.

Table 5. Distribution of gender and MM stage according to the outcome of COVID-19 infection.

| | | Outcome | | | | | |
| | | Mild | | Moderate | | Severe | |
		N	%	N	%	N	%
Gender	MALES (N = 95)	82	86.3%	10	10.5%	3	3.2%
	FEMALES (N = 74)	67	90.5%	5	6.8%	2	2.7%
Stage	RISS I (N = 52)	45	86.5%	6	11.6%	1	1.9%
	RISS II (N = 62)	56	90.3%	4	6.5%	2	3.2%
	RISS III (N = 55)	48	87.3%	5	9.1%	2	3.6%

Table 6. Differences in COVID-19 outcomes according to medical history, BMI, and age.

| | Outcome | | | | | | |
| | Mild | | Moderate | | Severe | | |
	N	%	N	%	N	%	*p*
Diabetes Mellitus	10	6.7%	3	20%	1	20%	0.099
Copd	15	10.1%	1	6.7%	0	0%	0.697
Cad	6	4%	0	0%	0	0%	1.000
Hypertension	63	42.3%	7	46.7%	1	20%	0.593
	Mean	SD	Mean	SD	Mean	SD	
Bmi	26.80	4.46	28.27	6.76	25.88	3.36	0.456
Age	64.34	10.60	65.00	9.30	65.00	12.29	0.966

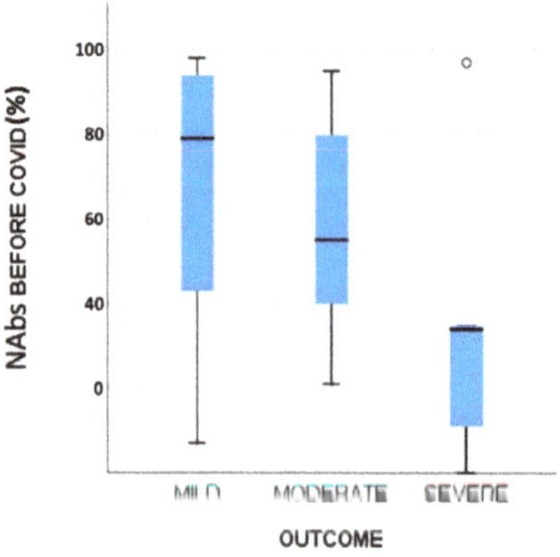

Figure 1. Boxplot of the differences in NAbs before SARS-CoV-2 infection based on COVID-19 outcomes.

Table 7. Differences in COVID-19 outcomes depending on performance status, response to treatment for MM, and NAbs before SARS-CoV-2 infection (SD: stable disease, PR: partial response, VGPR: very good partial response, CR: complete response, sCR: stringent complete response).

		Outcome						
		Mild		Moderate		Severe		
		N	%	N	%	N	%	*p*
Performance Status	0	122	81.9%	9	60%	2	40%	0.017
	1	2	18.1%	6	40%	3	60%	
RESPONSE	SD	15	10.5%	5	33.3%	2	40%	0.126
	PR	33	23.1%	1	6.7%	0	0%	
	VGPR	52	36.4%	4	26.7%	2	40%	
	CR	32	22.4%	3	20%	1	20%	
	sCR	11	7.7%	2	13.3%	0	0%	
		Mean	SD	Mean	SD	Mean	SD	
NAbs before COVID (%)		67.49	28.85	59	24.69	35.4	37.57	0.040

Regarding treatments, despite the various drug types included in the study and their combinations, it was observed that treatment with belantamab mafodotin was associated with adverse COVID-19 outcomes. A total of 47% of patients receiving belantamab mafodotin were male, with a median age of 66.2 years, and a median Nab measurement before the infection of 29.8%. As far as their response to treatment for MM was concerned, five patients had a partial response to therapy (29.4%), five patients had a very good partial response (29.4%), and seven patients had a complete response (41.2%). Seven of these patients received molnupiravir (41.2%) and ten received ritonavir-nirmatrelvir (58.8%). As shown in Figure 2, there was a significantly higher risk for severe SARS-CoV-2 infection in patients that received belantamab mafodotin ($p < 0.001$) according to the Fisher's exact test.

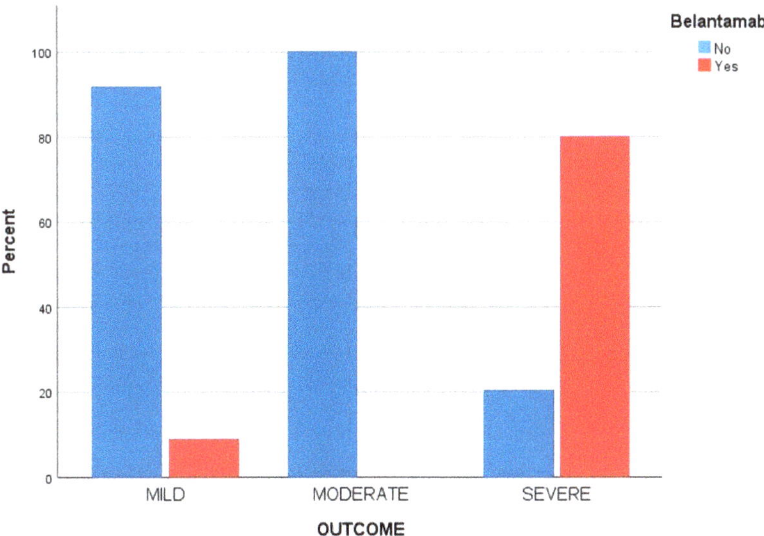

Figure 2. Bar chart of the differences in COVID-19 outcomes according to type of active treatment.

4. Discussion

The SARS-CoV-2 pandemic is a major cause of morbidity and mortality worldwide. Patients with MM are at high risk for severe infection, breakthrough infection, and they present suboptimal humoral responses to COVID-19 vaccination [16–20]. Unfortunately, in some patients with cancer, the infection cannot be completely controlled with antiviral drugs and supportive care, and they ultimately develop severe disease and need hospitalization [8]. The majority of immunocompromised patients may not be fully protected from severe infection after vaccination and the early use of oral antiviral drugs in cases of infection seems to be effective in preventing severe COVID-19 [3,5,21,22]. The use of specific antiviral treatment with molnupiravir and ritonavir-boosted nirmatrelvir aims to reduce the risk of severe infection and hospitalization in patients with MM who undergo mild SARS-CoV-2 infection and do not require supplemental oxygen [12,23,24]. Molnupiravir is the prodrug of the ribonucleoside analogue β-D-N4-hydroxycytidine inhibiting the SARS-CoV-2 replication and reducing viral load [23,25–27]. In phase I/II/III studies, it has shown both safety and efficacy, reducing the risk of hospitalization and mortality by approximately 50% in non-hospitalized adults with mild-to-moderate SARS-CoV-2 disease who are at risk for poor COVID-19-related outcomes [22]. Ritonavir-nirmatrelvir is an oral protease inhibitor that has also reduced the risk for hospitalization and death by approximately 90% in clinical trials [11]. Both molnupiravir and ritonavir-nirmatrelvir are administrated within the first five days of symptoms [12].

Our analysis included patients with MM who were receiving antimyeloma therapy and were infected with SARS-CoV-2. Patients with MM have a high risk of severe infection because of their immunosuppression and their defective response to vaccination [28]. Thus, the patients were eligible to receive either ritonavir-nirmatrelvir or molnupiravir according to the national guidelines. We compared the demographic characteristics of the patients, such as their age, health problems, vaccination status, levels of neutralizing antibodies, antimyeloma therapies, response to therapy, and the severity of their SARS-CoV-2 infection as well as their mortality. No significant differences were revealed between the two antiviral treatment groups in terms of both baseline characteristics and infection outcomes.

No statistically significant difference was observed in the infection outcome regarding the demographic characteristics of the patients including BMI, age, and past medical history defined as the presence of DM, COPD, CAD, or hypertension. Regarding PS, a statistically significant difference of mild cases versus severe cases was observed, while the difference between mild and moderate cases was borderline non-significant. The PS of each patient could be affected by age, BMI, comorbidities, antimyeloma therapy, tolerance to this therapy, and severity of the underlying MM [29]. A longer follow up of our data would enable the characterization of long-term outcomes related to COVID-19, such as the persistence of symptoms related to long COVID.

A difference between severe and mild cases was observed concerning NAbs response levels before SARS-CoV-2 infection. Patients with severe disease had lower neutralizing antibody levels before SARS-CoV-2 infection compared to patients with mild disease. Regarding the role of active treatment, it was observed that treatment with belantamab mafodotin was associated with severe COVID-19 outcomes. Belantamab mafodotin is a humanized IgG1 k monoclonal antibody against the B-cell maturation antigen (BCMA) conjugated with a cytotoxic agent, maleimidocaproyl monomethyl auristatin F (mcMMAF). The antibody–drug conjugate binds to BCMA on myeloma cell surfaces causing cell-cycle arrest and inducing antibody-dependent cellular cytotoxicity [30,31].

Previous studies have consistently shown that patients with MM who receive anti-BCMA therapies, including conjugated monoclonal antibodies and bispecific antibodies, have inferior humoral responses following initial and booster vaccination compared with other treatment regimens [30,32]. Pertinent data have also shown that treatments with anti-CD38 monoclonal antibodies or anti-BCMA bispecific T-cell engagers have been associated with inferior CD4+ T-cell responses against SARS-CoV-2as well [5,33]. These patients may be more susceptible to SARS-CoV-2 infection and severe COVID-19. However, it should be noted that only a few patients were receiving anti-BCMA drugs in our study; therefore, the significant results of the univariate analysis should be interpreted with caution. Furthermore, anti-BCMA agents are currently administered in the relapsed/refractory disease setting. The degree of immune dysfunction probably increases with multiple lines of treatment due to exposure to different drug classes and myeloma burden. A larger study sample would be required to perform a robust multivariate analysis and control for potential confounders.

Lasagna et al. collected data from patients with solid tumors on active treatment and examined the effectiveness of oral antivirals in preventing severe SARS-CoV-2 infection and mortality. The majority of treated patients showed a reduction in the duration of symptoms and only one patient required hospitalization. This study confirmed the efficacy of oral antivirals in patients with solid tumors who receive antineoplastic therapy and underlined the need for the early management of patients with cancer who are infected with SARS-CoV-2, in order to avoid both hospitalization and death [6].

Real-world data are important to evaluate both the effectiveness and safety of COVID-19 antivirals in patients with cancer [8]. These data suggest that outpatient therapies for mild SARS-CoV-2 infection may reduce the duration of symptoms, and the risk of hospitalization and morbidity. It seems that the main benefit of the early use of antivirals lies in the prevention of severe infection that requires hospitalization. Our data suggest that the effect of antivirals may be limited when a patient develops severe infection. However, a main limitation of this study pertains to the absence of a placebo or a non-treatment group of patients with myeloma and COVID-19. Therefore, we were not able to precisely determine the benefit of antivirals administration compared to supportive care alone. Furthermore, our study was not randomized in order to minimize any potential bias in the comparisons between the two antivirals. The outcomes may vary depending on the availability of other supportive care, which could be a major factor in determining patients' disease course. The dominant SARS-CoV-2 variants in the community may also differentiate patient outcomes. During the study period, the Omicron strain prevailed, which has been associated with less severe disease compared to previous strains. A longitudinal analysis during different SARS-CoV-2 waves, including control groups, would be necessary to reach firm conclusions about the precise benefit of antivirals. Although the study sample size was small, our study provides useful real-world data that may be used as a basis for the design of larger randomized studies in the field.

In conclusion, ritonavir-nirmatrelvir and molnupiravir may be beneficial in preventing severe disease from SARS-CoV-2 infection in MM patients who are under anti-myeloma treatment. This prospective study indicated the comparable effects of the two treatment options, providing an important background for further research.

Author Contributions: Conceptualization, methodology, software, data Curation, formal analysis, validation, supervision and project administration: M.G., M.A.D. and E.T.; investigation, writing—review and editing, visualization: V.S., I.N.-S., P.M., F.T., D.F., M.M., M.R., E.E.-P., E.K. and E.T. All authors have read and agreed to the published version of the manuscript.

Funding: This research received no external funding.

Institutional Review Board Statement: The study was conducted in accordance with the Declaration of Helsinki, and approved by the Ethics Committee of "Alexandra" General Hospital (Reference No. 15/23 December 2020), Athens, Greece.

Informed Consent Statement: Informed consent was obtained from all subjects involved in the study.

Data Availability Statement: Data are available upon request from the corresponding author.

Conflicts of Interest: The authors declare no conflict of interest.

References

1. Chari, A.; Samur, M.K.; Martinez-Lopez, J.; Cook, G.; Biran, N.; Yong, K.; Hungria, V.; Engelhardt, M.; Gay, F.; García Feria, A.; et al. Clinical Features Associated with COVID-19 Outcome in Multiple Myeloma: First Results from the International Myeloma Society Data Set. *Blood* **2020**, *136*, 3033–3040. [CrossRef] [PubMed]
2. Stampfer, S.D.; Goldwater, M.-S.; Jew, S.; Bujarski, S.; Regidor, B.; Daniely, D.; Chen, H.; Xu, N.; Li, M.; Green, T.; et al. Response to MRNA Vaccination for COVID-19 among Patients with Multiple Myeloma. *Leukemia* **2021**, *35*, 3534–3541. [CrossRef] [PubMed]
3. Rosati, M.; Terpos, E.; Bear, J.; Burns, R.; Devasundaram, S.; Ntanasis-Stathopoulos, I.; Gavriatopoulou, M.; Kastritis, E.; Dimopoulos, M.-A.; Pavlakis, G.N.; et al. Low Spike Antibody Levels and Impaired BA.4/5 Neutralization in Patients with Multiple Myeloma or Waldenstrom's Macroglobulinemia after BNT162b2 Booster Vaccination. *Cancers* **2022**, *14*, 5816. [CrossRef] [PubMed]
4. Abdallah, A.-O.; Mahmoudjafari, Z.; Atieh, T.; Ahmed, N.; Cui, W.; Shune, L.; Mohan, M.; McGuirk, J.; Remker, C.; Foss, M.; et al. Neutralizing Antibody Responses against SARS-CoV-2 in Patients with Plasma Cell Disorders Who Are on Active Treatment after Two Doses of MRNA Vaccination. *Eur. J. Haematol.* **2022**, *109*, 458–464. [CrossRef] [PubMed]
5. Ramasamy, K.; Sadler, R.; Jeans, S.; Weeden, P.; Varghese, S.; Turner, A.; Larham, J.; Gray, N.; Carty, O.; Barrett, J.; et al. Immune Response to COVID-19 Vaccination Is Attenuated by Poor Disease Control and Antimyeloma Therapy with Vaccine Driven Divergent T-Cell Response. *Br. J. Haematol.* **2022**, *197*, 293–301. [CrossRef]
6. Lasagna, A.; Cassaniti, I.; Lilleri, D.; Quaccini, M.; Ferrari, A.; Sacchi, P.; Bruno, R.; Baldanti, F.; Pedrazzoli, P. Effectiveness of the Available Early Therapies in Reducing Severe COVID-19 in Non-Hospitalized Patients with Solid Tumors on Active Treatment. *Front. Med.* **2022**, *9*, 1036473. [CrossRef]
7. Wang, L.; Berger, N.A.; Xu, R. Risks of SARS-CoV-2 Breakthrough Infection and Hospitalization in Fully Vaccinated Patients with Multiple Myeloma. *JAMA Netw. Open* **2021**, *4*, e2137575. [CrossRef]
8. Song, Q.; Bates, B.; Shao, Y.R.; Hsu, F.-C.; Liu, F.; Madhira, V.; Mitra, A.K.; Bergquist, T.; Kavuluru, R.; Li, X.; et al. Risk and Outcome of Breakthrough COVID-19 Infections in Vaccinated Patients with Cancer: Real-World Evidence From the National COVID Cohort Collaborative. *J. Clin. Oncol.* **2022**, *40*, 1414–1427. [CrossRef]
9. Wen, W.; Chen, C.; Tang, J.; Wang, C.; Zhou, M.; Cheng, Y.; Zhou, X.; Wu, Q.; Zhang, X.; Feng, Z.; et al. Efficacy and Safety of Three New Oral Antiviral Treatment (Molnupiravir, Fluvoxamine and Paxlovid) for COVID-19: A Meta-Analysis. *Ann. Med.* **2022**, *54*, 516–523. [CrossRef]
10. Saravolatz, L.D.; Depcinski, S.; Sharma, M. Molnupiravir and Nirmatrelvir-Ritonavir: Oral COVID Antiviral Drugs. *Clin. Infect. Dis.* **2022**, *76*, 165–171. [CrossRef]
11. Hammond, J.; Leister-Tebbe, H.; Gardner, A.; Abreu, P.; Bao, W.; Wisemandle, W.; Baniecki, M.; Hendrick, V.M.; Damle, B.; Simón-Campos, A.; et al. Oral Nirmatrelvir for High-Risk, Nonhospitalized Adults with Covid-19. *N. Engl. J. Med.* **2022**, *386*, 1397–1408. [CrossRef] [PubMed]
12. Jayk Bernal, A.; Gomes da Silva, M.M.; Musungaie, D.B.; Kovalchuk, E.; Gonzalez, A.; Delos Reyes, V.; Martín-Quirós, A.; Caraco, Y.; Williams-Diaz, A.; Brown, M.L.; et al. Molnupiravir for Oral Treatment of Covid-19 in Nonhospitalized Patients. *N. Engl. J. Med.* **2022**, *386*, 509–520. [CrossRef] [PubMed]
13. Cao, Z.; Gao, W.; Bao, H.; Feng, H.; Mei, S.; Chen, P.; Gao, Y.; Cui, Z.; Zhang, Q.; Meng, X.; et al. VV116 versus Nirmatrelvir–Ritonavir for Oral Treatment of COVID-19. *N. Engl. J. Med.* **2023**, *388*, 406–417. [CrossRef] [PubMed]
14. Pang, N.Y.-L.; Pang, A.S.-R.; Chow, V.T.; Wang, D.-Y. Understanding Neutralising Antibodies against SARS-CoV-2 and Their Implications in Clinical Practice. *Mil. Med. Res.* **2021**, *8*, 47. [CrossRef]
15. Upton, G.J.G. Fisher's Exact Test. *J. R. Stat. Soc. Ser. A (Stat. Soc.)* **1992**, *155*, 395. [CrossRef]
16. Storti, P.; Marchica, V.; Vescovini, R.; Franceschi, V.; Russo, L.; Notarfranchi, L.; Raimondi, V.; Toscani, D.; Burroughs Garcia, J.; Costa, F.; et al. Immune Response to SARS-CoV-2 MRNA Vaccination and Booster Dose in Patients with Multiple Myeloma and Monoclonal Gammopathies: Impact of Omicron Variant on the Humoral Response. *Oncoimmunology* **2022**, *11*, 2120275. [CrossRef]
17. Chuleerarux, N.; Manothummetha, K.; Moonla, C.; Sanguankeo, A.; Kates, O.S.; Hirankarn, N.; Phongkhun, K.; Thanakitcharu, J.; Leksuwankun, S.; Meejun, T.; et al. Immunogenicity of SARS-CoV-2 Vaccines in Patients with Multiple Myeloma: A Systematic Review and Meta-Analysis. *Blood Adv.* **2022**, *6*, 6198–6207. [CrossRef]
18. Cook, G.; John Ashcroft, A.; Pratt, G.; Popat, R.; Ramasamy, K.; Kaiser, M.; Jenner, M.; Henshaw, S.; Hall, R.; Sive, J.; et al. Real-world Assessment of the Clinical Impact of Symptomatic Infection with Severe Acute Respiratory Syndrome Coronavirus (COVID-19 Disease) in Patients with Multiple Myeloma Receiving Systemic Anti-cancer Therapy. *Br. J. Haematol.* **2020**, *190*, e83. [CrossRef]
19. Blennow, O.; Salmanton-García, J.; Nowak, P.; Itri, F.; Van Doesum, J.; López-García, A.; Farina, F.; Jaksic, O.; Pinczés, L.I.; Bilgin, Y.M.; et al. Outcome of Infection with Omicron SARS-CoV-2 Variant in Patients with Hematological Malignancies: An EPICOVIDEHA Survey Report. *Am. J. Hematol.* **2022**, *97*, E312–E317. [CrossRef]
20. Pagano, L.; Salmanton-García, J.; Marchesi, F.; Blennow, O.; Gomes da Silva, M.; Glenthøj, A.; van Doesum, J.; Bilgin, Y.M.; López-García, A.; Itri, F.; et al. Breakthrough COVID-19 in Vaccinated Patients with Hematologic Malignancies: Results from the EPICOVIDEHA Survey. *Blood* **2022**, *140*, 2773–2787. [CrossRef]
21. Najjar-Debbiny, R.; Gronich, N.; Weber, G.; Khoury, J.; Amar, M.; Stein, N.; Goldstein, L.H.; Saliba, W. Effectiveness of Paxlovid in Reducing Severe COVID-19 and Mortality in High Risk Patients. *Clin. Infect. Dis.* **2022**, *76*, e342–e349. [CrossRef]

22. Mahase, E. Covid-19: Molnupiravir Reduces Risk of Hospital Admission or Death by 50% in Patients at Risk, MSD Reports. *BMJ* **2021**, *375*, n2422. [CrossRef] [PubMed]

23. Fischer, W.A.; Eron, J.J.; Holman, W.; Cohen, M.S.; Fang, L.; Szewczyk, L.J.; Sheahan, T.P.; Baric, R.; Mollan, K.R.; Wolfe, C.R.; et al. A Phase 2a Clinical Trial of Molnupiravir in Patients with COVID-19 Shows Accelerated SARS-CoV-2 RNA Clearance and Elimination of Infectious Virus. *Sci. Transl. Med.* **2022**, *14*, eabl7430. [CrossRef] [PubMed]

24. Fischer, W.; Eron, J.J.; Holman, W.; Cohen, M.S.; Fang, L.; Szewczyk, L.J.; Sheahan, T.P.; Baric, R.; Mollan, K.R.; Wolfe, C.R.; et al. Molnupiravir, an Oral Antiviral Treatment for COVID-19. *medRxiv* **2021**. *preprint*. [CrossRef]

25. Vitiello, A.; Troiano, V.; La Porta, R. What Will Be the Role of Molnupiravir in the Treatment of COVID-19 Infection? *Drugs Ther. Perspect.* **2021**, *37*, 579–580. [CrossRef] [PubMed]

26. Agostini, M.L.; Pruijssers, A.J.; Chappell, J.D.; Gribble, J.; Lu, X.; Andres, E.L.; Bluemling, G.R.; Lockwood, M.A.; Sheahan, T.P.; Sims, A.C.; et al. Small-Molecule Antiviral β-d-N4-Hydroxycytidine Inhibits a Proofreading-Intact Coronavirus with a High Genetic Barrier to Resistance. *J. Virol.* **2019**, *93*, e01348-19. [CrossRef]

27. Zhou, S.; Hill, C.S.; Sarkar, S.; Tse, L.V.; Woodburn, B.M.D.; Schinazi, R.F.; Sheahan, T.P.; Baric, R.S.; Heise, M.T.; Swanstrom, R. β-d-N4-Hydroxycytidine Inhibits SARS-CoV-2 Through Lethal Mutagenesis but Is Also Mutagenic To Mammalian Cells. *J. Infect. Dis.* **2021**, *224*, 415–419. [CrossRef] [PubMed]

28. Terpos, E.; Engelhardt, M.; Cook, G.; Gay, F.; Mateos, M.-V.; Ntanasis-Stathopoulos, I.; van de Donk, N.W.C.J.; Avet-Loiseau, H.; Hajek, R.; Vangsted, A.J.; et al. Management of Patients with Multiple Myeloma in the Era of COVID-19 Pandemic: A Consensus Paper from the European Myeloma Network (EMN). *Leukemia* **2020**, *34*, 2000–2011. [CrossRef]

29. Jensen, C.E.; Vohra, S.N.; Nyrop, K.A.; Deal, A.M.; LeBlanc, M.R.; Grant, S.J.; Muss, H.B.; Lichtman, E.I.; Rubinstein, S.M.; Wood, W.A.; et al. Physical Function, Psychosocial Status, and Symptom Burden Among Adults with Plasma Cell Disorders and Associations with Quality of Life. *Oncologist* **2022**, *27*, 694–702. [CrossRef]

30. Tzogani, K.; Penttilä, K.; Lähteenvuo, J.; Lapveteläinen, T.; Lopez Anglada, L.; Prieto, C.; Garcia-Ochoa, B.; Enzmann, H.; Gisselbrecht, C.; Delgado, J.; et al. EMA Review of Belantamab Mafodotin (Blenrep) for the Treatment of Adult Patients with Relapsed/Refractory Multiple Myeloma. *Oncologist* **2021**, *26*, 70–76. [CrossRef]

31. Lassiter, G.; Bergeron, C.; Guedry, R.; Cucarola, J.; Kaye, A.M.; Cornett, E.M.; Kaye, A.D.; Varrassi, G.; Viswanath, O.; Urits, I. Belantamab Mafodotin to Treat Multiple Myeloma: A Comprehensive Review of Disease, Drug Efficacy and Side Effects. *Curr. Oncol.* **2021**, *28*, 640–660. [CrossRef] [PubMed]

32. Ntanasis-Stathopoulos, I.; Karalis, V.; Gavriatopoulou, M.; Malandrakis, P.; Sklirou, A.D.; Eleutherakis-Papaiakovou, E.; Migkou, M.; Roussou, M.; Fotiou, D.; Alexopoulos, H.; et al. Second Booster BNT162b2 Restores SARS-CoV-2 Humoral Response in Patients with Multiple Myeloma, Excluding Those Under Anti-BCMA Therapy. *Hemasphere* **2022**, *6*, e764. [CrossRef] [PubMed]

33. Aleman, A.; Upadhyaya, B.; Tuballes, K.; Kappes, K.; Gleason, C.R.; Beach, K.; Agte, S.; Srivastava, K.; Van Oekelen, O.; Barcessat, V.; et al. Variable Cellular Responses to SARS-CoV-2 in Fully Vaccinated Patients with Multiple Myeloma. *Cancer Cell* **2021**, *39*, 1442–1444. [CrossRef] [PubMed]

Article

A Novel Mathematical Model That Predicts the Protection Time of SARS-CoV-2 Antibodies

Zhaobin Xu [1],*, Dongqing Wei [2], Hongmei Zhang [1] and Jacques Demongeot [3],*

[1] Department of Life Science, Dezhou University, Dezhou 253023, China
[2] School of Life Sciences and Biotechnology, Shanghai Jiao Tong University, Shanghai 200030, China
[3] Laboratory AGEIS EA 7407, Team Tools for e-Gnosis Medical, Faculty of Medicine, University Grenoble Alpes (UGA), 38700 La Tronche, France
* Correspondence: xuzhaobin@dzu.edu.cn (Z.X.); jacques.demongeot@univ-grenoble-alpes.fr (J.D.)

Abstract: Infectious diseases such as SARS-CoV-2 pose a considerable threat to public health. Constructing a reliable mathematical model helps us quantitatively explain the kinetic characteristics of antibody-virus interactions. A novel and robust model is developed to integrate antibody dynamics with virus dynamics based on a comprehensive understanding of immunology principles. This model explicitly formulizes the pernicious effect of the antibody, together with a positive feedback stimulation of the virus–antibody complex on the antibody regeneration. Besides providing quantitative insights into antibody and virus dynamics, it demonstrates good adaptivity in recapturing the virus-antibody interaction. It is proposed that the environmental antigenic substances help maintain the memory cell level and the corresponding neutralizing antibodies secreted by those memory cells. A broader application is also visualized in predicting the antibody protection time caused by a natural infection. Suitable binding antibodies and the presence of massive environmental antigenic substances would prolong the protection time against breakthrough infection. The model also displays excellent fitness and provides good explanations for antibody selection, antibody interference, and self-reinfection. It helps elucidate how our immune system efficiently develops neutralizing antibodies with good binding kinetics. It provides a reasonable explanation for the lower SARS-CoV-2 mortality in the population that was vaccinated with other vaccines. It is inferred that the best strategy for prolonging the vaccine protection time is not repeated inoculation but a directed induction of fast-binding antibodies. Eventually, this model will inform the future construction of an optimal mathematical model and help us fight against those infectious diseases.

Keywords: SARS-CoV-2; antibody dynamics; vaccine; protection time; mathematical modeling

Citation: Xu, Z.; Wei, D.; Zhang, H.; Demongeot, J. A Novel Mathematical Model That Predicts the Protection Time of SARS-CoV-2 Antibodies. *Viruses* **2023**, *15*, 586. https://doi.org/10.3390/v15020586

Academic Editors: Jason Yiu Wing Kam and Hernan Garcia-Ruiz

Received: 30 January 2023
Revised: 10 February 2023
Accepted: 17 February 2023
Published: 20 February 2023

1. Introduction

The COVID-19 epidemic has caused more than 630 million infections and over 6.5 million deaths worldwide by the end of 2022, and it can hardly disappear in a short time based on observation [1–3]. Until now, vaccination has been the only approach to fighting SARS-CoV-2 infections. As the global promotion of the SARS-CoV-2 vaccine continues, we are gaining a better understanding of the antibodies triggered by vaccines and natural infections. It is gradually becoming apparent that antibodies against SARS-CoV-2, whether acquired from natural infection or triggered by vaccination, decay over time, with a concomitant decrease in protective efficiency. Unlike antibodies from vaccinations such as smallpox [4], antibodies against SARS-CoV-2 do not provide durable protection [5].

A significant portion of the population is reluctant to be vaccinated, according to a public poll survey [6,7]. The first concern comes from the side effects of vaccination, such as blood clots [8,9]. The second concern is this deficiency of the long-lasting protection effect brought by vaccination since many clinical reports indicate that the vaccine's protection effect is declining over time [10–13]. Based on the clinical data, a significant decrease in the immune effect of the vaccine against reinfection has been noticed. The SARS-CoV-2 vaccine still shows an overall positive effect on protection currently. However, more people are concerned about how long the protection can be provided by the antibodies triggered by their immune response [5,14]. There is an urgent need to evaluate the protection time of neutralizing antibodies quantitatively.

Constructing a reliable mathematical model helps us quantitatively explain the virus dynamics in the host body, which could provide a reasonable prediction toward those sensitive concerns faced by the public. Researchers have conducted many studies on virus infections using computational approaches [15,16].

Specifically, depending on whether the immune response is integrated into the models, traditional cellular models can be divided into two categories. The first set contains models that do not include the immune response [17–21], that is, traditional viral kinetic models, whose following equation can be classically expressed as:

$$\frac{dT}{dt} = sT - dT - \beta TV \tag{1}$$

$$\frac{dI}{dt} = \beta TV - \delta I \tag{2}$$

$$\frac{dV}{dt} = pI - cV \tag{3}$$

where T denotes the target cells or the number of susceptible cells; I denotes the number of cells where the infection has occurred; and V denotes the number of viruses. This simplified model is wildly used to study the virus dynamics upon infection. While admitting its advantages in recapturing the viral dynamics in virus infection, concerns about this model are proposed. The limitation of this model originates from the absence of a connection to the body's immune response. The main driving force behind eliminating the virus is activating the adaptive immune response by lymphocyte B and T cells, not the depletion of the susceptible cells. Therefore, the estimation based on those depletion mechanisms is skeptical, as it would return a much smaller initial susceptible cell number. The number of susceptible cells should be higher than the number estimated by several orders. Mandating the estimation would inevitably lead to underestimating T_0, where T_0 represents the number of initial susceptible cells. The T_0 term is similar to the initial number of susceptible people when the *SIR* (susceptible, infection, recovery) model is used to predict the epidemic trend [22,23]. The estimated T_0 values vary greatly due to the differences in the fitting data, which significantly reduces their reliability.

The second set consists of models that consider the immune response, where the classical expression of the model considering antibody binding is displayed as follows [24–30]:

$$\frac{dT}{dt} = \chi D - \beta TV \tag{4}$$

$$\frac{dI_1}{dt} = \beta TV - kI_1 \tag{5}$$

$$\frac{dF}{dt} = \omega V - \alpha F \tag{6}$$

$$\frac{dD}{dt} = \delta I_2 - \chi D \tag{7}$$

$$\frac{dI_2}{dt} = kI_1 - \delta I_2 \tag{8}$$

$$\frac{dA}{dt} = fV - hA \tag{9}$$

$$\frac{dV}{dt} = \frac{p}{1 + \varepsilon_1 F} I_2 - cV - \gamma TV - \kappa AV \tag{10}$$

where T denotes the overall number of susceptible cells; A denotes the number of antibodies; V denotes the amount of virus; D denotes the number of dead cells; F denotes the amount of interferon in nonspecific immunity; I_1 denotes the number of cells infected in class 1 (cells infected without viral multiplication); and I_2 denotes the number of cells infected in class 2 (cells infected with viral multiplication). This model effectively combines antibody dynamics with viral changes. It performs well in recapturing and explaining the dynamic interaction between the virus and the host immune system. However, the kinetic properties of antibodies have not been explicitly integrated into this model. Different antigen-binding antibodies will exhibit different kinetics, and this diversity in binding behavior is an essential factor influencing the host-pathogen interaction. The difference in the antibody binding capacity would lead to a significant variation in the virus dynamics. However, this classical model does not explicitly formulize the antibody term.

Although mathematical modeling of antibody kinetics has been significantly advanced and developed in recent years, there is still great potential to be further improved. Therefore, a new mathematical model of antibody dynamics is proposed by us. This model displays good fitness in explaining some puzzling phenomena, especially in the case of SARS-CoV-2 infection, with improvements to previous models in several aspects.

First: New immunology responses are formulized and integrated into this model based on biochemistry principles. We explicitly formulize the virtual effect of antibodies on eliminating pathogenic microorganisms while noticing the stimulation effect of pathogenic antigens on antibodies [31]. This positive feedback stimulation is explicitly represented in a mathematical function as well.

Second: The dynamic model developed hypothesizes that environmental factors maintain memory cell levels. The natural attenuation of antibody levels can be recaptured in this model. Furthermore, their discontinuous decay trend can also be captured with the introduction of environmental antigens. Antigen-like substances in the environment can maintain a specific concentration of B cells or T cells at a certain level. This can explain why some vaccines can provide lifelong protection.

Third: The antigen–antibody binding process is represented as reversible in our model. Instead of using the equilibrium constant, the binding dynamic is described in both the binding and reverse dissociation reactions.

Based on the basic principles of immunology, we established the theoretical hypothesis of antibody kinetics. The results and methods sections illustrate specific principles and rationales in detail. In the end, this antibody kinetic model proposes some possible mechanisms underlying some real-world scenarios:

I.   How are memory cells maintained?
II.   How does our immune system screen for antibodies with a strong binding affinity?
III.   Why do people who get influenza and other vaccines have a lower mortality rate from SARS-CoV-2?
IV.   How can we effectively calculate the duration of protection of a specific antibody?
V.   Why are some recovered patients retested as positive cases without infections from other people?
VI.   Why do vaccinations show considerable differences in protection efficiency?
VII.   How can we improve the protective efficiency and duration of vaccines?

2. Materials and Methods

2.1. Mathematical Representation of the Antibody Production Process

A simple mathematical representation of the immune response is described in the diagram in Figure 1 below.

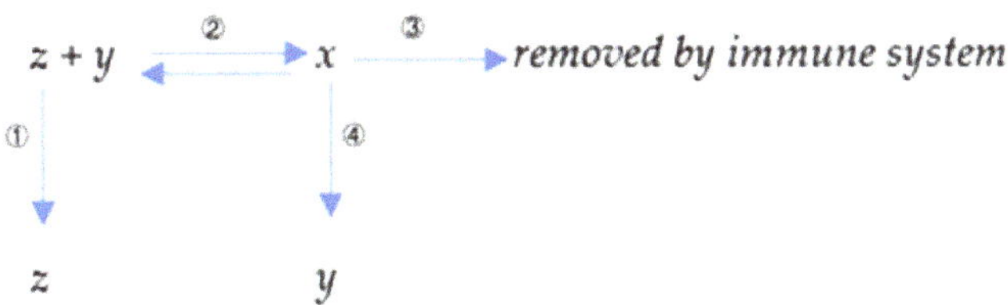

Figure 1. A simple diagram of host-virus interaction.

Here, x denotes the number of antibody–antigen (virus) complexes, y denotes the total number of antibodies, and z denotes the number of viruses. Six processes are displayed in our model. The first reaction represents the virus's proliferation or replication with a reaction constant k_1. The second reaction represents the binding reaction between virus and antibody, with a forward reaction constant k_2 and reverse constant k_{-2}. The third reaction represents removing the antibody–virus complex with a reaction constant k_3 with the help of natural killer (NK) cells [32]. The fourth reaction represents the induction of a new antibody by the antibody–virus complex with a kinetic constant k_4. In immunology, those virus–antibody complexes are on the surface of B-cells since the antibodies are initially produced by B-cells and will attach to the plasma membrane of B-cells. Those complexes would further bind to the helper cells because the antibody has another structure binding region toward those receptors. Those helper cells will present the antigen part, a virus in this case, to the T-cells. The physical placement should be such that B-cells bind to those helper cells and further present themselves close to T-cells. The T-cells will handle those antigen substances; if those substances are not self-originated, they will secrete signal molecules to promote the proliferation or division of B-cells that attach to them. Therefore, B-cells finally proliferate, along with the antibodies their B-cells generate. The fifth reaction represents the degradation of the virus with a constant k_5. The sixth reaction represents the degradation of antibodies with a rate constant k_6.

The differential equations based on those reactions are derived as below:

We established the following equations to represent the proliferation process of antibodies:

$$\frac{dx}{dt} = k_2yz - k_{-2}x - k_3x, \tag{11}$$

$$\frac{dy}{dt} = k_{-2}x - k_2yz + k_4x - k_6y, \tag{12}$$

$$\frac{dz}{dt} = k_{-2}x - k_2yz - k_5z + k_1z. \tag{13}$$

2.2. Mathematical Modeling including Environmental Antigens

The equations introduced in Section 2.1 describe antibody dynamics in the presence of the associated virus. In reality, other environmental antigens can induce antibody coupling and production. To further account for this possibility, we add a new set of equations as shown in Figure 2 below:

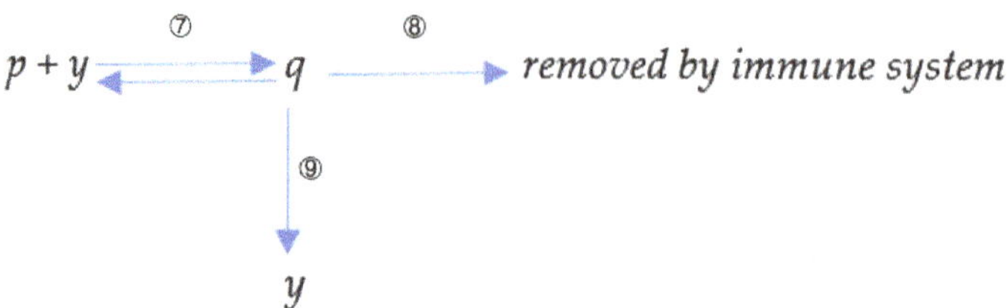

Figure 2. The role of environmental antigens in immune response.

Where p denotes antigen-like substances in the environment that are supposed to be constant; q denotes antibodies bound to antigen-like substances in the environment. Reaction 7 represents the binding reaction between antibodies and environmental antigenic substances with a forward constant k_7 and a reverse constant k_{-7}. Reaction 8 represents the removal of the antibody–antigen complex q with a reaction rate k_3. Reaction 9 represents the induction of a new antibody by q. Therefore, a new set of equations is derived as follows:

$$\frac{dx}{dt} = k_2yz - k_{-2}x - k_3x, \tag{14}$$

$$\frac{dy}{dt} = k_{-2}x - k_2yz + k_4x - k_6y - k_7py + k_{-7}q + k_4q, \tag{15}$$

$$\frac{dz}{dt} = k_{-2}x - k_2yz - k_5z + k_1z, \tag{16}$$

$$\frac{dp}{dt} = 0, \tag{17}$$

$$\frac{dq}{dt} = k_7py - k_{-7}q - k_3q. \tag{18}$$

The antibody level will eventually drop to zero based on the first model described in Equations (11)–(13). Since the decaying coefficient k_6 is more significant than zero, the antibody will finally fade away quickly. However, as we all know, some antibodies can persist in the human body for a long time and provide lifelong protection. It forms the basis for a vaccine. Therefore, "environmental antigen-like substances" are introduced into the second model. They could be food-resourced, air-resourced, or even self-resourced. The presentation of those substances to T-cells would give weak signals to proliferate the memory B-cells or T-cells. The detailed physical background is introduced in Section 3.1.

2.3. A Simplified Model Simulating the Proliferation of Antibodies with Different Binding Kinetic Characteristics by the Immune System

$$\frac{dx_i}{dt} = Ki_{on}y_iz - Ki_{off}x_i - k_3x_i, \tag{19}$$

$$\frac{dy_i}{dt} = Ki_{off}x_i - Ki_{on}y_iz + k_4x_i - k_6(y_i - w_i), \tag{20}$$

$$\frac{dz}{dt} = \sum_{i=1}^{n}(Ki_{off}x_i - Ki_{on}y_iz) - k_5z + k_1z. \tag{21}$$

Here, x_i denotes the amount of the i-th antibody that binds to the viral antigen, y_i denotes the total number of the i-th antibody, w_i is a constant that indicates the threshold for maintaining the i-th antibody by the antigen-like substances in the environment. Ki_{on} denotes the binding constant of the free antibody i and the viral antigen. Ki_{off} denotes the reaction coefficient for the dissociation reaction of the antibody i viral complex. After we finish the second model, we want to show how our immune system selects good-binding antibodies. Therefore, we introduce the third model. The third model is a combination of the first model and second model. The difference between the third and second models is that they do not explicitly integrate environmental factors. Since each kind of antibody will have a corresponding environmental counterpart, it would be too complicated if we explicitly integrated all of them. Therefore, we use a simplified term w_i to represent the minimal antibody level threshold maintained by its corresponding environmental-antigenic substances. Each antibody has different binding kinetics toward the virus. The modeling result demonstrates the dynamic process by which our immune system selects those fast-binding antibodies over slower ones.

The above three models are numerically solved by MATLAB using the ode15s function [33].

3. Results

3.1. Physical Mechanism behind This Approach and the Underlying Relationships among the Three Models

Our antibody model is based on the following four rationales.

The first is the stimulus of the antibody–antigen complex on the regeneration of its corresponding antibody. The T-cells will induce the proliferation of the B-cells through the antigen-presenting process. Eventually, the corresponding antibody can be reproduced following the magnification of antibody-generating B-cells [34]. Although this regulation is critical in initializing the adaptive immune response, it was not explicitly represented in previous studies [24–30]. The $k_4 x$ term in Equation (12) expresses this positive feedback regulation. Integrating feedback regulation into the classical representation of biological reactions allows us to simulate reactions more comprehensively. Specifically, when we refer to the concentration alternation of antibody and virus, the change of antibody in Equation (12) should be $k_{-2}x - k_2yz - k_6y$, if we use chemical reactions to describe it. However, in this way, we ignore the virus–antibody complex's stimulation of the regeneration of antibodies. In immunology, those virus–antibody complexes will proliferate the corresponding antibody with the help of T-cells after an antigen presentation [35]. Therefore, we use an external $k_4 x$ term to model this positive feedback effect.

The second is the kinetic relationship between antibodies and antigens. This antigen–antibody binding is represented in the second process in the first diagram. The model is endowed with more dynamic characteristics with the introduction of a reversible reaction.

The third is the pernicious effect of antibodies on antigens, which is expressed as the $-k_3 x$ term in Equation (11). It has been recognized that the antibody–antigen complex can be efficiently removed with the help of functional immune cells such as NK (natural killer) cells [32].

The fourth is the introduction of environmental antigenic substances. We hypothesize that the environmental antigen contributes to the maintenance of memory cells. The declination of induced neutralizing antibodies is ubiquitous in almost all virus infections, such as Zika [36], Dengue [29], and SARS-CoV-2 [37]. However, this declination is always discontinuous after their concentration drops to a stable level. This is contributed by the existence of memory cells [38]. Immunologists regarded the long-lasting B-cells or T-cells as "memory cells." Experiments gradually demonstrated that although the so-called "memory cells" are some specific forms of immune cells [39,40], they have similar half-lives as normal CD8+ cells [41]. Therefore, maintaining antibody levels in "memory cells" should be explained as a state of equilibrium between decay and regeneration. We suppose this final equilibrium state derives from environmental antigen-like substances. Due to the presence of cross-interaction [42], the neutralizing antibody, no matter how specific it is, could have weak binding with other macromolecules in the solution. Those weak binding partners are defined as "environmental antigen-like substances" in our model; they could be food-resourced, air-resourced, or self-resourced. The presentation of those substances to T cells would give weak proliferation signals to neighboring cells. Therefore, a lower bound for specific neutralizing antibodies is present due to the positive feedback regulation of environmental antigenic substances. This hypothesis could explain why some antibodies persist in the body at detectable levels for a long time. We propose that the presence of environmental antigenic substances allows us to produce lifelong immunity to some pathogens.

The antibody level will eventually drop to zero based on the first model described in Equations (11)–(13). This implies that the antibody would finally fade in a short time. However, some antibodies can persist in the human body for a long time and provide lifelong protection, which is why we use vaccines. Following the last rationale, a second model is constructed to integrate environmental antigenic substances. The protein sequence of environmental antigen-like substances is ordinarily close to our own body and less antigenic, leading to a weak T-cell proliferation signal. An equilibrium state can be reached when the decay of memory cells is equal to their new synthesis. A good fit between the two scenarios further improves the reliability of this model. The first is that those environmental antigen-like molecules would help maintain the antibody at a certain level. This scenario can be easily recaptured in our second model. Our model can also explain the second scenario: environmental antigen-like substances cannot naturally elevate the antibody level to the final equilibrium concentration. Environmental antigenic substances can only help maintain the antibody level instead of triggering the increment. An exception is the allergy reaction, which happens under extreme conditions with the massive presence of environmental antigenic substances, especially when they have a strong binding affinity. Our second model can also recapture this phenomenon by increasing p and k_7 values.

Although the absolute equilibrium of specific antibodies is challenging to simulate mathematically, the presence of environmental antigen-like substances dramatically attenuates the antibody decay rate. The simulation results are shown in Figure 3. The decay rate of antibodies is closely related to the concentration of the antigen-like substance in the environment. The high concentration of the environmental antigenic substance would lead to a slow decay speed. The antibody decay rate is significantly slower (shown in the solid yellow curve) when there are massive environmental antigen-like substances, as shown in Figure 3A. The antibody level attenuation is also significantly influenced by the binding kinetics between environmental antigen and its corresponding antibody. The antibody decay rate is significantly slower (shown in the dashed yellow line) when k_3 sets a considerable value in Figure 3B. It represents a better binding kinetics between environmental antigen-like substance and its corresponding antibody.

It is common sense that environmental antigenic substances alone can hardly trigger antibody proliferation without pathogenic antigens. Numerical simulations also indicate that antigen-like substances in the environment hardly stimulate antibody proliferation. One scenario in which environmental antigen-like substances do not trigger antibody growth is recaptured in Figure 4A. The insufficient capacity of environmental antigens to induce antibody magnification derives from their poor binding affinity toward antibodies. In most cases, the presence of antigen-like substances in the environment will not directly stimulate antibody proliferation but will significantly attenuate the decay rate after antibody proliferation, as shown in Figure 4A. The situation can be altered when the antigen-like substances are sufficient or their binding affinities toward some antibodies are powerful. Antigen-like substances in the environment can also stimulate antibody levels, as shown in Figure 4B. This might help elucidate the underlying mechanism of allergy. A significant difference between environmental antigen-like substances and pathogenic antigens is that the former are deficient in self-replication. However, their concentrations are maintained at a relatively stable level due to constant replenishment from the environment.

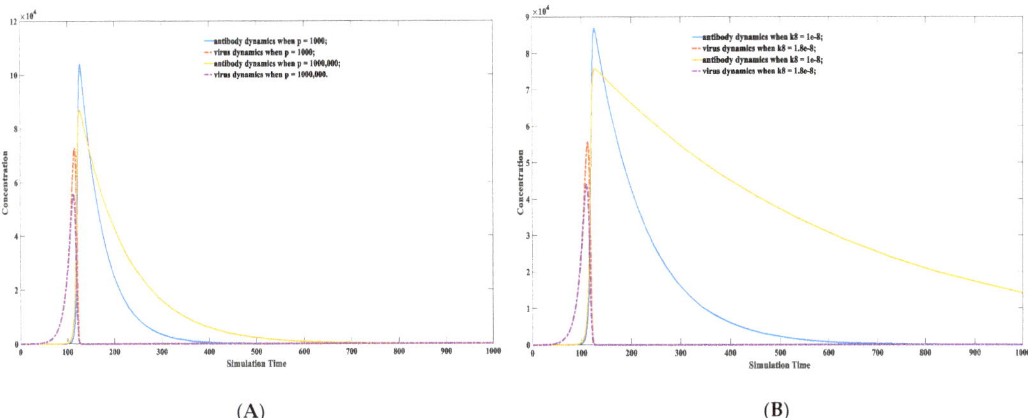

(**A**)   (**B**)

Figure 3. Antibody and virus dynamics modeling using different $p(0)$ value. $p(0)$ represents the concentration of environmental antigens. The virus-antibody dynamics are modeled under different environmental antigen concentrations (**A**) and different environmental antigen attributes (**B**). As shown in (**A**), the antibody decay rate is significantly slower (shown in solid yellow curve) when there is a large amount of environmental antigen-like substances. The parameter set we used is: $x(0) = 0$, $y(0) = 100$, $z(0) = 10$, $p(0) = 1000$ or $p(0) = 1{,}000{,}000$, $k_1 = 0.1$, $k_2 = 1 \times 10^{-5}$, $k_{-2} = 1 \times 10^{-14}$, $k_3 = 1$, $k_4 = 2$, $k_5 = 0.02$, $k_6 = 0.02$, $k_7 = 1 \times 10^{-8}$, $k_{-7} = 1 \times 10^{-14}$. As shown in (**B**), the antibody decay rate is significantly slower (shown in solid yellow curve) when k_8 sets a large value which corresponds to a better binding kinetics between environmental antigen-like stuff and its corresponding antibody. The parameter set we used is: $x(0) = 0$, $y(0) = 100$, $z(0) = 10$, $p(0) = 1{,}000{,}000$, $k_1 = 0.1$, $k_2 = 1 \times 10^{-5}$, $k_{-2} = 1 \times 10^{-14}$, $k_3 = 1$, $k_4 = 2$, $k_5 = 0.02$, $k_6 = 0.02$, $k_7 = 1 \times 10^{-8}$ or $k_7 = 1.8 \times 10^{-8}$, $k_{-7} = 1 \times 10^{-14}$.

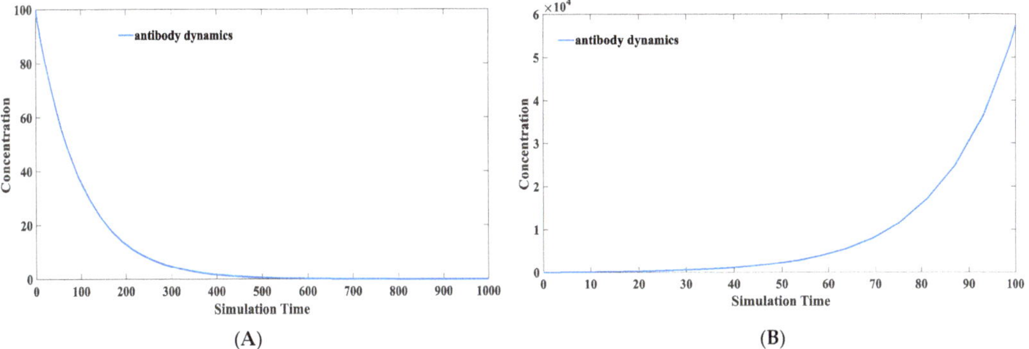

(**A**)   (**B**)

Figure 4. Antibody dynamics modeling with different k_7 values. (**A**) One scenario where environmental antigen-like substances do not trigger antibody growth. As shown in (**A**), the antibody does not engage proliferation due to the presence of environment antigen-like molecules. The parameter set we used is: $x(0) = 0$, $y(0) = 100$, $z(0) = 0$, $p(0) = 1{,}000{,}000$, $k_1 = 0.1$, $k_2 = 1 \times 10^{-5}$, $k_{-2} = 1 \times 10^{-14}$, $k_3 = 1$, $k_4 = 2$, $k_5 = 0.02$, $k_6 = 0.02$, $k_7 = 1 \times 10^{-8}$, $k_{-7} = 1 \times 10^{-14}$. (**B**) One scenario where environmental antigen-like substances do trigger antibody proliferation. As shown in (**B**), the antibody does engage proliferation due to the presence of environment antigen-like molecules. The parameter set we used is: $x(0) = 0$, $y(0) = 100$, $z(0) = 0$, $p(0) = 1{,}000{,}000$, $k_1 = 0.1$, $k_2 = 1 \times 10^{-5}$, $k_{-2} = 1 \times 10^{-14}$, $k_3 = 1$, $k_4 = 2$, $k_5 = 0.02$, $k_6 = 0.02$, $k_7 = 1 \times 10^{-7}$, $k_{-7} = 1 \times 10^{-14}$. The antibodies might significantly increase when the environmental antigenic substances bind strongly with them. This always induces allergic reactions.

3.2. Characteristics of Immune Response after Infected with Different Virus Strains

For the same type of virus, highly virulent strains have high replication activity [43]. Infection with highly virulent variants will lead to a prominent peak viral load and can exhibit strong transmissibility per time unit. However, its infection cycle is typically short. These infection dynamics are defined as acute infections in clinical performance [44]. The cytokine storm induced by a quickly elevated antibody level is responsible for the acute response [45]. Weak virulent strains have the opposite behaviors, which could lead to chronic infection in clinical settings. It does not typically induce a quick immune response, resulting in a more extended infection cycle. All those situations can be recaptured using our model, which is displayed in Figure 5.

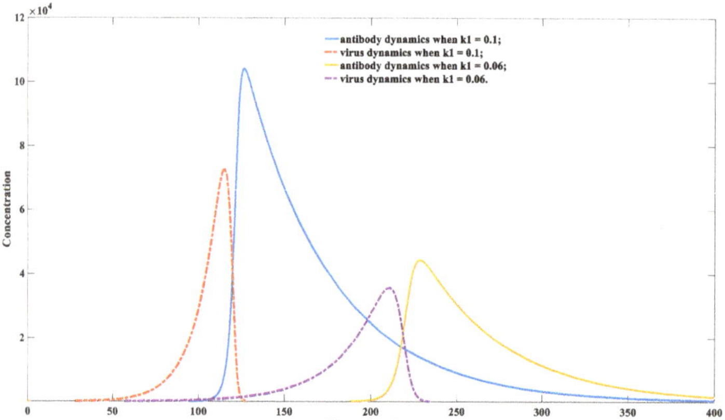

Figure 5. Different immune behaviors toward variants with different replication activities. The parameter set we used is: $x(0) = 0$, $y(0) = 100$, $z(0) = 10$, $p(0) = 1000$, $k_1 = 0.06$ or 0.1, $k_2 = 1 \times 10^{-5}$, $k_{-2} = 1 \times 10^{-14}$, $k_3 = 1$, $k_4 = 2$, $k_5 = 0.02$, $k_6 = 0.02$, $k_7 = 1 \times 10^{-8}$, $k_{-7} = 1 \times 10^{-14}$. As shown in this figure, the antibody response is milder when the host is infected by a less toxic strain (smaller k_1 value 0.06). The peak viral load is also less compared to its high toxic counterpart (bigger k_1 value 0.1). This indicates that even for the same virus infection, the immune response would vary greatly due to the differences in viral replication capacity of different variants.

It can be noticed from Figure 5 that, under the exact initial viral invasion dosage, the infection caused by a less virulent strain has a more extended latency phase. Its peak viral load is significantly lower than a more virulent strain. The maximal level of the yellow line is significantly lower than that of the dashed blue line, indicating that the antibody production induced by the weak virus infection is significantly lower than that induced by a virulent strain. It is also reflected in Figure 5 that the infection cycle caused by the weak virus is remarkably longer than its strong counterpart.

3.3. How the Immune System Screens for Highly Binding Antibodies

Our immune system can automatically detect suitable binding antibodies and selectively proliferate them. Simulating the antibody screening process by mathematical modeling remains challenging for computational biology researchers. This study provides a preliminary attempt to elucidate this process. Our model suggests a strong correlation between the proliferation potential of a specific antibody and its binding kinetics. Antibodies with fast binding rates would obtain rapid proliferation, while those with low binding rates proliferate slower. Thus, the antibodies with fast binding rates can grow faster and dominate in the final antibodies' composition, although various antibodies are triggered to proliferate during infection. Generally, antibodies with fast binding capacity have a more robust binding affinity. However, this relationship is not absolute. Binding affinity is the ratio of forward binding to reverse dissociation [46]. Infected individuals would have antibodies with good binding kinetics after infection [47,48]. Those antibodies display individual differences in binding affinity, with some antibodies having relatively stronger binding affinity [49]. We propose that the dominant factor affecting the efficiency of antibodies is not their absolute binding affinity K_d but the binding reaction constant k_2. Both k_2 and k_2 have physical meaning in Equation (12): k_2 stands for the forward reaction constant, while k_2 represents the reaction constant of the reverse one. The dissociation constant $K_d = k_2/k_{-2}$ can be used to evaluate the concentration of each component in an equilibrium state. K_d can also be transformed into the binding energy between those two macromolecules and thus is an important indicator that describes the binding affinity between two molecules [50].

As shown in Figure 6, the antibody with the fastest binding rate (solid blue line in this figure) demonstrates a maximal proliferation magnitude. This selection will eventually lead to a high proportion of fast-binding antibodies in the final antibody composition. Although these five antibodies have the same binding affinity of $K_d = 1 \times 10^{-9}$, their proliferations vary because of the differences in their binding speeds, and the antibody with the faster binding rate will gradually dominate. This example illustrates that the driving force in antibody reproduction is its binding speed, not its binding affinity. A different k_2 value and k_{-2} value is applied to keep the dissociation constant K_d as a fixed number. It is subversive to claim that the immune system would like to select those fast-binding antibodies rather than tightly binding antibodies. Experimental researchers always seek an antibody with an excellent binding affinity [47–49]. However, our model indicates that a solid-binding antibody might not perform well in preventing reinfection. Its kinetic behavior is more critical than its static binding affinity.

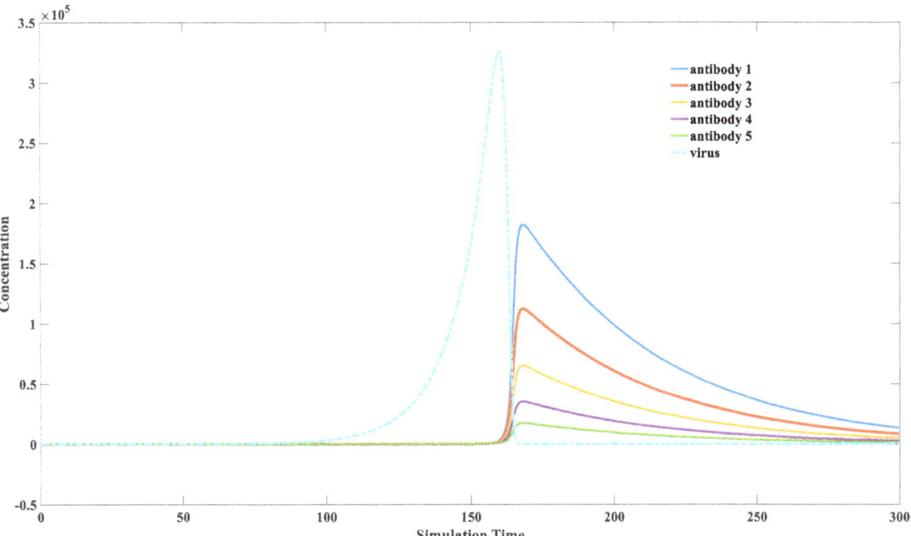

Figure 6. Dynamics of different antibodies with different kinetic attributes. The parameter sets we used are: $x(0) = 0$, $y(0) = 1$, $z(0) = 1$, $w = 1$, $k_1 = 0.1$, $k_2 = 1 \times 10^{-5}$, $k_{-2} = 1 \times 10^{-14}$, $k_3 = 1$, $k_4 = 2$, $k_5 = 0.02$, $k_6 = 0.02$, for antibody 1; $k_2 = 9 \times 10^{-6}$, $k_{-2} = 9 \times 10^{-15}$ for antibody 2; $k_2 = 8 \times 10^{-6}$, $k_{-2} = 8 \times 10^{-15}$ for antibody 3; $k_2 = 7 \times 10^{-6}$, $k_{-2} = 7 \times 10^{-15}$ for antibody 4; $k_2 = 6 \times 10^{-6}$, $k_{-2} = 6 \times 10^{-15}$ for antibody 5. It is demonstrated in this figure that the faster-binding antibodies engage amplification at a greater magnitude. In this way, the immune system selects those specific neutralizing antibodies.

3.4. High Concentrations of Weakly Binding Antibodies Can Provide Effective Protection

It has been statistically discovered that vaccination with other vaccines, such as the influenza vaccine, can also provide some degree of protection against SARS-CoV-2 [51–54]. The third model is utilized to explain this phenomenon. Those parameters used in Figure 7 are defined in Section 2.3.

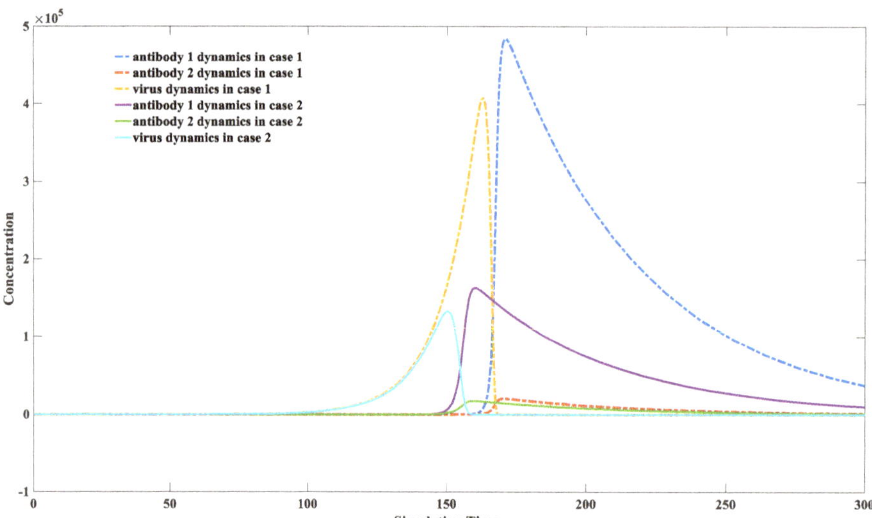

Figure 7. High concentrations of weakly binding antibodies can provide effective protection. A plot of the inhibitory capacity of a specific concentration of weakly binding antibodies against infection is presented. Two types of antibodies are presented in this figure: antibody 1 has a strong binding capacity ($K1_{on} = 1 \times 10^{-5}$) while antibody 2 has a relatively weak binding capacity ($K2_{on} = 5 \times 10^{-6}$). Two scenarios are simulated: both antibodies have low initial concentrations in case 1; weakly binding antibody has a high initial level in case 2. The parameter sets we used are: $x_1(0) = x_2(0) = 0$, $y_1(0) = y_2(0) = 1$, $z(0) = 1$, $w_1 = w_2 = 1$, $K1_{on} = 1 \times 10^{-5}$, $K1_{off} = 1 \times 10^{-14}$, $K2_{on} = 5 \times 10^{-6}$, $K2_{off} = 1 \times 10^{-14}$, $k_3 = 1$, $k_4 = 2$, $k_5 = k_6 = 0.02$, $k_1 = 0.1$ for case 1; $y_2(0) = 100$ for case 2. It can be seen that the peak viral load and antibody level are significantly lower in case 2, which corresponds to a milder immune response. It indicates that the elevated weakly binding antibodies could also provide protection against severe infection.

Figure 7 illustrates that the presence of weakly binding antibodies can produce an inhibitory capacity against the virus. The concentration of the weakly binding antibody increases from one in case 2 to a thousand in case 1. The maximal value of the dashed yellow line is smaller than that of the solid blue line, indicating a more substantial inhibition effect on virus reproduction. It can be noticed that the maximal level of the solid-binding antibody also decreases with the enhanced initial level of the weakly binding antibody. A particular concentration of weakly binding antibodies can simultaneously inhibit the proliferation of strongly binding, efficient antibodies in the host.

This example suggests that vaccinated people will always have better protection against reinfection than non-vaccinated people, regardless of the mutation of the virus. However, when we aim to stimulate the production of high levels of firmly bound antibodies through vaccination, the presence of weakly bound antibodies might interfere with this process. The vaccination effect on each individual might vary due to the weak antibody inference. Some individuals can induce sufficient fast-binding antibodies, while others cannot, due to differences in their initial antibody library. The simulation results indicate antibody interferences might be greatly enhanced in people vaccinated against other viruses. This explains why some vaccines may occasionally lead to a higher overall mortality rate [14,55]. Keeping a relatively high antibody level for a specific pathogen, such as SARS-CoV-2, will shelter more people from infection. However, it might also have the side effect of decreasing the efficiency of other vaccines. The long-term effect of repeated booster vaccinations on SARS-CoV-2 needs a comprehensive study.

3.5. Calculation of the Protection Time Brought by Natural Infection

The prediction of protection time remains to be elucidated using a computational approach. The calculation of protection time is presented in this section with the application of a numerical approach. Figure 8 illustrates how we calculate the protection time upon natural infection. Details about how we calculate the protection time are described below:

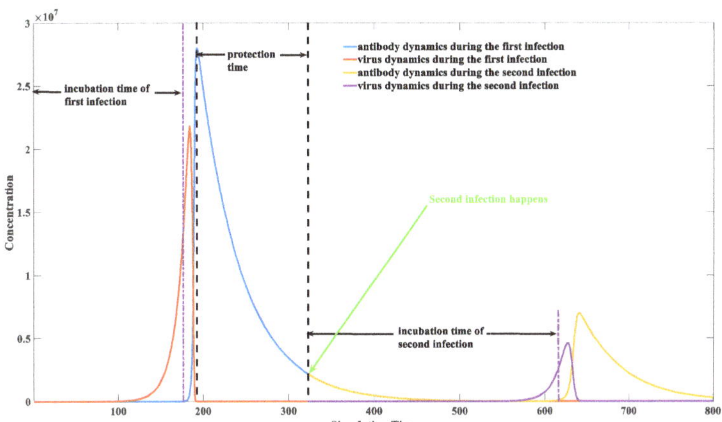

Figure 8. An illustration of protection time calculation. The parameter set we used is: $x(0) = 0$, $y(0) = 100$, $z(0) = 10$, $p(0) = 10{,}000$, $k_1 = 0.1$, $k_2 = 1 \times 10^{-7}$, $k_{-2} = 1 \times 10^{-14}$, $k_3 = 1$, $k_4 = 2$, $k_5 = 0.02$, $k_6 = 0.02$, $k_7 = 1 \times 10^{-8}$, $k_{-7} = 1 \times 10^{-14}$. Incubation time is calculated as the time interval between virus entrance and the production of antibodies. The viruses would engage a proliferation earlier than the antibodies. The patient is still asymptomatic even though the virus has reached a high level. Symptoms such as fever would appear when the antibody–virus complexes reach beyond a specific threshold. The second infection is marked with a green arrow. It can be seen in this figure that a second infection could occur when the IgG level drops below a certain level. It does not necessarily require a zero IgG level when a breakthrough infection happens.

First, we model the dynamic behaviors of all components (virus, virus–antibody complex, antibody). An initial number is assigned to each component before the first infection. For instance, $z(0) = 1$ stands for one invading virus. The virus number would vanish to zero soon after the antibody booms. The virus number dropped to zero after the 166th simulation time point, as illustrated in Figure 8.

After the antibody completely removes the virus, the second invasion is performed. The initial virus number would be manipulated to a non-zero integer at each following time unit, starting from 166 to 800. The early invasion would not lead to virus proliferation because the antibody level at that time was still high. However, the protection effect will gradually fade as the antibody level decreases. The earliest invasion that could lead to a virus proliferation is at the 326th time point in Figure 8. It is a time point when a breakthrough infection can happen. The time between the first infection and this time point is the protection duration brought by the first infection. Therefore, the protection time can be roughly estimated from 166th to 326th. We can also notice that the maximal viral load in the second infection is significantly lower than in the first infection. This indicates that the reinfection or breakthrough infection typically has a milder symptom than the first, although it will occur with a higher chance as time goes on. The protective effect will decrease as time increases, as shown in Figure 9 in the next section.

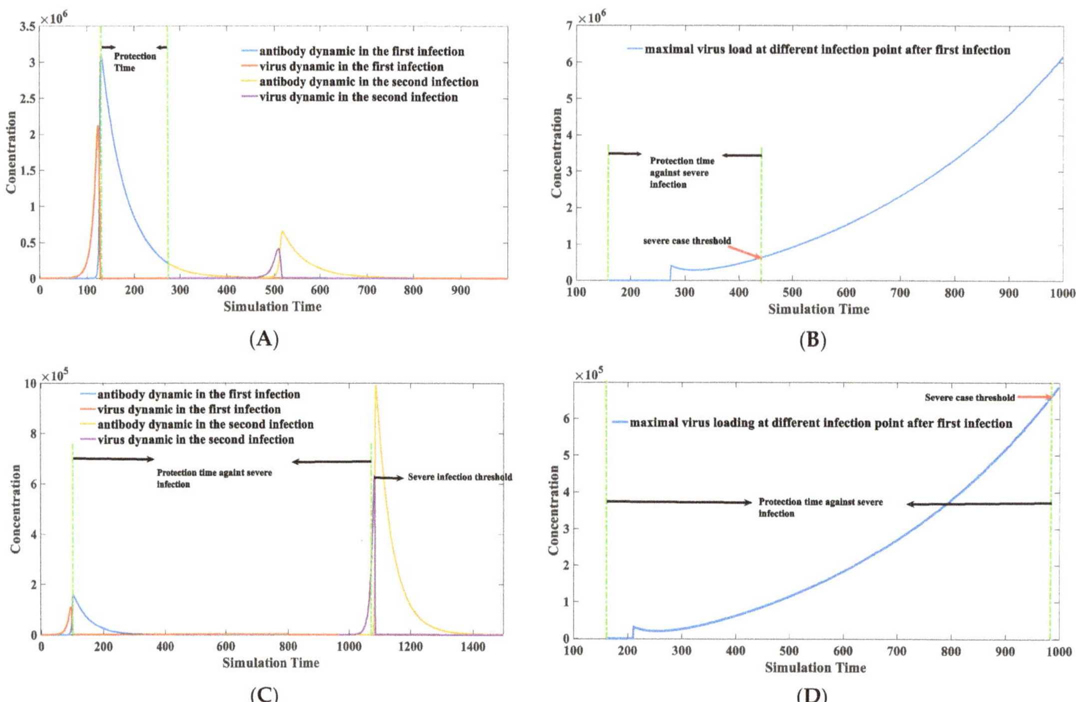

(A)　　　　　　　　　　　　　　　**(B)**

(C)　　　　　　　　　　　　　　　**(D)**

Figure 9. Protection time calculation when the antibody has a specific binding kinetic constant k_2. It can be inferred from this figure that a fast-binding neutralizing antibody would provide a longer protection time when we compared (**A**) with (**C**). The protection time against severe infection can also be prolonged, given the faster binding kinetics when comparing (**B**) to (**D**). (**A**) Protection time calculation when the antibody has a weak binding kinetic constant $k_2 = 1 \times 10^{-6}$. The parameter set we used is: $x(0) = 0$, $y(0) = 100$, $z(0) = 10$, $q(0) = 1 \times 10^6$, $k_1 = 0.1$, $k_2 = 1 \times 10^{-6}$, $k_{-2} = 1 \times 10^{-14}$, $k_3 = 1$, $k_4 = 2$, $k_5 = 0.02$, $k_6 = 0.02$, $k_7 = 1 \times 10^{-9}$, $k_{-7} = 1 \times 10^{-14}$. (**B**) Maximal virus load at different infection points when the antibody has a binding kinetic constant $k_1 = 1 \times 10^{-6}$. The parameter set used is the same as (**A**). (**C**) Protection time calculation when the antibody has a strong binding kinetic constant $k_2 = 1 \times 10^{-5}$. The parameter set we used is: $x(0) = 0$, $y(0) = 100$, $z(0) = 10$, $q(0) = 1 \times 10^6$, $k_1 = 0.1$, $k_2 = 1 \times 10^{-5}$, $k_{-2} = 1 \times 10^{-14}$, $k_3 = 1$, $k_4 = 2$, $k_5 = 0.02$, $k_6 = 0.02$, $k_7 = 1 \times 10^{-9}$, $k_{-7} = 1 \times 10^{-14}$. (**D**) Maximal virus load at different infection points when the antibody has a binding kinetic constant $k_2 = 1 \times 10^{-5}$. The parameter set used is the same as (**C**).

3.6. Factors Affecting the Duration of Antibody Protection: Concentration of the Environmental Antigen-like Substance, Viral Replication Capacity, and Antibody Binding Kinetics

Based on Equation (16), the virus will magnify itself when the $k_{-2}x - k_2yz - k_5z + k_1z$, term is larger than zero. The analytic solution on the reinfection possibility is difficult to derive, although the antibody level and the virus replication capacity directly influence this term. A high concentration of environmental antigenic substances p and an excellent binding kinetic k_7 can help maintain the antibody at a relatively high level, thus providing an extended protection time. Meanwhile, the antibody kinetic features could also greatly influence the protection time, which is shown in Figure 9. A lifelong protective effect can happen when a virus has a weak replication capacity (small k_1 value), a high concentration of environmental antigen-like substances (large p-value), and a fast-binding antibody with an enormous k_2 value. Typical examples of this category are the smallpox virus [4], tetanus bacillus [56], and so on. In contrast, other viruses, such as influenza and SARS-CoV-2, may not be able to trigger a super-long protection time [5]. Meanwhile, for influenza and SARS-CoV-2 infections, there is a considerable variation in the protective time after natural infection or vaccination [57,58]. This variation is mainly attributed to the difference in their corresponding antibody kinetics.

As claimed in the previous section, the binding kinetic constant k_2 is much more important than the binding affinity K_d. Figure 9 illustrates how k_2 influences the protection time in different infection cases.

The second infection time is marked in Figure 9A, which starts around the 275th time unit. The protection time can be roughly estimated to be between 125th and 275th time unit. The protection time is significantly extended (100th to 1080th time unit) in Figure 9C when a better binding antibody is introduced. A virus loading threshold for the severe case is arbitrarily set as 6.4×10^5 marked by the red arrow in Figure 9B. The protection time against severe infection can be derived based on Figure 9B,D, which is the time interval between those two green lines. Figure 9D demonstrates that the severe infection would happen after the 975th time unit with the same severe threshold. The protection time can be roughly estimated to range from 125th to 975th time unit. This protection phase is much longer compared to Figure 9B. This example demonstrates that an excellent binding antibody will protect people longer than the weak one. It indicates that the protection time brought by infection or vaccination could significantly vary for different individuals. The primary reason lies in the differences in the binding kinetics of antibodies–virus interaction. Another important discovery in this modeling is that people who generate suitable antibodies with fast binding kinetics rarely display significant symptoms in the first infection, which can be reflected in the smaller virus peak loading amount in the first infection in Figure 9C compared to Figure 9A. It is suggested that those fast-binding antibodies typically exist in patients without intense infection symptoms. Repeated vaccination might be limited in reshaping the composition of the final antibody reservoir, i.e., it does not significantly alter the binding activity of the antibodies. Therefore, targeted improvement of antibody binding activity against certain viruses is the key to extending vaccine protection in the long term. The heterogeneity of the individual antibody library might cause significant differences in the binding activity of antibodies to the same viral infection or vaccination, leading to a significant variation in protection duration. Therefore, we may need to change the vaccination strategy. Vaccination should induce the production of antibodies with high binding activity instead of a random incitation. The targeted induction of specific antibodies using gene editing provides a potential solution [59–61].

3.7. Parameter Estimation in Real Scenario

According to the data from clinical experiments, we can further fit the specific parameters to obtain the dynamic characteristics of antibodies within certain populations. The core content of this model is to accurately simulate the trend of mutual change between antibodies and viruses, so it is not required to add units to the mathematical model. By adjusting different parameters, especially the mean, and variance of the parameters, we can get the characteristics of the antibody dynamics behavior of a population. By comparing the available statistical data, we can estimate the distribution of antibody dynamics parameters of the whole population. Although these parameters have no direct physical meaning, they can more accurately reflect the decline of immunity in different individuals and the possibility of reinfection at different times.

We treat the concentration of the environmental antigenic substance as a constant value. The primary alternation is the difference in the binding capacity of the antibody in different individuals, which is defined as k_2 in our model. The binding kinetic between the environmental antigenic substance and the antibody, defined as k_7 in our model, varies among different people. The antibody kinetic performance characteristics of the population obtained using the above parameter combination are shown in Figure 10A.

It was demonstrated that IgG antibody levels to the SARS-CoV-2 nucleocapsid waned within months, according to six months of data from a longitudinal seroprevalence study of 3217 UK healthcare workers [62]. The simulation result has a good match with this clinical report. The main factor determining the peak antibody concentration is k_2, while the driving force determining the antibody waning speed comes from k_7. The change in the overall antibody level of the population over time is shown in Figure 10A, and the change in the overall protective efficacy is shown in Figure 10B. According to this model, we can judge that in the absence of virus mutation, the overall protective efficacy of an initial 100% efficacy of the COVID-19 vaccine in a human population (10,000 human simulations) would drop to 97.21% after 100 days, 65.44% after 150 days, 39.28% after 200 days, and 28% after 240 days. The simulation results in Figure 8B are consistent with the clinical data on vaccine effectiveness. Based on a cohort study among US veterans, for the period 1 February 2021 to 1 October 2021, vaccine effectiveness against infection (VE-I) declined over time ($p < 0.01$ for time dependence), even after adjusting for age, sex, and comorbidity. VE-I declined for all vaccine types, with the most significant declines for Janssen, followed by Pfizer-BioNTech and Moderna. Specifically, in March, VE-I was 86.4% for Janssen, 89.2% for Moderna, and 86.9% for Pfizer-BioNTech. By September, VE-I had declined to 13.1% for Janssen, 58.0% for Moderna, and 43.3% for Pfizer-BioNTech [63]. According to a retrospective cohort study on the effectiveness of the mRNA BNT162b2 vaccine, effectiveness against infections declined from 88% (95% CI 86–89) during the first month after complete vaccination to 47% after five months. Among sequenced infections, vaccine effectiveness against infections of the delta variant was high during the first month after complete vaccination (93%) but declined to 53% after four months [64]. Similar trends were observed in the cohort study conducted in Qatar [65].

3.8. Recovered Patients with Retest Positive for SARS-CoV-2

In SARS-CoV-2 infections, we noticed some rare occurrences of reinfections by patients themselves [66]. This recurrent positivity is not caused by the invasion of environmental viruses but by the re-proliferation of internal viruses [67]. The self-reinfection may lead to a second epidemic outbreak, making it more challenging to prevent and control the epidemic. The scenario of self-reinfection is a frequent phenomenon in virus infection. Many people suffer from recurrent respiratory infections [68]. It is also typical when HBV or HCV infection occurs [69]. Our model suggests a possible mechanism underlying this phenomenon. The patient will experience reinfection under certain parameter sets, as shown in Figure 11.

This example above shows that it is difficult to completely eliminate a specific pathogen in the presence of fast-binding antibodies. However, the peak viral load during the infection is small, as shown in the solid red line in Figure 11A. This will lead to a relatively mild symptom. The peak viral load in Figure 11A is significantly lower than the typical infection cases illustrated in Figures 8 and 9.

The low virus load does not have an intense stimulation on antibody production. Therefore, it is impossible to eliminate the virus from the body due to insufficient antibody quantity. The virus will regain the opportunity to proliferate when the antibody level decreases later. The patient may display a positive nucleic acid test result again. Multiple virus resurgences are marked in the dashed green cycle after the first infection, as displayed in Figure 11A. However, the infections in those cases are generally less symptomatic and even asymptomatic. A more extreme case is a long-term chronic infection, such as the appearance of a long-positive patient [70]. Like self-reinfection, an equilibrium state can be reached if the antibody–antigen interaction is moderate in those long-positive patients. In this case, pathogens would not be eliminated but maintained at a relatively stable level, forming a chronic infection. The low concentration of pathogenic antigens only provides a limited driving force for promoting antibody reproduction. The antibody and the virus will remain relatively low for a long time, as reflected in Figure 11B.

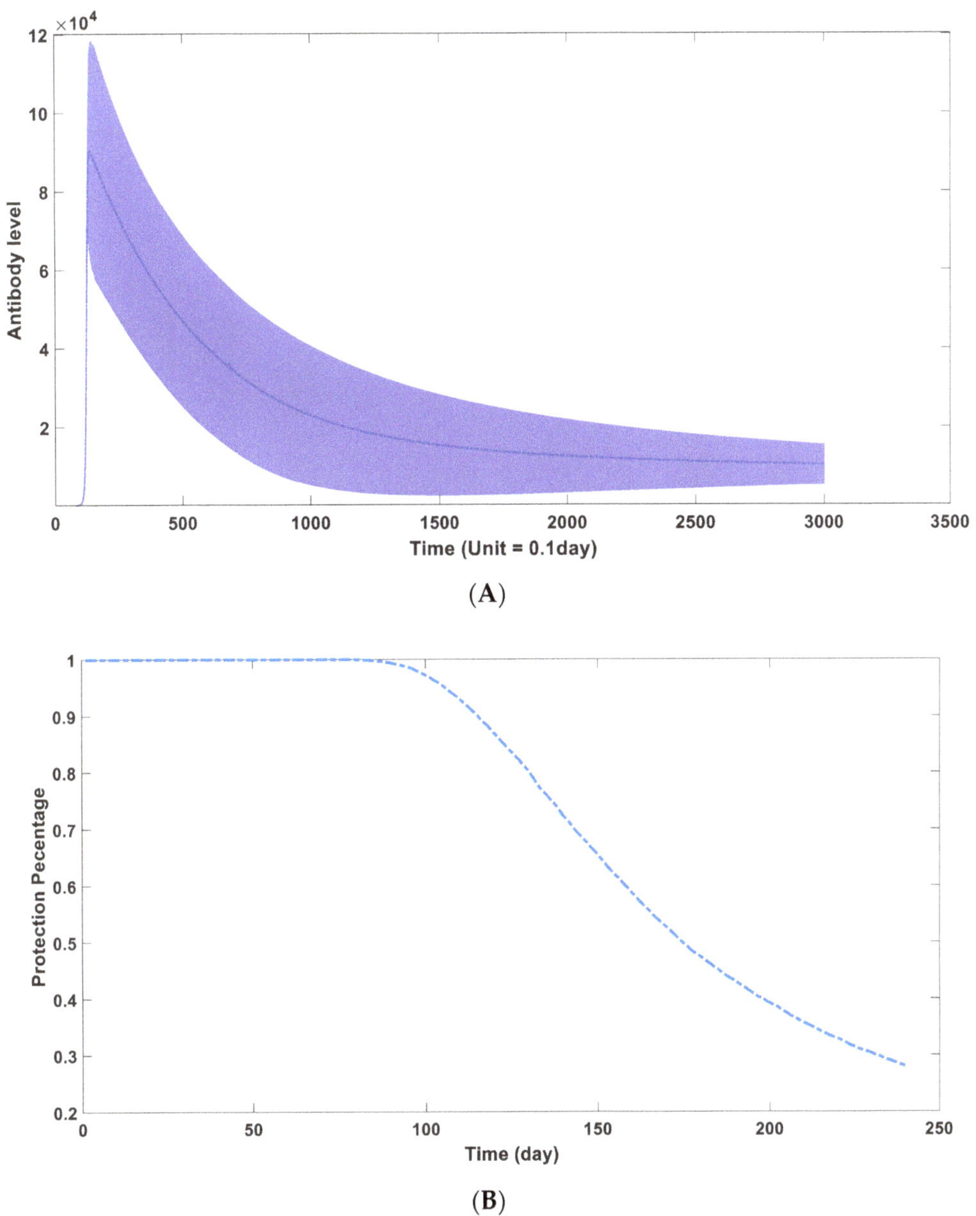

Figure 10. (**A**) the dynamic behavior of antibodies in the overall population through time. (The blue zone around mean curve stands for 95% confidence interval). It can be seen that the IgG level would significantly decline after reaching a peak level. However, its degradation does not follow a simple mathematical formula. Its descent rate would gradually decline and be maintained at a relatively stable level after 200 days. (**B**) The protection performance of antibodies in the overall population through time. It can be seen in this figure that the protection efficiency of induced neutralizing antibodies would be maintained at a relatively high level in the first 100 days. Its protection efficiency would engage a rapid decline after the first three months.

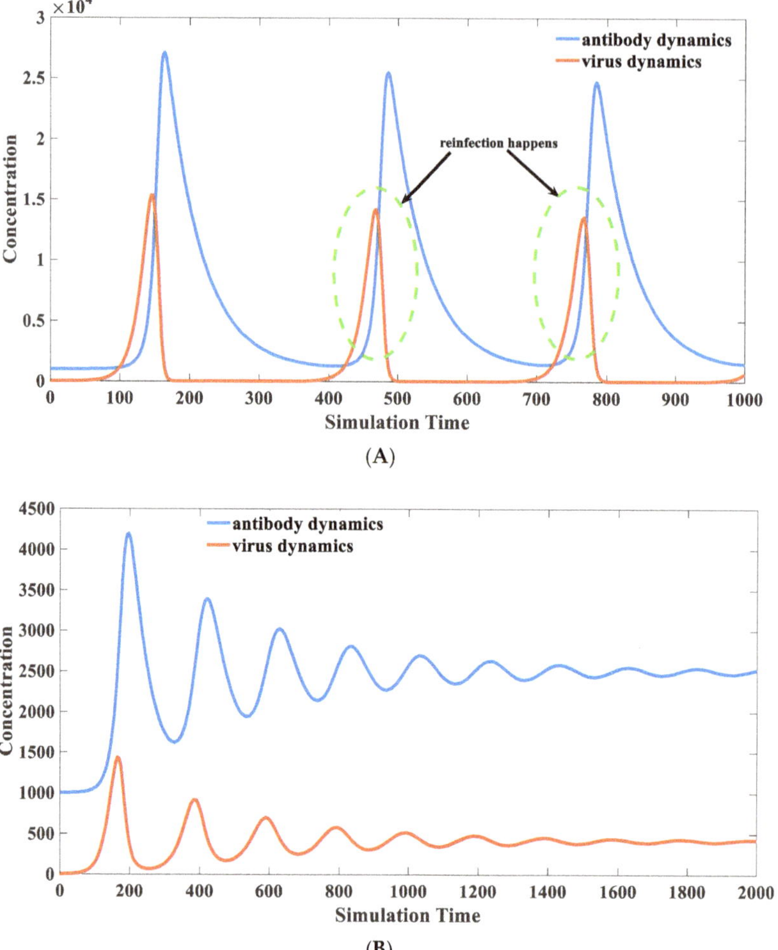

Figure 11. (**A**) Self-reinfection scenario. The parameter sets we used are: $x(0) = 0$, $y(0) = 1000$, $z(0) = 1$, $w = 1000$, $k_1 = 0.1$, $k_2 = 1 \times 10^{-5}$, $k_{-2} = 1 \times 10^{-14}$, $k_3 = 1$, $k_4 = 2$, $k_5 = 0.02$, $k_6 = 0.02$. Reinfections are represented as repeated waves in this figure. It indicates that the viruses could re-proliferate when the antibodies cannot completely eliminate them. The viruses start to proliferate when the antibodies drop to a certain level, leading to self-reinfection in this case. (**B**) Scenario of chronic infection. The parameter sets we used are: $x(0) = 0$, $y(0) = 1000$, $z(0) = 1$, $w = 1000$, $k_1 = 0.1$, $k_2 = 3 \times 10^{-5}$, $k_{-2} = 1 \times 10^{-5}$, $k_3 = 1$, $k_4 = 2$, $k_5 = 0.02$, $k_6 = 0.02$. In this case, pathogens would not be eliminated but maintained at a relatively stable level, forming a chronic infection. The low concentration of pathogenic antigens only provides a limited driving force for promoting antibody reproduction.

Pathogenic antigens and environmental antigenic substances all contribute to chronic inflammation. For chronic infections caused by pathogenic microorganisms, a short-term boost in antibody levels can be used to accomplish a complete clearance of pathogenic microorganisms [71]. Chronic infections could permanently disappear or significantly improve after they become acute infections in the clinic. The chronic symptom will disappear or become invisible after the healing from the acute infection [72]. Conversely, chronic inflammation caused by environmental factors can be removed by shielding environmental antigens for a certain period [73]. The blockade of antibody–antigen interaction will decrease the corresponding antibody level, thereby attenuating or eliminating the immune response. An allergic reaction can be significantly alleviated in the upcoming contact after this treatment.

4. Discussion

The application of mathematical modeling displays good epidemic forecast capacity at the population level [22,23,74,75]. Mathematical models are also helpful in quantitatively elucidating the immune process's dynamics on an individual level. We are motivated to develop a new mathematical model that can explicitly model the antibody dynamics based on those pioneering modeling attempts [17–21,24–30]. A computational antibody dynamic model is finally proposed, which can help us explain some phenomena mentioned in the introduction.

I. How are memory cells maintained?

The environmental antigen-like substances maintain memory cells in our model. It was demonstrated that, in the absence of a pathogen, the environmental antigen could not directly stimulate antibody proliferation in most cases except in allergic reactions. The concentration of memory cells is closely correlated to the property and concentration of its corresponding environmental antigenic substances. Some memory cells against a specific pathogen can be maintained at a high level for decades. The persistent stimulation by the environmental antigenic substances, rather than the eternal lifespan, leads to the long-term existence of those immune memory cells. The existence of super-binding antibodies and high concentrations of environmental antigens can also trigger allergic reactions.

II. How does our immune system screen for antibodies with solid binding affinity?

Based on the model, antibodies with faster binding kinetics would increase proliferation. The binding constant is not linearly correlated to the binding affinity. The immune system tends to select antibodies with fast binding kinetics rather than solid binding affinity. The underlying mechanism comes from the positive feedback regulation of the virus–antibody complex. A good binding antibody would lead to a faster generation of the virus–antibody complex. This complex would further promote the proliferation of its corresponding antibody.

III. Why do people vaccinated by the influenza vaccine or other vaccines have a lower mortality rate from SARS-CoV-2 infection?

An interesting phenomenon during the COVID-19 pandemic is that the mortality rate is significantly lower in people who have received the influenza vaccine and other vaccines than in their unvaccinated counterparts [51–54]. The non-specific binding antibody level can be elevated after non-specific vaccination. Although weakly binding antibodies inhibit the proliferation of firmly binding antibodies in vivo, they can significantly inhibit virus proliferation and reduce peak viral load. This would lead to a lower mortality rate and milder symptoms. The potential concerns are also discussed in Section 3.4. A particular concentration of weakly binding antibodies can inhibit the proliferation of strongly binding antibodies, as revealed in Figure 7. The rationality of repeated booster vaccination needs a comprehensive evaluation because it might weaken the immune response to other pathogens.

IV. How could we effectively calculate the protection duration of a specific antibody?

Concerns are remarkably booming when people realize the protection of the SARS-CoV-2 vaccine has declined over time. Given the experimental data on viruses and antibodies, the kinetic parameters of this model can be inferred. The protection time of the vaccine or natural infection can be further deduced. Personalized prediction is also feasible with the availability of individual antibody behavior. The protection times of different individuals can be immensely varied, ranging from extraordinarily durable to very transient. Three factors that influence the protection time are discussed in Section 3.6. The virus's attributes, such as the replication speed, influence the antibody production time. A faster-replicating virus, for instance, an RNA virus, inclines to have a shorter antibody protection time after natural infection. Virus immunogenicity against T-cells influences the duration of antibody protection. Viruses with strong T-cell immunogenicity provide a strong stimulus for antibody proliferation, exhibiting extended and even lifelong protection. The antibody binding kinetics also contribute to the difference in its protection performance, as illustrated in Section 3.6.

V.   Why are there cases of self-reinfection?

Self-reinfection prevails in less virulent strains such as the Omicron variant. Our model provides a plausible mechanism underlying this unusual phenomenon. The individual might exhibit self-reinfection with the presence of a good binding antibody; repeated infection can also be displayed when a less virulent strain invades the patient. This can help explain why Omicron infections are inclined to be asymptomatic and self-reinfect.

An extreme case of self-reinfection is a chronic status in which the individual displays positive nucleic acid results for a long time. The chronic infection of SARS-CoV-2, together with other chronic infections, can also be explained by this model. It was demonstrated that a sub-equilibrium state could be maintained under specific parameter combinations with a relatively low antibody level, thus exhibiting long-term chronic inflammation. The long-term chronic infection is maintained by the constantly active (even weakly) antigens, which can come (i) from the non-coding human genome, where there are, for example, proteins of ancient viruses such as HER-V that produce antibodies a long time after their passage in the host genome, providing cross-protection against other infectious agents [76–79]; (ii) from the V(D)J mechanism of the innate immunity which can also give birth to antibodies with a large spectrum of actively participating to non-specific defenses against pathogens [80–83]; and (iii) from other vaccines (it is for example known that BCG reprograms the innate immunity [84]).

VI. Why do vaccinations show considerable differences in protection?

According to our antibody kinetic model, the antibody interference effect is the most important contributing factor to the differences in vaccine protection time, apart from viral mutations. Large amounts of interfering antibodies can inhibit the production of high concentrations of fast-binding antibodies after vaccination, causing a decrease in protection efficiency and protection period. The antibody reservoir's heterogeneity in the human population also influences the generation of those fast-binding antibodies. The presence of good templates could promote the production of fast-binding mature antibodies after a few rounds of somatic hypermutation. Those fast-binding antibodies could provide extended protection. Environmental antigenic substances also influence the protection duration by influencing the antibodies' decay rate. Clinical reports [85] have demonstrated that age and gender are statistically associated with differences in antibody response after vaccination. The IgG antibodies triggered by the SARS-CoV-2 BNT162b2 vaccine significantly varied among ages. Young people tend to generate more neutralizing antibodies compared to their elder counterparts. This mainly reflects a variation in k_4 values in our model. A faster antibody generation capacity, equal to a larger k_4 value, would confer a stronger immunity to younger people. However, it does not guarantee that the protection duration in young people would be longer than that in older people for each individual. As discussed in Section 3.6, the antibody attributes, together with environmental antigenic substances, strongly impact the protection time. Suitable antibodies with fast binding constant k_2 and high concentrations of environmental antigenic substances would prolong the protection time. The clinical report also indicates that females have stronger immunity than males. However, this trend is as insignificant as the influences of age. In this study, the forecast is performed to predict the population's behavior, which does not consider the influences of age and gender. A more accurate and specific forecast can be performed in a future study when we integrate age and gender differences into the model. In this case, a different k_4 value would be assigned to each subgroup.

VII. How can we improve the protective efficiency and duration of vaccines?

Persistent exposure to massive antigenic substances can boost the neutralizing antibody level in a short period. However, the repeated booster can hardly magnify the proportion of solid-binding antibodies in the final antibody composition. The targeted induction of antibodies with excellent binding kinetics could provide prolonged protection against reinfection. Researchers are paving the way in this direction through gene-editing techniques [59–61]. Another concern with the constant booster shots is that the antibody interference might diminish the effectiveness of other vaccines, including the expected vaccines against massively mutated strains of SARS-CoV-2. Therefore, instead of pursuing a short-term antibody surge, we suggest that researchers aim to induce antibodies with fast-binding activity.

Besides providing a quantitative explanation of virus–host dynamics during the infection cycle, this model has several practical applications. One application is to predict the evolutionary direction of SARS-CoV-2 mathematically. The relationship between virulence and transmissibility can be simulated. A delicate equilibrium point that optimizes the transmissibility can be numerically obtained. Based on this model, we predict that the virulence of SARS-CoV-2 might further decrease, accompanied by an enhancement of transmissibility [86]. The second application is to optimize the vaccine inoculum dose mathematically. The antigen level, which can be represented as the initial inoculum dose, significantly influences the vaccine's efficiency. A low dose might not be able to induce sufficient IgG to provide durable protection. In contrast, a high dose is more inclined to trigger a high level of IgG. However, it also has a more substantial adverse effect, representing a high antibody–virus complex level in this model. The high dose also increases its production cost. Therefore, a mathematical optimization can be performed to evaluate the optimal inoculum dose before clinical trials. Thirdly, this study could theoretically pave the way for future vaccine development. The protection duration of antibodies generated by natural infection is explicitly modeled in this study. The modeling of vaccines is different because the vaccine, independent of its type, does not replicate but is injected with an extremely high initial viral concentration. Different vaccines would have a different protection duration against breakthrough infection. vaccination efficiency is very complicated and generally weaker than the protection brought by a natural infection. The inactivation process for the inactivated vaccine would truncate its original structure, especially the epitope spots of spikes and nucleocapsid proteins. Its efficiency might be less than the mRNA vaccine. The vaccine based on the viral vector also generated deficient antigens compared to the real virus. After all, two major factors influence the protection duration of different vaccines: the peak level of neutralizing antibodies and viral mutations. A large dose of inoculum would trigger more neutralizing antibodies, leading to more extended protection and stronger adverse effects caused by the immune response. All vaccines would display weakened or even zero protection against variants with tremendous mutations due to a reduced k_2 value in this model. mRNA vaccine is better than the traditional vaccine because it can gradually generate realistic antigens. The traditional concept for vaccine development, with the application of an inactivated virus, viral vector, or mRNA, strictly forbids the use of a live virus. Our model demonstrated that the attenuated virus with proliferation constraints (smaller k_1 value in our model) could induce the immune response and antibody level in a milder way (shown in Figure 5). Those attenuated live viruses form a solid basis for asymptomatic infections [87]. There are at least two remarkable advantages of attenuated live viruses in future vaccine development. Firstly, since our model demonstrates that the peak antibody level is independent of the initial virus concentration, a minimal dose of live viruses could elevate specific IgG to a high level. It is also more infectious than traditional vaccines. This means a large number of populations can be covered with a few vaccination attempts. Secondly, despite the unlikely inactivation process, the attenuated virus is structurally identical to the original virus. Therefore, the specificity of neutralizing antibodies triggered by those attenuated viruses is the same as that of those induced by natural infections. As demonstrated by our model, an excellent binding capacity could induce a more durable protection time. The safety of those attenuated live viruses can also be guaranteed based on the principle of RNA replication: truncated mRNA cannot produce fixed offspring (full genome size).

Nevertheless, we also have to admit that our model has many hypotheses and uncertainties. Furthermore, it still lacks an in vivo experiment data fitting process, while many of the predictions and theories in the article remain to be confirmed experimentally and statistically in the future.

Author Contributions: Conceptualization, Z.X.; methodology, Z.X. and J.D.; writing—original draft preparation, Z.X.; writing—review and editing, D.W. and H.Z.; funding acquisition, Z.X. All authors have read and agreed to the published version of the manuscript.

Funding: This research was funded by DeZhou University, grant number 30101418.

Data Availability Statement: Matlab source codes are available at https://github.com/zhaobinxu23/antibody_dynamics, accessed on 1 January 2022.

Acknowledgments: We thank Zuyi Huang from Villanova University. Yushan Zhu from Tsinghua University, Jian Song, Peiyan Guan, XiangYong Li, and Zhenlin Wei from Dezhou University for helpful conversations, comments, and clarifications.

Conflicts of Interest: The authors declare no conflict of interest.

References

1. Skegg, D.; Gluckman, P.; Boulton, G.; Hackmann, H.; Karim, S.S.A.; Piot, P.; Woopen, C. Future scenarios for the COVID-19 pandemic. *Lancet* **2021**, *397*, 777–778. [CrossRef] [PubMed]
2. Xu, Z.; Zhang, H. If we cannot eliminate them, should we tame them? Mathematics underpinning the dose effect of virus infection and its application on COVID-19 virulence evolution. *medRxiv* **2021**. [CrossRef]
3. Xu, Z.; Zhang, H.; Huang, Z. A Continuous Markov-Chain Model for the Simulation of COVID-19 Epidemic Dynamics. *Biology* **2022**, *11*, 190. [CrossRef] [PubMed]
4. Parrino, J.; Graham, B.S. Smallpox vaccines: Past, present, and future. *J. Allergy Clin. Immunol.* **2006**, *118*, 1320–1326. [CrossRef] [PubMed]
5. Sadoff, J.; Gray, G.; Vandebosch, A.; Cardenas, V.; Shukarev, G.; Grinsztejn, B.; Goepfert, P.A.; Truyers, C.; Fennema, H.; Spiessens, B.; et al. Safety and efficacy of single-dose Ad26. COV2. S vaccine against COVID-19. *N. Engl. J. Med.* **2021**, *384*, 2187–2201. [CrossRef] [PubMed]
6. Allen, J.D.; Feng, W.; Corlin, L.; Porteny, T.; Acevedo, A.; Schildkraut, D.; King, E.; Ladin, K.; Fu, Q.; Stopka, T.J. Why are some people reluctant to be vaccinated for COVID-19? A cross-sectional survey among US Adults in May–June 2020. *Prev. Med. Rep.* **2021**, *24*, 101494. [CrossRef]
7. Ruiz, J.B.; Bell, R.A. Predictors of intention to vaccinate against COVID-19: Results of a nationwide survey. *Vaccine* **2021**, *39*, 1080–1086. [CrossRef]
8. Klugar, M.; Riad, A.; Mekhemar, M.; Conrad, J.; Buchbender, M.; Howaldt, H.P.; Attia, S. Side effects of mRNA-based and viral vector-based COVID-19 vaccines among German healthcare workers. *Biology* **2021**, *10*, 752. [CrossRef]
9. Riad, A.; Pokorna, A.; Attia, S.; Klugarova, J.; Koscik, M.; Klugar, M. Prevalence of COVID-19 Vaccine Side Effects among Healthcare Workers in the Czech Republic. *J. Clin. Med.* **2021**, *10*, 1428. [CrossRef]
10. Rzymski, P.; Camargo, C.A., Jr.; Fal, A.; Flisiak, R.; Gwenzi, W.; Kelishadi, R.; Leemans, A.; Nieto, J.J.; Ozen, A.; Perc, M.; et al. COVID-19 Vaccine Boosters: The good, the bad, and the ugly. *Vaccines* **2021**, *9*, 1299. [CrossRef]
11. Gupta, R.K.; Topol, E.J. COVID-19 vaccine breakthrough infections. *Science* **2021**, *374*, 1561–1562. [CrossRef] [PubMed]
12. Cevik, M.; Grubaugh, N.D.; Iwasaki, A.; Openshaw, P. COVID-19 vaccines: Keeping pace with SARS-CoV-2 variants. *Cell* **2021**, *184*, 5077–5081. [CrossRef] [PubMed]
13. Krause, P.R.; Fleming, T.R.; Peto, R.; Longini, I.M.; Figueroa, J.P.; Sterne, J.A.C.; Cravioto, A.; Rees, H.; Higgins, J.P.T.; Boutron, I.; et al. Considerations in boosting COVID-19 vaccine immune responses. *Lancet* **2021**, *398*, 1377–1380. [CrossRef] [PubMed]
14. Klein, S.L.; Shann, F.; Moss, W.J.; Benn, C.S.; Aabye, P. RTS, S malaria vaccine and increased mortality in girls. *Mbio* **2016**, *7*, e00514-16. [CrossRef]
15. Bocharov, G.; Volpert, V.; Ludewig, B.; Meyerhans, A. *Mathematical Immunology of Virus Infections*; Springer International Publishing: Cham, Switzerland, 2018.
16. Eftimie, R.; Gillard, J.J.; Cantrell, D.A. Mathematical models for immunology: Current state of the art and future research directions. *Bull. Math. Biol.* **2016**, *78*, 2091–2134. [CrossRef]
17. Smith, A.M.; Adler, F.R.; McAuley, J.L.; Gutenkunst, R.N.; Ribeiro, R.M.; McCullers, J.A.; Perelson, A.S. Effect of 1918 PB1-F2 expression on influenza A virus infection kinetics. *PLoS Comput. Biol.* **2011**, *7*, e1001081. [CrossRef]

18. Baccam, P.; Beauchemin, C.; Macken, C.A.; Hayden, F.G.; Perelson, A.S. Kinetics of influenza A virus infection in humans. *J. Virol.* **2006**, *80*, 7590–7599. [CrossRef]

19. Beauchemin, C.A.; McSharry, J.J.; Drusano, G.L.; Nguyen, J.T.; Went, G.T.; Ribeiro, R.M.; Perelson, A.S. Modeling amantadine treatment of influenza A virus in vitro. *J. Theor. Biol.* **2008**, *254*, 439–451. [CrossRef]

20. Handel, A.; Longini, I.M., Jr.; Antia, R. Neuraminidase inhibitor resistance in influenza: Assessing the danger of its generation and spread. *PLoS Comput. Biol.* **2007**, *3*, e240. [CrossRef]

21. Canini, L.; Perelson, A.S. Viral kinetic modeling: State of the art. *J. Pharmacokinet. Pharmacodyn.* **2014**, *41*, 431–443. [CrossRef]

22. Khajanchi, S.; Sarkar, K. Forecasting the daily and cumulative number of cases for the COVID-19 pandemic in India. *Chaos Interdiscip. J. Nonlinear Sci.* **2020**, *30*, 071101. [CrossRef]

23. Mondal, J.; Khajanchi, S. Mathematical modeling and optimal intervention strategies of the COVID-19 outbreak. *Nonlinear Dyn.* **2022**, *109*, 177–202. [CrossRef] [PubMed]

24. Hancioglu, B.; Swigon, D.; Clermont, G. A dynamical model of human immune response to influenza A virus infection. *J. Theor. Biol.* **2007**, *246*, 70–86. [CrossRef] [PubMed]

25. Handel, A.; Longini, I.M., Jr.; Antia, R. Towards a quantitative understanding of the within-host dynamics of influenza a infections. *J. R. Soc. Interface* **2010**, *7*, 35–47. [CrossRef] [PubMed]

26. Lee, H.Y.; Topham, D.J.; Park, S.Y.; Hollenbaugh, J.; Treanor, J.; Mosmann, T.R.; Jin, X.; Ward, B.M.; Miao, H.; Holden-Wiltse, J.; et al. Simulation and prediction of the adaptive immune response to influenza A virus infection. *J. Virol.* **2009**, *83*, 7151–7165. [CrossRef] [PubMed]

27. Miao, H.; Hollenbaugh, J.A.; Zand, M.S.; Holden-Wiltse, J.; Mosmann, T.R.; Perelson, A.S.; Wu, H.; Topham, D.J. Quantifying the early immune response and adaptive immune response kinetics in mice infected with influenza A virus. *J. Virol.* **2010**, *84*, 6687–6698. [CrossRef]

28. Tridane, A.; Kuang, Y. Modeling the interaction of cytotoxic T lymphocytes and influenza virus infected epithelial cells. *Math. Biosci. Eng.* **2010**, *7*, 171.

29. Clapham, H.E.; Quyen, T.H.; Kien, D.T.H.; Dorigatti, I.; Simmons, C.; Ferguson, N. Modelling virus and antibody dynamics during dengue virus infection suggests a role for antibody in virus clearance. *PLoS Comput. Biol.* **2016**, *12*, e1004951. [CrossRef]

30. Best, K.; Guedj, J.; Madelain, V.; de Lamballerie, X.; Lim, S.Y.; Osuna, C.E.; Whitney, J.B.; Perelson, A.S. Zika plasma viral dynamics in nonhuman primates provides insights into early infection and antiviral strategies. *Proc. Natl. Acad. Sci. USA* **2017**, *114*, 8847–8852. [CrossRef]

31. Clark, E.A.; Ledbetter, J.A. How B and T cells talk to each other. *Nature* **1994**, *367*, 425–428. [CrossRef]

32. Mujal, A.M.; Delconte, R.B.; Sun, J.C. Natural killer cells: From innate to adaptive features. *Annu. Rev. Immunol.* **2021**, *39*, 417–447. [CrossRef] [PubMed]

33. Shampine, L.F.; Reichelt, M.W. The matlab ode suite. *SIAM J. Sci. Comput.* **1997**, *18*, 1–22. [CrossRef]

34. Abbas, A.K.; Lichtman, A.H.; Pober, J.S. B cell activation and antibody production. *Cell. Mol. Immunol.* **2005**, *1*, 243–267.

35. Guermonprez, P.; Valladeau, J.; Zitvogel, L.; Théry, C.; Amigorena, S. Antigen presentation and T cell stimulation by dendritic cells. *Annu. Rev. Immunol.* **2002**, *20*, 621–667. [CrossRef]

36. Bonyah, E.; Okosun, K.O. Mathematical modeling of Zika virus. *Asian Pac. J. Trop. Dis.* **2016**, *6*, 673–679. [CrossRef]

37. Yamayoshi, S.; Yasuhara, A.; Ito, M.; Akasaka, O.; Nakamura, M.; Nakachi, I.; Koga, M.; Mitamura, K.; Yagi, K.; Maeda, K.; et al. Antibody titers against SARS-CoV-2 decline, but do not disappear for several months. *EClinicalMedicine* **2021**, *32*, 100734. [CrossRef]

38. Pennock, N.D.; White, J.T.; Cross, E.W.; Cheney, E.E.; Tamburini, B.A.; Kedl, R.M. T cell responses: Naive to memory and everything in between. *Adv. Physiol. Educ.* **2013**, *37*, 273–283. [CrossRef]

39. Kurosaki, T.; Kometani, K.; Ise, W. Memory B cells. *Nat. Rev. Immunol.* **2015**, *15*, 149–159. [CrossRef]

40. Inoue, T.; Moran, I.; Shinnakasu, R.; Phan, T.G.; Kurosaki, T. Generation of memory B cells and their reactivation. *Immunol. Rev.* **2018**, *283*, 138–149. [CrossRef]

41. Van den Berg, S.P.H.; Derksen, L.Y.; Drylewicz, J.; Nanlohy, N.M.; Beckers, L.; Lanfermeijer, J.; Gessel, S.N.; Vos, M.; Otto, S.A.; de Boer, R.J.; et al. Quantification of T-cell dynamics during latent cytomegalovirus infection in humans. *PLoS Pathog.* **2021**, *17*, e1010152. [CrossRef]

42. Aalberse, R.C.; Kleine, B.I.; Stapel, S.O.; van Ree, R. Structural aspects of cross-reactivity and its relation to antibody affinity. *Allergy* **2001**, *56*, 27–29. [CrossRef] [PubMed]

43. Bull, J.J.; Lauring, A.S. Theory and empiricism in virulence evolution. *PLoS Pathog.* **2014**, *10*, e1004387. [CrossRef] [PubMed]

44. Iqbal, F.M.; Lam, K.; Sounderajah, V.; Clarke, J.M.; Ashrafian, H.; Darzi, A. Characteristics and predictors of acute and chronic post-COVID syndrome: A systematic review and meta-analysis. *EClinicalMedicine* **2021**, *36*, 100899. [CrossRef] [PubMed]

45. Fajgenbaum, D.C.; June, C.H. Cytokine storm. *N. Engl. J. Med.* **2020**, *383*, 2255–2273. [CrossRef] [PubMed]

46. Kastritis, P.L.; Bonvin, A.M.J.J. On the binding affinity of macromolecular interactions: Daring to ask why proteins interact. *J. R. Soc. Interface* **2013**, *10*, 20120835. [CrossRef]

47. Seydoux, E.; Homad, L.J.; MacCamy, A.J.; Parks, K.R.; Hurlburt, N.K.; Jennewein, M.F.; Akins, N.R.; Stuart, A.B.; Wan, Y.H.; Feng, J.; et al. Analysis of a SARS-CoV-2-infected individual reveals development of potent neutralizing antibodies with limited somatic mutation. *Immunity* **2020**, *53*, 98–105.e5. [CrossRef] [PubMed]

48. Ju, B.; Zhang, Q.; Ge, J.; Wang, R.; Sun, J.; Ge, X.; Yu, J.; Shan, S.; Zhou, B.; Song, S.; et al. Human neutralizing antibodies elicited by SARS-CoV-2 infection. *Nature* **2020**, *584*, 115–119. [CrossRef]

49. Muecksch, F.; Weisblum, Y.; Barnes, C.O.; Schmidt, F.; Schaefer-Babajew, D.; Wang, Z.; Lorenzi, J.C.; Flyak, A.I.; DeLaitsch, A.T.; Huey-Tubman, K.E.; et al. Affinity maturation of SARS-CoV-2 neutralizing antibodies confers potency, breadth, and resilience to viral escape mutations. *Immunity* **2021**, *54*, 1853–1868.e7. [CrossRef]

50. Schneider, E.D.; Kay, J.J. Order from disorder: The thermodynamics of complexity in biology. In *What Is Life? The Next Fifty Years: Speculations on the Future of Biology*; Murphy, M.P., O'neill, L.A.J., Eds.; Cambridge University Press: Cambridge, UK, 1997; p. 161.

51. Marín-Hernández, D.; Schwartz, R.E.; Nixon, D.F. Epidemiological evidence for association between higher influenza vaccine uptake in the elderly and lower COVID-19 deaths in Italy. *J. Med. Virol.* **2021**, *93*, 64. [CrossRef]

52. Salem, M.L.; El-Hennawy, D. The possible beneficial adjuvant effect of influenza vaccine to minimize the severity of COVID-19. *Med. Hypotheses* **2020**, *140*, 109752. [CrossRef]

53. Fink, G.; Orlova-Fink, N.; Schindler, T.; Grisi, S.; Ferrer, A.P.S.; Daubenberger, C.; Brentani, A. Inactivated trivalent influenza vaccination is associated with lower mortality among patients with COVID-19 in Brazil. *BMJ Evid.-Based Med.* **2021**, *26*, 192–193. [CrossRef] [PubMed]

54. Escobar, L.E.; Molina-Cruz, A.; Barillas-Mury, C. BCG vaccine protection from severe coronavirus disease 2019 (COVID-19). *Proc. Natl. Acad. Sci. USA* **2020**, *117*, 17720–17726. [CrossRef] [PubMed]

55. Aaby, P.; Mogensen, S.W.; Rodrigues, A.; Benn, C.S. Evidence of increase in mortality after the introduction of diphtheria-tetanus-pertussis vaccine to children aged 6–35 months in Guinea-Bissau: A time for reflection? *Front. Public Health* **2018**, *6*, 79. [CrossRef] [PubMed]

56. Rhee, P.; Nunley, M.K.; Demetriades, D.; Velmahos, G.; Doucet, J.J. Tetanus and trauma: A review and recommendations. *J. Trauma Acute Care Surg.* **2005**, *58*, 1082–1088. [CrossRef] [PubMed]

57. Milne, G.; Hames, T.; Scotton, C.; Gent, N.; Johnsen, A.; Anderson, R.M.; Ward, T. Does infection with or vaccination against SARS-CoV-2 lead to lasting immunity? *Lancet Respir. Med.* **2021**, *9*, 1450–1466. [CrossRef]

58. Ferdinands, J.M.; Alyanak, E.; Reed, C.; Fry, A.M. Waning of influenza vaccine protection: Exploring the trade-offs of changes in vaccination timing among older adults. *Clin. Infect. Dis.* **2020**, *70*, 1550–1559. [CrossRef]

59. Huang, D.; Tran, J.T.; Olson, A.; Vollbrecht, T.; Tenuta, M.; Guryleva, M.V.; Fuller, R.P.; Schiffner, T.; Abadejos, J.R.; Couvrette, L.; et al. Vaccine elicitation of HIV broadly neutralizing antibodies from engineered B cells. *Nat. Commun.* **2020**, *11*, 5850. [CrossRef]

60. Voss, J.E.; Gonzalez-Martin, A.; Andrabi, R.; Fuller, R.P.; Murrell, B.; McCoy, L.E.; Porter, K.; Huang, D.; Li, W.; Sok, D.; et al. Reprogramming the antigen specificity of B cells using genome-editing technologies. *Elife* **2019**, *8*, e42995. [CrossRef]

61. Moffett, H.F.; Harms, C.K.; Fitzpatrick, K.S.; Tooley, M.R.; Boonyaratanakornkit, J.; Taylor, J.J. B cells engineered to express pathogen-specific antibodies protect against infection. *Sci. Immunol.* **2019**, *4*, eaax0644. [CrossRef]

62. Lumley, S.F.; Wei, J.; O'Donnell, D.; Stoesser, N.E.; Matthews, P.C.; Howarth, A.; Hatch, S.B.; Marsden, B.D.; Cox, S.; James, T.; et al. The duration, dynamics and determinants of SARS-CoV-2 antibody responses in individual healthcare workers. *Clin. Infect. Dis.* **2021**. [CrossRef]

63. Cohn, B.A.; Cirillo, P.M.; Murphy, C.C.; Krigbaum, N.Y.; Wallace, A.W. SARS-CoV-2 vaccine protection and deaths among US veterans during 2021. *Science* **2022**, *375*, 331–336. [CrossRef] [PubMed]

64. Tartof, S.Y.; Slezak, J.M.; Fischer, H.; Hong, V.; Ackerson, B.K.; Ranasinghe, O.N.; Frankland, T.B.; Ogun, O.A.; Zamparo, J.M.; Gray, S.; et al. Effectiveness of mRNA BNT162b2 COVID-19 vaccine up to 6 months in a large integrated health system in the USA: A retrospective cohort study. *Lancet* **2021**, *398*, 1407–1416. [CrossRef] [PubMed]

65. Chemaitelly, H.; Tang, P.; Hasan, M.R.; AlMukdad, S.; Yassine, H.M.; Benslimane, F.M.; Khatib, H.A.; Coyle, P.; Ayoub, H.H.; Kanaani, Z.; et al. Waning of BNT162b2 vaccine protection against SARS-CoV-2 infection in Qatar. *N. Engl. J. Med.* **2021**, *385*, e83. [CrossRef]

66. Dao, T.L.; Hoang, V.T.; Gautret, P. Recurrence of SARS-CoV-2 viral RNA in recovered COVID-19 patients: A narrative review. *Eur. J. Clin. Microbiol. Infect. Dis.* **2021**, *40*, 13–25. [CrossRef]

67. Azam, M.; Sulistiana, R.; Ratnawati, M.; Fibriana, A.I.; Bahrudin, U.; Widyaningrum, D.; Aljunid, S.M. Recurrent SARS-CoV-2 RNA positivity after COVID-19: A systematic review and meta-analysis. *Sci. Rep.* **2020**, *10*, 20692. [CrossRef]

68. Bush, A. Recurrent respiratory infections. *Pediatr. Clin. N. Am.* **2009**, *56*, 67–100. [CrossRef] [PubMed]

69. Midgard, H.; Weir, A.; Palmateer, N.; Lo Re, V., 3rd; Pineda, J.A.; Macias, J.; Dalgard, O. HCV epidemiology in high-risk groups and the risk of reinfection. *J. Hepatol.* **2016**, *65*, S33–S45. [CrossRef]

70. Yang, J.R.; Deng, D.T.; Wu, N.; Yang, B.; Li, H.J.; Pan, X.B. Persistent viral RNA positivity during the recovery period of a patient with SARS-CoV-2 infection. *J. Med. Virol.* **2020**, *92*, 1681–1683. [CrossRef]

71. Chan, C.E.; Chan, A.H.; Hanson, B.J.; Ooi, E.E. The use of antibodies in the treatment of infectious diseases. *Singap. Med. J.* **2009**, *50*, 663–672.

72. Rehermann, B.; Nascimbeni, M. Immunology of hepatitis B virus and hepatitis C virus infection. *Nat. Rev. Immunol.* **2005**, *5*, 215–229. [CrossRef]

73. Durham, S.R. Allergen avoidance measures. *Respir. Med.* **1996**, *90*, 441–445. [CrossRef] [PubMed]

74. Rai, R.K.; Khajanchi, S.; Tiwari, P.K.; Venturino, E.; Misra, A.K. Impact of social media advertisements on the transmission dynamics of COVID-19 pandemic in India. *J. Appl. Math. Comput.* **2022**, *68*, 19–44. [CrossRef] [PubMed]

75. Khajanchi, S.; Sarkar, K.; Mondal, J.; Nisar, K.S.; Abdelwahab, S.F. Mathematical modeling of the COVID-19 pandemic with intervention strategies. *Results Phys.* **2021**, *25*, 104285. [CrossRef] [PubMed]

76. Laderoute, M.P.; Larocque, L.J.; Giulivi, A.; Diaz-Mitoma, F. Further evidence that human endogenous retrovirus K102 is a replication competent foamy virus that may antagonize HIV-1 replication. *Open AIDS J.* **2015**, *9*, 112. [CrossRef] [PubMed]

77. Demongeot, J.; Seligmann, H. SARS-CoV-2 and miRNA-like inhibition power. *Med. Hypotheses* **2020**, *144*, 110245. [CrossRef] [PubMed]

78. Srinivasachar Badarinarayan, S.; Sauter, D. Switching sides: How endogenous retroviruses protect us from viral infections. *J. Virol.* **2021**, *95*, e02299-20. [CrossRef] [PubMed]

79. Liu, H.; Bergant, V.; Frishman, G.; Ruepp, A.; Pichlmair, A.; Vincendeau, M.; Frishman, D. Influenza A Virus Infection Reactivates Human Endogenous Retroviruses Associated with Modulation of Antiviral Immunity. *Viruses* **2022**, *14*, 1591. [CrossRef]

80. Pasqual, N.; Gallagher, M.; Aude-Garcia, C.; Loiodice, M.; Thuderoz, F.; Demongeot, J.; Ceredig, R.; Marche, P.N.; Jouvin-Marche, E. Quantitative and qualitative changes in VJ α rearrangements during mouse thymocytes differentiation: Implication for a limited T cell receptor α chain repertoire. *J. Exp. Med.* **2002**, *196*, 1163–1174. [CrossRef]

81. Baum, T.P.; Pasqual, N.; Thuderoz, F.; Hierle, V.; Chaume, D.; Lefranc, M.P.; Jouvin-Marche, E.; Marche, P.N.; Demongeot, J. IMGT/GeneInfo: Enhancing V (D) J recombination database accessibility. *Nucleic Acids Res.* **2004**, *32*, D51–D54. [CrossRef]

82. Baum, T.P.; Hierle, V.; Pasqual, N.; Bellahcene, F.; Chaume, D.; Lefranc, M.P.; Jouvin-Marche, E.; Marche, P.N.; Demongeot, J. IMGT/GeneInfo: T cell receptor gamma TRG and delta TRD genes in database give access to all TR potential V (D) J recombinations. *BMC Bioinform.* **2006**, *7*, 1–7. [CrossRef]

83. Thuderoz, F.; Simonet, M.A.; Hansen, O.; Pasqual, N.; Dariz, A.; Baum, T.P.; Hierle, V.; Demongeot, J.; Marche, P.N.; Jouvin-Marche, E. Numerical modelling of the VJ combinations of the T cell receptor TRA/TRD locus. *PLoS Comput. Biol.* **2010**, *6*, e1000682. [CrossRef] [PubMed]

84. O'Neill, L.A.J.; Netea, M.G. BCG-induced trained immunity: Can it offer protection against COVID-19? *Nat. Rev. Immunol.* **2020**, *20*, 335–337. [CrossRef] [PubMed]

85. Pellini, R.; Venuti, A.; Pimpinelli, F.; Abril, E.; Blandino, G.; Campo, F.; Conti, L.; De Virgilio, A.; De Marco, F.; Di Domenico, E.G.; et al. Initial observations on age, gender, BMI and hypertension in antibody responses to SARS-CoV-2 BNT162b2 vaccine. *EClinicalMedicine* **2021**, *36*, 100928. [CrossRef] [PubMed]

86. Xu, Z.; Wei, D.; Zeng, Q.; Zhang, H.; Sun, Y.; Demongeot, J. More or less deadly? A mathematical model that predicts SARS-CoV-2 evolutionary direction. *Comput. Biol. Med.* **2023**, *153*, 106510. [CrossRef]

87. Xu, Z.; Yang, D.; Wang, L.; Demongeot, J. Statistical analysis supports UTR (untranslated region) deletion theory in SARS-CoV-2. *Virulence* **2022**, *13*, 1772–1789. [CrossRef]

Review

Understanding the SARS-CoV-2 Virus Neutralizing Antibody Response: Lessons to Be Learned from HIV and Respiratory Syncytial Virus

Nigel J. Dimmock * and Andrew J. Easton

School of Life Sciences, University of Warwick, Coventry CV4 7AL, UK
* Correspondence: n.j.dimmock@warwick.ac.uk

Abstract: The SARS-CoV-2 pandemic commenced in 2019 and is still ongoing. Neither infection nor vaccination give long-lasting immunity and, here, in an attempt to understand why this might be, we have compared the neutralizing antibody responses to SARS-CoV-2 with those specific for human immunodeficiency virus type 1 (HIV-1) and respiratory syncytial virus (RSV). Currently, most of the antibodies specific for the SARS-CoV-2 S protein map to three broad antigenic sites, all at the distal end of the S trimer (receptor-binding site (RBD), sub-RBD and N-terminal domain), whereas the structurally similar HIV-1 and the RSV F envelope proteins have six antigenic sites. Thus, there may be several antigenic sites on the S trimer that have not yet been identified. The epitope mapping, quantitation and longevity of the SARS-CoV-2 S-protein-specific antibodies produced in response to infection and those elicited by vaccination are now being reported for specific groups of individuals, but much remains to be determined about these aspects of the host–virus interaction. Finally, there is a concern that the SARS-CoV-2 field may be reprising the HIV-1 experience, which, for many years, used a virus for neutralization studies that did not reflect the neutralizability of wild-type HIV-1. For example, the widely used VSV-SARS-CoV-2-S protein pseudotype has 10-fold more S trimers per virion and a different configuration of the trimers compared with the SARS-CoV-2 wild-type virus. Clarity in these areas would help in advancing understanding and aid countermeasures of the SARS-CoV-2 pandemic.

Keywords: human infection; repeated infection; SARS-CoV-2; HIV-1; RSV; neutralizing antibody; neutralization assay; epitope specificity; analysis of serum antibody specificity; monoclonal antibody

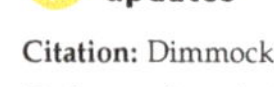
check for **updates**

Citation: Dimmock, N.J.; Easton, A.J. Understanding the SARS-CoV-2 Virus Neutralizing Antibody Response: Lessons to Be Learned from HIV and Respiratory Syncytial Virus. *Viruses* **2023**, *15*, 504. https://doi.org/10.3390/v15020504

Academic Editor: Jason Yiu Wing KAM

Received: 9 January 2023
Revised: 7 February 2023
Accepted: 9 February 2023
Published: 11 February 2023

1. Introduction

Severe acute respiratory syndrome coronavirus 2 (SARS-CoV-2) is a newly arrived human pathogen, and there is much to be learned about it and how it interacts with its new host. This respiratory pathogen emerged in 2019 and most likely originated from a bat via an unknown non-human animal source. The virus causes serious mortality in people rendered susceptible by age, comorbidities, immunodeficiency or combinations of these problems, but it is less severe in healthy individuals. Vaccines expressing the main SARS-CoV-2 surface (S) envelope protein were rapidly devised and deployed in wealthy countries, and this included the first global use of an mRNA vaccine. These vaccines successfully reduced SARS-CoV-2-associated mortality, but, as with natural disease, antibody-associated immunity to subsequent infection was short-lived; the longevity of T cell-based immunity and its potential role(s) in the protection from subsequent infection are not fully understood yet. While vaccination importantly provides amelioration of disease symptoms, successive immunizations (up to five, to date) did not improve the longevity of protection, and even the combination of post-infection immunity and vaccination gave a similarly short-lived immunity to infection [1–4]. Thus, the human population as a whole is still at risk from an infection that ranges from the subclinical to a variable morbidity that has serious personal and economic impacts by requiring time away from the workplace. The impacts also include a debilitating condition known as 'long COVID', the causes of which remain unclear, although it has been linked with the reactivation of latent viruses such as Epstein–Barr virus [5]. An additional concern is that the cycle of repeat infections can transmit the virus to those in the high-risk group, particularly if the infection is inapparent or if people no longer feel it necessary to self-isolate. Added to these problems is the continuing evolution of new virus strains which afford some degree of escape from existing immune defenses, rendering an inadequate antibody response even more ineffective.

All the concerns about a short-lived, poor or inadequate immunity following infection or vaccination emphasize the need to understand the nature of the protective immune response against SARS-CoV-2 to better arm ourselves in the future against this and other viruses. This article considers the SARS-CoV-2-specific immunity that results from infection or vaccination. This appears to be largely antibody-based: sterilizing immunity cannot be mediated through T-cells. Space constraints mean that we focus on the direct neutralization of the SARS-CoV-2 by antibody. This should not be taken to mean that we place no value on other mechanisms in which neutralizing and non-neutralizing antibodies act against viruses via complement or other components of the immune system or that Fc receptors play a role in preventing or ameliorating disease; doubtless others will expand on these areas in due course.

The cause of the short-lived nature of the immunity is not known, although in the case of infection, it seems likely that SARS-CoV-2-virus-mediated immunosuppression/immunomodulation is involved. The scenario of initial protection followed by its decline suggests that cells responsible for producing antibody are not renewed, lose the ability to respond properly and be amplified or are ablated. SARS-CoV-2 has a large genome and expresses many proteins, and there is no shortage of candidate immunosuppressors. However, the mRNA vaccine expresses only the S protein, which would have to possess both immunostimulatory and immunosuppressive elements.

2. The Nature of Neutralizing Antibodies

While much is said about the importance of virus-neutralizing antibodies, an often-unstated key point is that they are not all equal in their ability to cause a loss of virus infectivity. A plethora of neutralizing antibodies that recognize different epitopes, and hence have different specificities, can potentially be elicited by any infectious agent or vaccine, and, in addition, the antibody response can vary according to the genetics of the host and their past infection experience. However, for reasons not understood, only a small fraction of the potential range of antibodies with these specificities may be expressed. The neutralization titer of (usually) a serum or plasma sample is not an absolute but represents the sum of the activity of all the neutralizing antibodies present in a particular sample. For example, this could consist of antibodies that are specific to many different regions of the target protein and hence will cause neutralization by interfering with the several different mechanisms that make up the infectious process—not only by blocking attachment to the main virus receptor binding domain (RBD), as is often assumed. Alternatively, the same neutralization titer could be achieved by a population that comprises one or only a few different antibody specificities [6–8]. Without analysis of the antibody specificities, it is not possible to distinguish between these two examples. With a simple assessment of a neutralization titer, they appear identical, but biologically, they are very different. In addition, antibodies to a particular neutralization region of the target protein can vary in the amount of each antibody specificity that is produced and in the efficiency with which each mediates neutralization. Thus, there is a huge permutation capable of qualitative and quantitative variability in the antibody population between individuals, and understanding the detail of this variability is key to understanding the factors that underpin an effective and long-lasting immunity.

3. The SARS-CoV-2 Surface (S) Protein

The SARS-CoV-2 S cell-attachment protein is anchored by a trans-membrane region at its C-terminus and undergoes proteolytic cleavage and structural rearrangements to form a metastable structure that is able to catalyze the fusion-entry process [9]. This is common to all enveloped viruses [10,11]. In the SARS-CoV-2 S protein, cleavage at the S1–S2 junction gives rise to an outer S1 subunit and an inner subunit (S2) that has a new N terminus. S1 and S2 are not covalently bound to each other. The binding of the S1 RBD to its receptor, the angiotensin-converting enzyme-2 (ACE2), in the cell membrane ACE2 triggers a second proteolytic cleavage of S2 at S2' which exposes a new N terminus at the end of the 41-residue fusion peptide. Electron microscopy of SARS-CoV-2 virions shows that S1 exists in either a closed conformation ('RBD down') in which the RBDs are buried and unavailable for binding or an open conformation ('1 or more RBDs up') and appears to oscillate between the two positions [12,13]. The open conformation is required for the binding to ACE2 that leads to the fusion of the viral and cell membranes.

The number of fusion protein trimers on the surface of enveloped virus particles is very variable. SARS-CoV-2 (24+/−9 trimers of the S protein per virion) and HIV-1 (11–18 trimers of gp120-41 per virion) have a low number, whereas influenza virus A has an order of magnitude more (300–400 HA trimers per virion), all with similar-sized particles; respiratory syncytial virus (RSV) trimers are densely packed [14–17]. Curiously, influenza virus and SARS-CoV-2 trimers are not uniformly distributed and are not arranged in any discernible pattern [15,16]. Further, and perhaps uniquely, the SARS-CoV-2 trimers are hinged and can be tilted in any direction up to 50 degrees, rather than lying at right angles to the virion surface, as usually depicted. In addition, the number of S trimers present on a SARS-CoV-2 virion differs markedly from that of a VSV virion pseudotyped with a SARS-CoV-2 S protein which is commonly used to measure SARS-CoV-2-specific neutralizing antibodies, as VSV has 400 trimers per virion [18]. SARS-CoV-2 S protein-pseudotyped retroviruses, also used to measure SARS-CoV-2 neutralization, might more faithfully reproduce the SARS-CoV-2 S trimer distribution [12] and therefore be a more appropriate test system for evaluating SARS-CoV-2 neutralization.

A clear consequence of a small number of surface protein trimers per virion is fewer targets for neutralizing antibodies, as with HIV-1 [14]. How the distribution and/or number of trimers affect a SARS-CoV-2 neutralization event is not certain. It is not known for SARS-CoV-2 how many S trimers need to engage an antibody molecule to cause neutralization, and this is likely to vary with the function of the antigenic site in the infection process. If blocking the receptor binding domain (RBD) is the mechanism of neutralization, it seems unlikely that the binding of a single IgG molecule to one S trimer of the virion would be sufficient to cause neutralization, as this leaves up to 23 +/- 9 other trimers free to engage cell receptors. Indeed, rabies virus, closely related to VSV, could bind 130 molecules of a neutralizing monoclonal IgG (ca. 400 surface trimers per virion) without losing any infectivity [19]. The neutralization of a virion by a single antibody molecule would require interaction between all, or a sufficient number of, trimers and the triggering of some sort of cooperative event that renders the virion non-infectious—a situation for which there is, as of yet, no direct evidence. Additionally, it is not known whether a single SARS-CoV-2 S trimer can bind one, two or three molecules of IgG. The Mr of an S protein trimer ectodomain is approx. 540, and that of an IgG is approx. 160, so it might be possible to bind more than one IgG per trimer, depending on the position of the epitope and any spatial interference mediated by the mobility of the bound IgG. An S trimer construct is reported to simultaneously bind three Fabs of the monoclonal antibody (MAb) 4A8, but a Fab is approximately 30% of the mass of the whole antibody molecule and hence offers much less steric interference [20]. Unless trimers become clustered, they are spaced too far apart to be crosslinked by IgG, and the three repeated epitopes of a single trimer are too close together to allow for intramolecular crosslinking by an IgG molecule. However, the sparsity of S trimers on the virion surface and the flexibility of the S ectodomain allow for the maximum access of antibodies directed to any region of the S trimer. The sparsity of S trimers could also impinge on the efficiency of the fusion-entry process of SARS-CoV-2, as, for example, influenza virus requires the cooperation of three or more trimers for its fusion-entry process [21].

4. SARS-CoV-2-Specific Human Monoclonal Antibodies

Several groups have isolated SARS-CoV-2 S protein-specific monoclonal antibodies, e.g., [22,23]. Zost et al. isolated 389 SARS-CoV-2 S protein-specific monoclonal antibodies from two people who had been infected with the virus [22]. These were identified by a reaction with a stabilized prefusion form of the S protein ectodomain (1208 residues, with 6 residues substituted), an RBD construct (223 residues), or an S protein N-terminal domain construct (350 residues). Most MAbs, including those reacting with the N-terminal construct, were not neutralizing, but 67/70 neutralizing MAbs reacted with the RBD construct. In a further report, most of the 40 neutralizing MAbs tested blocked the interaction of the S trimer with the ACE-2 primary receptor [24]. Such data have led to the assumption that these MAbs neutralize by blocking the attachment of the virus to the host cell. However, blocking the interaction of individual proteins is a very different scenario from blocking a virion, which carries many copies of the target molecule and is many orders of magnitude larger, from attaching to the cell surface, and critical experiments clarifying the details of the neutralization process need to be conducted. One RBD-binding MAb (CR3022) has an entirely different mode of action and prematurely activates the post-fusion state of the S protein [25], although its epitope is distinct from the ACE-2 binding site [20]; a different MAb affects SARS-CoV-1 and MERS in the same way [26].

It would be strange if all SARS-CoV-2 MAbs were directed to the RBD region, as the S proteins of other enveloped viruses have several neutralization sites that are located on disparate regions of the protein. Other enveloped viruses, such as HIV-1 and influenza virus, with surface proteins that effect both attachment and fusion, stimulate antibodies which block the fusion-entry process [27,28]. For example, HIV-1 has six definably different antigenic sites with which neutralizing antibodies interact [29]. The apparent absence of SARS-CoV-2-specific neutralizing antibodies targetting regions other than the RBD is difficult to understand, as the selecting proteins, especially the prefusion form of the SARS-CoV-2 S ectodomain, contain the majority of the S protein trimer. However, there are several possible explanations: (a) it may be that the engineering required to stabilize the selecting protein for its use as an antigen alters its antigenicity. The conformational changes that result from cleavages that give rise to S1 and S2, and then S2', are likely to be accompanied by the formation of new epitopes or an increased exposure of existing epitopes and, provided they are available for a sufficient time to activate B cells, to generate their cognate antibodies. However, antibodies to the fusion peptide of SARS-CoV-1 and SARS-CoV-2 exist in people who have recovered from infection and may be neutralizing [30,31], although assays confirming that fusion is inhibited were not reported. (b) Infection may bias the antibody response to the RBD region as seen with HIV-1 [32]. (c) MAbs were obtained from blood that was withdrawn at 50 days after infection, as this gave a better antibody response than blood taken at 35 days: thus, the time at which the blood sample was taken may be relevant [22]. This has been seen elsewhere; for example, the antibody response in Ebola virus infections may take up to one year to fully develop. (d) There may indeed be no other antigenic regions: however, this seems unlikely, as other viruses, such as HIV-1, as noted above, have multiple antigenic sites involved in other aspects of the entry process.

A final, but crucial, factor when extrapolating to antibody protection in vivo is the nature of the neutralization assay. As already mentioned, there is a concern regarding the relevance of assays based on pseudoviruses, such as VSV, which carry an order of magnitude more S trimers per virion than those that are present on the genuine SARS-CoV-2 virus [19]. In addition, it would be more relevant to use a competitive in vitro assay that mimics the in vivo situation in allowing for the simultaneous interaction of the virus, antibody, and host cells, rather than the conventional test in which preincubated virus-antibody complexes are added to susceptible cells. This is underlined by the high affinity of the S protein trimer for its receptor (in the μM range, whereas that of influenza virus is 1000-fold lower), suggesting that SARS-CoV-2 can escape neutralization by rapidly binding to and entering into a host cell [15]. It would be interesting to compare neutralization titers by the two methods suggested.

5. Mapping S-Specific MAbs That Neutralize SARS-CoV-2

Several studies describe the identification of SARS-CoV-2-specific highly neutralizing MAbs (e.g., [20,22,24]). Individual laboratories have made efforts to map their MAbs, but the key residues needed for epitope binding have been determined only for a few, usually through structural studies [20]. Binding assays usually use very large structures such as trimers of the S ectodomain, the RBD, and the N-terminal domain, which comprise many different epitopes. There is an urgent need in the global SARS-CoV-2 community for a comprehensive, interlaboratory mapping collaboration. In essence, the following conclusions emerge:

1. Highly neutralizing MAbs (N_{50} < 150 μg/mL) representing the RBD region and the N-terminal domain region have been found, but the frequency of occurrence and the number of such antibodies in body fluids during/after infection are not known, nor do we know if their production is stimulated by vaccination.

2. Most, but not all, MAbs (e.g., 47D11: reviewed by [20]) identified so far are inhibitory in the S trimer-ACE2 protein-binding assay, although, as noted above, this does not necessarily imply that these antibodies neutralize by preventing the attachment of the SARS-CoV-2 virus particle to the cell. Testing with authentic SARS-CoV-2 virus, not pseudovirions, is required to determine the mechanism(s) of neutralization.

3. There is clear evidence that MAbs that inhibit in the S trimer-ACE2 binding assay, and therefore impact the RBD, are diverse and interact with a number of different epitopes.

4. MAb P2B-2F6 competes in the S trimer-ACE2 binding assay and is shown by electron microscopy to bind to the RBD region in both its open (up) and closed positions, whereas ACE-2 binds to the RBD only in its open position. MAb P2B-2F6 has a higher affinity for the RBD than the ACE-2 receptor.

5. Many MAbs bind the N-terminal construct, but few are neutralizing; an exception is the highly neutralizing 4A8 MAb.

6. There is one report of post-infection antibodies that are specific for a peptide in the fusion region (mentioned above) [31].

7. Most studies show that the same selecting S protein constructs that identify neutralizing MAbs also reveal many non-neutralizing MAbs. While some of the latter could be antiviral in vivo through antibody-dependent cell cytotoxicity or other mechanisms, there is a concern that the ubiquitous non-neutralizing MAbs could block the binding and action of neutralizing antibodies and permit infection.

6. Comparison with Other Virus Systems

Here, we compare the SARS-CoV-2-specific antibody response with those found in other human virus systems—notably, human immunodeficiency virus type 1 (HIV-1) and respiratory syncytial virus (RSV). The HIV-1 field has been concerned with neutralizing antibodies for many years, and it now arguably represents the best studied and understood system of all. At first glance, it may seem strange to compare SARS-CoV-2 with HIV-1, as they cause such different infections and diseases, but both are enveloped viruses, and their major surface proteins, like those of all enveloped viruses, have a very similar structure. These are type 1 transmembrane homotrimer proteins with an external N-terminus and a C-terminus inside the virion, and they are cleaved to give an outer domain bound covalently or non-covalently, depending on the virus, to the inner domain. They all undergo profound structural rearrangements on contact with their primary cell receptor as part of the fusion-entry process.

7. What Can the SARS-CoV-2 Field Learn from HIV-1?

Like SARS-CoV-2, the HIV-1 and RSV fields are noted for the diversity of virus variants that differ in the sequence of their main surface protein, and they are classified into a number of clades, with multiple strains in each. It is an immense problem to devise a vaccine that can stimulate an immune response broad enough to protect against all variations of the virus. Particularly for HIV-1, this needs to be a sterilizing immunity, since, once infection is initiated, the HIV-1 genome is integrated into the DNA of the cell and cannot be removed. T-cell immunity is of little value, as it acts only on infected cells, so reliance has to be placed on the neutralizing antibody response. The HIV-1 field has trialled an unprecedented variety of neutralization immunogens and experimental hosts, mostly with modest results. Antibodies are made, but, as with SARS-CoV-2, these mostly exert an ineffective neutralizing response that comprises narrowly specific antibodies that can be directed against highly mutable epitopes; these antibodies may not prevent infection and are relatively short-lived.

Another issue for HIV-1 was the assay system which, for convenience, used virus strains that had been adapted to grow in the laboratory—this adaptation requires the virus to mutate but, at the same time, renders it far easier to neutralize than a wild-type virus [33]. Thus, the laboratory-adapted strains are a poor indicator of protection in vivo. It is not clear if the SARS-CoV-2 system has the same problem, but it is suggested that adaptation to Vero-hACE2-TMPRSS2 cells may modify the S protein so that antibodies no longer neutralize as efficiently [23]. Second, there is concern over pseudotype assays that, for the convenience of scale and automation, are used as surrogates for the neutralization of SARS-CoV-2, as the distributions of S trimers on VSV and authentic SARS-CoV-2 are very different, as mentioned above [14,15]. The S protein-VSV pseudotype was variously shown to give neutralization that was statistically close to the neutralization of an infectious SARS-CoV-2 laboratory-adapted virus of that time [34] or was 21-fold more sensitive [22].

Eventually, the HIV-1 studies led to the discovery of broadly neutralizing monoclonal antibodies, which, as their name indicates, neutralize a plethora of different wild-type strains (31–99% of known strains) in vitro to a high titer (i.e., requiring from 3.7 to 0.002 µg/mL) [29]. Crucially, they also protect in vivo. The authors cite the macaque model, in which animals are given a natural mucosal challenge and develop an HIV-like illness, to show that 100% sterilizing protection could be achieved using a mixture of broadly neutralizing monoclonal antibodies and that this is related to the neutralization titer. Furthermore, they calculate that, for the HIV-1 infection of humans, a combination of four broadly neutralizing monoclonal antibodies at a serum concentration of 30 µg/mL could provide sterilizing immunity.

In summary, (a) experience with HIV-1 warns the SARS-CoV-2 field to question the validity of the pseudotyped virus used routinely in SARS-CoV-2 neutralization tests and to ask if the tests faithfully represent wild-type circulating virus; (b) like the SARS-CoV-2 vaccines, HIV-1 vaccines, in general, stimulate sub-optimal neutralizing antibody responses and fail to elicit the broadly neutralizing antibodies that are really needed for effective protection from infection; and (c) it needs to be determined if the human antibody repertoire is capable of responding to SARS-CoV-2 to produce the sort of broadly neutralizing antibodies that are so effective against HIV-1, and, if it is, a vaccine that can elicit them needs to be devised. Such antibodies have recently been found using a bioinformatic approach and include one that neutralizes the S2′ cleavage site [35].

8. What Can the SARS-CoV-2 Field Learn from the Respiratory Syncytial Virus?

Since its first description in 1956, RSV has been studied extensively to try to lessen the burden of the respiratory disease that it generates in seasonal epidemics, especially in the very young and the elderly. Several features of RSV infection and disease are similar to those found in SARS-CoV-2 infections, and the approaches to tackling these for RSV may provide useful insights into studies of SARS-CoV-2 with a view to developing therapeutic interventions. As with SARS-CoV-2, the robust antibody and cellular immune response to RSV infection that is detected following infection is short-lived. This, coupled with the appearance of genetic variants, means that repeat infections occur throughout life [36]. While the underlying processes leading to the short-lived immunity following natural infection are not entirely understood, it is known that RSV directly interferes with several aspects of the innate immune system, including the inflammatory and type 1 interferon responses as well as the intracellular processes involved in antigen presentation. These effects have consequences for the adaptive immune response which receives signals from the innate immune system, and the result is a reduction in or loss of B cell memory and T cell function, as reviewed in [37].

RSV has two envelope proteins: an attachment protein (G) and a fusion protein (F) that is present as a trimer in the prefusion state on the surface of infected cells and the virus particle; virus particles expressing only F are infectious, indicating that F is able to attach to host cells [38]. Initial studies of the neutralizing antibodies generated following infection with RSV were directed against the fusion (F) protein. However, neutralizing antibodies directed against both the pre-and post-fusion forms of the F protein are also present in human serum, and these interact with epitopes that are retained after the conformational change from pre- to post-fusion states [39,40]. No vaccine is currently available for RSV, though several are in clinical trials and have shown promising early results [41,42]. The only currently used treatment for RSV infection is a series of monthly injections of palivizumab, which is approved for prophylactic treatment in high-risk infants. Palivizumab is an F protein-specific, humanized mouse monoclonal antibody that prevents virus entry into cells. In recent years, modified versions of palivizumab have been generated to extend its half-life in the body following administration in order to reduce the need for monthly injections [43]. While palivizumab provides protection in approximately 50% of previously uninfected infants, it is not used for the treatment of infections in adults, where it is less effective. In recent years, it has become clear that it is necessary to better understand the immune response to infection to try to generate new approaches for prophylaxis and treatment [44].

Recent approaches have focused on investigation of the nature of the antibodies produced following natural infection. A study of 364 RSV F protein-specific monoclonal antibodies derived from memory B cells taken from three adult volunteers showed that these recognized a total of six different antigenic sites, including one not previously described [45]. The neutralizing anti-F protein antibodies with the greatest activity recognized the pre-fusion form of the F protein, and several showed 100-fold higher neutralizing activity than palivizumab, emphasizing the value of such studies for identifying potential prophylactic and therapeutic candidates. A similar study of 23 monoclonal antibodies isolated from human memory B cells confirmed this observation [46].

In a separate study, Andreano et al. investigated the repertoire of antibodies produced by single-cell sorted memory B cells from four adult volunteers. The data show that, while three subjects had generated antibodies specific for the pre-fusion form of the F protein, the majority of neutralizing antibodies were directed against the F protein bound to both the pre- and post-fusion forms. However, 78 of the 82 antibodies with dual-binding capacity were bound significantly more strongly to the pre-fusion form, which is found most predominantly on virus particles [47]. The data from these studies emphasize that the neutralization of virus infectivity in vivo is a potentially complex combination of a spectrum of different antibodies with different neutralizing activities. In the case of anti-RSV F protein antibodies, these are all targeted to epitopes closely located to each other on the F protein, so the relative concentrations of each, as well as their inherent affinity for their target epitope, will determine the competitive outcome of the interaction. In addition, these studies showed that many of the monoclonal antibodies have neutralizing activity against different genetic subtypes of RSV and, indeed, to the more distantly related human metapneumovirus, suggesting that they may exert broad range protection. The data also suggested that the genetic lineages of neutralizing antibodies seen in young children may differ from the lineages seen in adults. Such a bias in the immune response would be an important consideration for the design of vaccines for different target populations. It will be important to establish whether a similar situation exists with SARS-CoV-2 infections.

9. How to Better Understand the SARS-CoV-2 Neutralizing Antibody Response

To improve approaches for developing a strong, potentially long-lasting protective antibody response against SARS-CoV2, we need a measure of the individual antibody specificities that make up the neutralizing response produced to infection and vaccination—qualitatively in terms of their epitope specificity and quantitatively to determine their SARS-CoV-2-specific individual neutralizing titers, the breadth of neutralization, and the longevity of synthesis [48]. Crudely put, we need to know if the serum or plasma sample contains one or a few antibodies of high-titer specificities, antibodies of many different low-titer specificities, or a combination of both. In this way, we can analyze the response to a SARS-CoV-2 infection and determine the efficacy of a vaccine. This is needed to allow us to understand why the SARS-CoV-2 neutralizing antibody response is short-lived and ineffective in preventing infection. Only with this information can the SARS-CoV-2 neutralizing antibody response be properly evaluated, and through this, we will gain a better understanding of the deficiencies of the SARS-CoV-2 infection- and vaccine-induced neutralizing antibody responses and of why people can be infected repeatedly.

One way to analyze the SARS-CoV-2 neutralizing antibody response would be to assemble as extensive a library of SARS-CoV-2 S protein-specific monoclonal antibodies as possible, with emphasis on their specific neutralizing ability (µg required for 50% neutralization) and their capacity to neutralize a broad range of SARS-CoV-2 virus strains. Progress has already been made, but there is a worrying lack of MAbs that do not block the RBD. The main focus would then be to interrogate human blood/nasal wash antibody samples to see if these have the ability to block the binding of individual monoclonal antibodies to the authentic S protein. The blocking of binding would indicate the presence of a cognate MAb in the human sample or of one that has an overlapping footprint.

Alternatively, an epitope library for SARS-CoV-2 based on the sequence/structure of the S protein could be constructed. A linear peptide library is easily made, but peptides locked in a relevant conformation would likely be more useful. This could be approached using AI systems such as AlphaFold, which, in recent times, have revolutionized the determination of the 3D structures of proteins [49]. Each peptide would be synthesized with an anchor sequence to attach it to a substrate, and a library of such peptides would interrogate the antibody sample using an automated system. Basic quantitation could be provided by applying serial dilutions of the sample.

Electron- and cryo-electron microscopy are other methods that have been used successfully to map HIV-1 [50] and SARS-CoV-2 [20] surface protein-specific antibodies present in polyclonal sera after infection or vaccination. The traditional method of selecting neutralizing antibody escape mutants and determining which amino acid substitutions had occurred is available to those with biosecure facilities.

Other, but related, questions include why SARS-CoV-2 vaccines do not stimulate long-lived broadly neutralizing antibodies, and we need to know if such antibodies can be found in the human antibody repertoire. As for therapy, we need to know if such antibodies are to be found in other (possibly unconventional) animal systems and hence become a source for humanized monoclonal antibody reagents. Thinking outside the box led to the immunization of cows, which proved to be a surprising source of HIV-1-specific broadly neutralizing antibodies [51]. These have an ultralong CDRH3, which is a key feature of HIV-1-specific broadly neutralizing antibodies [29].

We have the space to quote only a limited number of papers; we apologize to the authors whose excellent work has been omitted.

Author Contributions: Conceptualization, N.J.D., writing—original draft preparation, review and editing, N.J.D. and A.J.E. All authors have read and agreed to the published version of the manuscript.

Funding: The authors received no funding.

Viruses **2023**, 15, 504

Institutional Review Board Statement: Not applicable.

Informed Consent Statement: Not applicable.

Data Availability Statement: Not applicable.

Acknowledgments: Not applicable.

Conflicts of Interest: The authors have an interest in a company and patents relating to anti-respiratory virus measures using defective interfering viruses.

References

1. Wajnberg, A.; Amant, F.; Firpo, A.; Altman, D.R.; Bailey, M.J.; Mansour, M.; McMahon, M.; Meade, P.; Mendu, D.R.; Muellers, K.; et al. Robust neutralizing antibodies to SARS-CoV-2 infection persist for months. *Science* **2020**, *370*, 1227–1230. [CrossRef]
2. Kim, Y.; Bae, J.Y.; Kwon, K.; Chang, H.H.; Lee, W.K.; Park, H.; Choi, I.; Park, M.S.; Kim, S.W. Kinetics of neutralizing antibodies against SARS-CoV-2 infection according to sex, age, and disease severity. *Sci. Rep.* **2022**, *12*, 13491. [CrossRef]
3. Phakaratsakul, S.; Manopwisedjaroen, S.; Boonarkart, C.; Kupatawintu, P.; Chaiwanichsiri, D.; Roytrakul, T.; Thitithanyanont, A. Dynamics of neutralizing antibodies and binding antibodies to domains of SARS-CoV-2 spike protein in COVID-19 survivors. *Viral Imunol.* **2022**, *35*, 545–552. [CrossRef]
4. Willyard, C. How quickly does COVID immunity fade? What Scientists know. *Nature* **2023**. [CrossRef]
5. Apostolou, E.; Rizwan, M.; Moustardas, P.; Sjögren, P.; Bertilson, B.C.; Bragee, B.; Polo, O.; Rosen, A. Saliva antibody-fingerprint of reactivated latent viruses after mild/asymptomatic COVID-19 is unique in patients with myalgicencephalomyelitis/chronic fatigue syndrome. *Front. Immunol.* **2022**, *13*, 949787. [CrossRef]
6. Lambkin, R.; McLain, L.; Jones, S.E.; Aldridge, S.L.; Dimmock, N.J. Neutralization escape mutants of type A influenza virus are readily selected by antiserum from mice immunized with whole virus: A possible mechanism for antigenic drift. *J. Gen. Virol.* **1994**, *75*, 3493–3502. [CrossRef] [PubMed]
7. Lambkin, R.; Dimmock, N.J. Longitudinal study of an epitope-biased haemagglutination-inhibition antibody response in rabbits immunized with type A influenza virus. *Vaccine* **1996**, *14*, 212–218. [CrossRef]
8. Lambkin, R.; Dimmock, N.J. All rabbits immunized with type A influenza virions have a serum antibody haemagglutination-inhibition antibody response biased to a single epitope in antigenic site B. *J. Gen. Virol.* **1995**, *76*, 889–897. [CrossRef] [PubMed]
9. Koppisetti, R.K.; Fulcher, Y.G.; van Doren, S.R. Fusion peptide of SARS-CoV-2 spike rearranges into a wedge inserted into bilayered micelles. *J. Am. Chem. Soc.* **2021**, *143*, 13205–13211. [CrossRef]
10. Harrison, S.C. Viral membrane fusion. *Virology* **2015**, *479–480*, 498–507. [CrossRef]
11. Barrett, C.T.; Dutch, R.E. Viral membrane fusion and the transmembrane domain. *Viruses* **2020**, *12*, 693. [CrossRef] [PubMed]
12. Walls, A.C.; Park, Y.-J.; Tortorici, M.A.; Wall, A.; McGuire, A.T.; Veesler, D. Structure, function, and antigenicity ofthe SARS-CoV-2 spike protein. *Cell* **2020**, *180*, 281–292. [CrossRef]
13. Wrapp, D.; Wang, N.; Corbett, K.S.; Goldsmith, J.A.; Hsieh, C.-H.; Abiona, O.; Graham, B.S.; McLellan, J.S. Cryo-EM structure of the 2019-nCoV spike in the prefusion conformation. *Science* **2020**, *367*, 1260–1263. [CrossRef] [PubMed]
14. Klein, J.S.; Bjorkman, P.J. Few and far between: How HIV may be evading antibody avidity. *PloS Path.* **2010**, *27*, e1000908. [CrossRef]
15. Ke, Z.; Oton, J.; Qu, K.; Cortese, M.; Zila, V.; McKeane, L.; Nakane, T.; Zivanov, J.; Neufeldt, C.J.; Cerikan, B.; et al. Structures and distributions of SARS-CoV-2 spike proteins on intact virions. *Nature* **2020**, *588*, 498–502. [CrossRef] [PubMed]
16. Harris, A.; Cardone, G.; Winkler, D.C.; Heymann, J.B.; Brecher, M.; White, J.M.; Steven, A.C. Influenza virus pleiomorphy characterized by cryoelectron tomography. *Proc. Nat. Acad. Sci. USA* **2006**, *103*, 19123–19127. [CrossRef] [PubMed]
17. Conley, M.J.; Short, J.M.; Burns, A.M.; Streetley, J.; Hutchings, J.; Bakker, S.E.; Power, B.J.; Jaffery, H.; Haney, J.; Zanetti, G.; et al. Helical ordering of envelope-associated proteins and glycoproteins in respiratory syncytial virus. *EMBO J.* **2022**, *41*, e109728. [CrossRef]
18. Thomas, D.; Newcomb, W.W.; Brown, J.C.; Wall, J.S.; Hainfeld, J.F.; Trus, B.L.; Steven, A.C. Mass and molecular composition of vesicular stomatitis virus: A scanning transmission electron microscopy analysis. *J. Virol.* **1985**, *54*, 598–607. [CrossRef]
19. Flamand, A.; Raux, H.; Gaudin, Y.; Ruigrok, R.W.H. Mechanisms of rabies virus neutralization. *Virology* **1993**, *194*, 302–313. [CrossRef]
20. Wang, M.-Y.; Zhao, R.; Gao, L.-J.; Gao, X.-F.; Wang, D.-P.; Cao, J.-M. SARS-CoV-2: Structure, biology, and structure-based therapeutics development. *Front. Cell. Infect. Microbiol.* **2020**, *10*, 587269. [CrossRef]
21. Danieli, T.; Pelletier, S.; Henis, Y.I.; White, J.M. Membrane fusion mediated by the influenza virus hemagglutinin requires the concerted action of at least three hemagglutinin trimers. *J. Cell Biol.* **1996**, *133*, 559–569. [CrossRef]
22. Zost, S.J.; Gilchuk, P.; Chen, R.E.; Case, J.B.; Reidy, J.X.; Trivette, A.; Nargi, R.S.; Sutton, R.E.; Suryadevara, N.; Chen, E.C.; et al. Rapid isolation and profiling of a diverse panel of human monoclonal antibodies targeting the SARS-CoV-2 spike protein. *Nat. Med.* **2020**, *26*, 1422–1427. [CrossRef]
23. Chen, R.E.; Zhang, X.; Case, J.B.; Winkler, E.S.; Liu, Y.; VanBlargan, L.A.; Liu, J.; Errico, J.M.; Xie, X.; Suryadevara, N.; et al. Resistance of SARS-CoV-2 variants to neutralization by monoclonal and serum-derived polyclonal antibodies. *Nat. Med.* **2021**, *27*, 717–726. [CrossRef]

24. Zost, S.J.; Gilchuk, P.; Case, J.B.; Binshtein, E.; Chen, R.E.; Nkolola, J.P.; Schäfer, A.; Reidy, J.X.; Trivette, A.; Nargi, R.S.; et al. Potently neutralizing and protective human antibodies against SARS-CoV-2. *Nature* **2020**, *584*, 443–449. [CrossRef]

25. Huo, J.; Zhao, Y.; Ren, J.; Zhou, D.; Duyvesteyn, H.M.E.; Ginn, H.M.; Carrique, L.; Malinauskas, T.; Ruza, R.R.; Shah, P.N.M.; et al. Neutralization of SARS-CoV-2 by destruction of the prefusion spike. *Cell Host Microbe* **2020**, *28*, 445–454. [CrossRef] [PubMed]

26. Walls, A.C.; Xiong, X.; Park, Y.-J.; Tortorici, M.A.; Snijder, J.; Quispe, J.; Cameroni, E.; Gopal, R.; Dai, M.; Lanzavecchia, A.; et al. Unexpected receptor functional mimicry elucidates activation of coronavirus fusion. *Cell* **2019**, *176*, 1024–1039. [CrossRef] [PubMed]

27. Edwards, M.J.; Dimmock, N.J. Hemagglutinin 1-specific immunoglobulin G and Fab molecules mediate postattachment neutralization of influenza A virus by inhibition of an early fusion event. *J. Virol.* **2001**, *75*, 10208–10218. [CrossRef]

28. Armstrong, S.J.; McInerney, T.L.; McLain, L.; Wahren, B.; Hinkula, J.; Levi, M.; Dimmock, N.J. Two neutralizing anti-V3 monoclonal antibodies act by affecting different functions of human immunodeficiency virus type 1. *J. Gen. Virol.* **1996**, *77*, 2931–2941. [CrossRef] [PubMed]

29. Sok, D.; Burton, D.R. Recent progress in broadly neutralizing antibodies to HIV. *Nat. Immunol.* **2018**, *19*, 1179–1188. [CrossRef] [PubMed]

30. Zhang, H.; Wang, G.; Li, J.; Nie, Y.; Shi, X.; Lian, G.; Wang, W.; Yin, X.; Zhao, Y.; Qu, X. Identification of an antigenic determinant on the S2 domain of the severe acute respiratory syndrome coronavirus spike glycoprotein capable of inducing neutralizing antibodies. *J. Virol.* **2004**, *78*, 6938–6945. [CrossRef]

31. Poh, C.M.; Guillaume, C.; Wang, B.; Naqiah, A.S.; Lee, C.Y.-P.; Chee, R.S.-L.; Fong, S.-W.; Yeo, N.K.-W.; Lee, W.-H.; Torres-Ruesta, A.; et al. Two linear epitopes on the SARS-CoV-2 spike protein that elicit neutralizing antibodies in COVID-19 patients. *Nat. Commun.* **2020**, *11*, 2806. [CrossRef]

32. Nara, P.L.; Garrity, R.R.; Goudsmit, J. Neutralization of HIV-1: A paradox of humoral proportions. *FASEB J.* **1991**, *5*, 2437–2455. [CrossRef]

33. Moore, J.P.; Cao, Y.; Quing, L.; Sattentau, Q.J.; Pyati, J.; Koduri, R.; Robinson, J.; Barbas III, C.F.; Burton, D.R.; Ho, D.D. Primary isolates of human immunodeficiency virus type 1 are relatively resistant to neutralization by monoclonal antibodies to gp120, and their neutralization is not predicted by studies with monomeric gp120. *J. Virol.* **1995**, *69*, 101–109. [CrossRef] [PubMed]

34. Bianchini, F.; Crivelli, V.; Abernathy, M.E.; Guerra, C.; Palus, M.; Muri, J.; Marcotte, H.; Piralla, A.; Redotti, M.; De Gasparo, R.; et al. Human neutralizing antibodies to cold linear epitopes and subdomain 1 of the SARS-CoV-2 spike glycoprotein. *Sci. Immunol.* **2023**. [CrossRef] [PubMed]

35. Bewley, K.R.; Coombes, N.S.; Gagnon, L.; McInroy, L.; Baker, N.; Shaik, I.; St-Jean, J.R.; St-Amant, N.; Buttigieg, K.R.; Humphries, H.E.; et al. Quantification of SARS-CoV-2 neutralizing antibody by wild-type plaque reduction neutralization, microneutralization and pseudotyped virus neutralization assays. *Nat. Protoc.* **2021**, *16*, 3114–3140. [CrossRef] [PubMed]

36. Hall, C.B.; Walsh, E.E.; Long, C.E.; Schnabel, K.C. Immunity to and frequency of reinfection with respiratory syncytial virus. *J. Infect. Dis.* **1991**, *163*, 693–698. [CrossRef]

37. Ascough, S.; Paterson, S.; Christopher Chiu, C. Induction and subversion of human protective immunity: Contrasting influenza and respiratory syncytial virus. *Front. Immunol.* **2018**, *9*, 323. [CrossRef] [PubMed]

38. Karron, R.A.; Buonagurio, D.A.; Georgiu, A.F.; Whitehead, S.S.; Adamus, J.E.; Clements-Mann, M.L.; Harris, D.O.; Randolph, V.B.; Udem, S.A.; Murphy, B.R.; et al. Respiratory syncytial virus (RSV) SH and G proteins are not essential for viral replication in vitro: Clinical evaluation and molecular characterization of a cold-passaged, attenuated RSV subgroup B mutant. *Proc. Nat. Acad. Sci. USA* **1997**, *94*, 13961–13966. [CrossRef]

39. Sastre, P.; Melero, J.A.; García-Barreno, B.; Palomo, C. Comparison of affinity chromatography and adsorption to vaccinia virus recombinant infected cells for depletion of antibodies directed against respiratory syncytial virus glycoproteins present in a human immunoglobulin preparation. *J. Med. Virol.* **2005**, *76*, 248–255. [CrossRef] [PubMed]

40. Magro, M.; Mas, V.; Chappell, K.; Vázquez, M.; Cano, O.; Luque, D.; Terrón, M.C.; Melero, J.A.; Palomo, C. Neutralizing antibodies against the preactive form of respiratory syncytial virus fusion protein offer unique possibilities for clinical intervention. *Proc. Nat. Acad. Sci. USA* **2012**, *109*, 3089–3094. [CrossRef]

41. Baber, J.; Arya, M.; Moodley, Y.; Jaques, A.; Jiang, Q.; Swanson, K.A.; Cooper, D.; Maddur, M.S.; Loschko, J.; Gurtman, A.; et al. A phase 1/2 study of a respiratory syncytial virus prefusion F vaccine with and without adjuvant in healthy older adults. *J. Infect. Dis.* **2022**, *226*, 2054–2063. [CrossRef] [PubMed]

42. Walsh, E.E.; Falsey, A.R.; Scott, D.A.; Gurtman, A.; Zareba, A.M.; Jansen, K.U.; Gruber, W.C.; Dormitzer, P.R.; Swanson, K.A.; Radley, D.; et al. A randomized phase 1/2 Study of a respiratory syncytial virus prefusion F vaccine. *J. Infect. Dis.* **2022**, *225*, 1357–1366. [CrossRef]

43. Venkatesan, P. Nirsevimab: A promising therapy for RSV. *Lancet Microbe* **2022**, *3*, e335. [CrossRef]

44. Graham, B.S. Immunological goals for respiratory syncytial virus vaccine development. *Curr. Opin. Immunol.* **2019**, *59*, 57–64. [CrossRef]

45. Gilman, M.S.A.; Castellanos, C.A.; Chen, M.; Ngwuta, J.O.; Goodwin, E.; Moin, S.M.; Mas, V.; Melero, J.A.; Wright, P.F.; Graham, B.S.; et al. Rapid profiling of RSV antibody repertoires from the memory B cells of naturally infected adult donors. *Sci. Immunol.* **2016**, *1*, eaaj1879. [CrossRef]

46. Xiao, X.; Tang, A.; Cox, K.S.; Wen, Z.; Callahan, C.; Sullivan, N.L.; Nahas, D.D.; Cosmi, S.; Galli, J.D.; Minnier, M.; et al. Characterization of potent RSV neutralizing antibodies isolated from human memory B cells and identification of diverse RSV/hMPV cross-neutralizing epitopes. *Mabs* **2019**, *11*, 1415–1427. [CrossRef]
47. Andreano, E.; Paciello, I.; Bardelli, M.; Tavarini, S.; Sammicheli, C.; Frigimelica, E.; Guidotti, S.; Torricelli, G.; Biancucci, M.; D'Oro, U.; et al. The respiratory syncytial virus (RSV) prefusion F-protein functional antibody repertoire in adult healthy donors. *EMBO Mol. Med.* **2021**, *13*, e14035. [CrossRef] [PubMed]
48. Dimmock, N.J. COVID-19 infection, vaccines, and immunity—The antibody response requires detailed analysis. *Microbiol. Res.* **2021**, *12*, 626–629. [CrossRef]
49. Callaway, E. What's next for Alphafold and the AI protein-folding revolution. *Nature* **2022**, *604*, 234–238. [CrossRef] [PubMed]
50. Bianchi, M.; Turner, H.L.; Nogal, B.; Cottrell, C.A.; Oyen, D.; Pauthner, M.; Bastidas, R.; Nedellec, R.; McCoy, L.E.; Wilson, I.A. Electron-microscopy-based epitope mapping defines specificities of polyclonal antibodies elicited during HIV-1 BG505 envelope trimer Immunization. *Immunity* **2018**, *49*, 288–300. [CrossRef]
51. Sok, D.; Le, K.M.; Vadnais, M.; Saye-Francisco, K.I.; Jardine, J.G.; Torres, J.L.; Berndsen, Z.T.; Kong, L.; Stanfield, R.; Jennifer Ruiz, J.; et al. Rapid elicitation of broadly neutralizing antibodies to HIV by immunization of cows. *Nature* **2017**, *548*, 108–111. [CrossRef] [PubMed]

Article

Safety and Efficacy of Outpatient Treatments for COVID-19: Real-Life Data from a Regionwide Cohort of High-Risk Patients in Tuscany, Italy (the FEDERATE Cohort)

Tommaso Manciulli [1], Michele Spinicci [1,2], Barbara Rossetti [3], Roberta Maria Antonello [1], Filippo Lagi [1,2], Anna Barbiero [1], Flavia Chechi [1,4], Giuseppe Formica [1], Emanuela Francalanci [1,5], Mirco Alesi [1,5], Samuele Gaggioli [1], Giulia Modi [1,6], Sara Modica [7], Riccardo Paggi [1,4], Cecilia Costa [6], Alessandra Morea [8], Lorenzo Paglicci [9], Ilaria Rancan [1,10], Francesco Amadori [11], Agnese Tamborrino [12], Marta Tilli [1,6], Giulia Bandini [1], Alberto Moggi Pignone [1], Beatrice Valoriani [9], Francesca Montagnani [10,13], Mario Tumbarello [10,13], Pierluigi Blanc [4], Massimo Di Pietro [6], Luisa Galli [12,14], Donatella Aquilini [5], Antonella Vincenti [11], Spartaco Sani [8], Cesira Nencioni [3], Sauro Luchi [7], Danilo Tacconi [9], Lorenzo Zammarchi [1,2] and Alessandro Bartoloni [1,2,*]

1   Dipartimento di Medicina Sperimentale e Clinica, Università degli Studi di Firenze, 50121 Firenze, Italy
2   SOD Malattie Infettive e Tropicali, Azienda Ospedaliero-Universitaria Careggi, 50134 Firenze, Italy
3   UOC Malattie Infettive, Ospedale Misericordia, 58100 Grosseto, Italy
4   SOC Malattie Infettive, Ospedale San Jacopo, 51100 Pistoia, Italy
5   UO Malattie Infettive, Ospedale Santo Stefano, 59100 Prato, Italy
6   UO Malattie Infettive, Ospedale Santa Maria Annunziata, 50012 Firenze, Italy
7   SOC Malattie Infettive ed Epatologia, Ospedale San Luca, 55100 Lucca, Italy
8   UO Malattie Infettive, Ospedali Riuniti di Livorno, 57124 Livorno, Italy
9   UO Malattie Infettive, Ospedale San Donato, 52100 Arezzo, Italy
10   Dipartimento di Biotecnologie Mediche, Università degli Studi di Siena, 53100 Siena, Italy
11   UO Malattie Infettive, Ospedale Apuane, 54100 Massa, Italy
12   Dipartimento Scienze della Salute, Università degli Studi di Firenze, 50139 Firenze, Italy
13   UOC Malattie Infettive e Tropicali, Azienda Ospedaliero Universitaria Senese, 53100 Siena, Italy
14   UO Malattie Infettive, Azienda Ospedaliero-Universitaria "Meyer", 50139 Firenze, Italy
*   Correspondence: alessandro.bartoloni@unifi.it

check for
updates

Citation: Manciulli, T.; Spinicci, M.; Rossetti, B.; Antonello, R.M.; Lagi, F.; Barbiero, A.; Chechi, F.; Formica, G.; Francalanci, E.; Alesi, M.; et al. Safety and Efficacy of Outpatient Treatments for COVID-19: Real-Life Data from a Regionwide Cohort of High-Risk Patients in Tuscany, Italy (the FEDERATE Cohort). *Viruses* **2023**, *15*, 438. https://doi.org/10.3390/v15020438

Academic Editor: Jason Yiu Wing KAM

Received: 31 December 2022
Revised: 28 January 2023
Accepted: 28 January 2023
Published: 5 February 2023

Abstract: Early COVID-19 treatments can prevent progression to severe disease. However, real-life data are still limited, and studies are warranted to monitor the efficacy and tolerability of these drugs. We retrospectively enrolled outpatients receiving early treatment for COVID-19 in 11 infectious diseases units in the Tuscany region of Italy between 1 January and 31 March 2022, when Omicron sublineages BA.1 and BA.2 were circulating. Eligible COVID-19 patients were treated with sotrovimab (SOT), remdesivir (RMD), nirmatrelvir/ritonavir (NRM/r), or molnupiravir (MOL). We gathered demographic and clinical features, 28-day outcomes (hospitalization or death), and drugs tolerability. A total of 781 patients (median age 69.9, 66% boosted for SARS-CoV-2) met the inclusion criteria, of whom 314 were treated with SOT (40.2%), 205 with MOL (26.3%), 142 with RMD (18.2%), and 120 with NRM/r (15.4%). Overall, 28-day hospitalization and death occurred in 18/781 (2.3%) and 3/781 (0.3%), respectively. Multivariable Cox regression showed that patients receiving SOT had a reduced risk of meeting the composite outcome (28-day hospitalization and/or death) in comparison to the RMD cohort, while no significant differences were evidenced for the MOL and NRM/r groups in comparison to the RMD group. Other predictors of negative outcomes included cancer, chronic kidney disease, and a time between symptoms onset and treatment administration > 3 days. All treatments showed good safety and tolerability, with only eight patients (1%) whose treatment was interrupted due to intolerance. In the first Italian multicenter study presenting real-life data on COVID-19 early treatments, all regimens demonstrated good safety and efficacy. SOT showed a reduced risk of progression versus RMD. No significant differences of outcome were observed in preventing 28-day hospitalization and death among patients treated with RMD, MOL, and NRM/r.

Keywords: SARS-CoV-2; COVID-19; sotrovimab; nirmatrelvir/ritonavir; molnupiravir; remdesivir

1. Introduction

Therapeutic options for the early phase of coronavirus disease 2019 (COVID-19) have been sought since the start of the pandemic. In the last year, several compounds have been licensed for the treatment of patients with recent symptoms onset [1]. These include monoclonal antibodies (mAbs); the oral antivirals nirmatrelvir/ritonavir (NRM/r; an inhibitor of the main protease, also called 3-chymotrypsin-like protease, of severe acute respiratory syndrome coronavirus 2 (SARS-CoV-2)) and molnupiravir (MOL; an inhibitor of the RNA-dependent RNA polymerase of SARS-CoV-2); and the intravenous antiviral remdesivir (RMD, another inhibitor of the RNA-dependent RNA polymerase of SARS-CoV-2) [2]. Based on the results of clinical trials, most international guidelines issued recommendations which prioritize the use of NRM/r (relative risk [RR] reduction 88% for hospitalization or death) or RMD (RR reduction 87%) over MOL (RR reduction 30%) for the treatment of COVID-19 patients who do not require hospitalization or supplemental oxygen [1,3]. The mAbs group has included several molecules, often used in combination to achieve viral neutralization. The clinical trials testing mAbs use in outpatients have been carried out in patients carrying the Alpha variant [4–6], and over time some of these drugs became inefficacious as SARS-CoV-2 mutated, generating new variants [7]. Sotrovimab (SOT), a recombinant human monoclonal antibody against SARS-CoV-2, which obtained US Food and Drug Administration Emergency Use Authorization for the treatment of high-risk outpatients with mild-to-moderate COVID-19 in May 2021, has been shown to have lower neutralizing activity against Omicron BA.1 than against the ancestral strain and other variants of concern, even less neutralizing activity against Omicron BA.2, and lost inhibitory capability against BA.4 and BA.5 [8,9]. Monoclonal antibodies that have maintained activity against BA.4/5 include bebtelovimab [1,7,10], currently not approved in Europe, and the combination of tigaxevimab and cilgavimab [7,11], approved for both pre-/post-exposure prophylaxis and early treatment of immunocompromised patients at increased risk of severe COVID. However, the clinical trials for most of these drugs were carried out in the pre-Omicron era, and among non-vaccinated subjects [4,12]. The introduction of anti-SARS-CoV-2 vaccines caused a drastic reduction in COVID-19 severity and mortality [13–15], and it also changed the susceptibility of the target population to the virus. As such, continued surveillance of the efficacy and effectiveness of the compounds used to treat patients at high risk of severe disease is needed. To date, few studies on the efficacy and tolerability of these drugs under real-world conditions, characterized by new variants of concern (VOCs) and large vaccination coverage, have been published. For instance, a recent observational study from Israel showed that NRM/r was able to reduce hospitalization rates and deaths in treated versus untreated subjects [16,17]. Early treatment with either MOL or NRM/r was confirmed to reduce the risks of mortality and in-hospital disease progression in comparison with untreated controls in a large cohort of patients (mostly unvaccinated) in Hong Kong, during the wave of SARS-CoV-2 Omicron subvariant BA.2.2, as NRM/r was additionally associated with a reduced risk of hospitalization. Other observational studies have confirmed the promising results of NRM/r [18–20], while data on the performance of MOL appears to be less clear-cut, as one study has shown that untreated patients had similar outcomes to those receiving MOL [19,21]. However, available studies suffer from limitations arising from their retrospective, observational nature. Since the start of January 2022, different regimens for early treatment of COVID-19 have been available in Italy (SOT, RMD, and MOL), while NRM/r was made available at the start of February 2022. As such, these drugs have been employed in the "Omicron era" of COVID-19 and used on subjects deemed at risk of severe disease by the Italian National Drug Agency (AIFA), with at-risk conditions including chronic diseases such as hypertension with organ damage, chronic kidney disease, chronic heart disease, cancers, chronic lung disease, and immunosuppression, as well as an age over 65 years [22].

We present data on the safety and efficacy of the four outpatient regimens available in Italy (SOT, RMD, NRM/r, and MOL), obtained from a multicenter study conducted in 11 infectious diseases units operating in the Tuscany region of Italy.

2. Materials and Methods

2.1. Patient Population

We retrospectively retrieved data on patients treated at the outpatient services of 11 infectious diseases units in Tuscany, Italy, between 1 January 2022 and 31 March 2022. Patients were considered eligible if: (i) they had received SOT, RMD, NRM/r, or MOL; (ii) they were treated in an outpatient setting; (iii) they had at least one risk factor according to AIFA criteria; and (iv) they were classified as having mild or moderate COVID-19 infection according to WHO criteria. Symptoms that allowed for treatment were defined by AIFA criteria and included fever, malaise, smell or taste disturbances, chills, dyspnea, sore throat, headache, myalgia, and GI symptoms. On the other hand, patients were excluded if they were: (i) hospitalized for reasons other than COVID-19 at the time of treatment, (ii) without a risk factor for severe COVID-19 according to AIFA criteria, and/or (iii) asymptomatic or suffering from a severe or critical disease. Children from a pediatric infectious diseases center were included, in light of a recent position paper by the Italian Society of Pediatrics recommending early treatment options for children at risk of COVID-19 progression [23].

2.2. Data Collection

Collected data included demographic information (sex at birth, age), data on risk factors for COVID-19 progression according to the AIFA criteria: age > 65 years, hypertension with organ damage, chronic heart disease, chronic kidney disease (CKD), chronic lung disease, chronic liver disease, immunosuppression (either congenital or iatrogenic), oncological patients including those with blood and solid cancers undergoing active treatment, and obesity, defined by a body mass index [the weight in kilograms divided by the square of the height in meters] $\geq$30. Among these conditions, no weights were attributed by AIFA to regulate the prescription of anti-SARS-CoV-2 early treatments. We also collected information on vaccination status (defined as one-dose, full-cycle, or boosted regardless of the type of vaccine used, as information on vaccine type was not readily available), date of symptoms onset, date of treatment administration, latency between symptoms onset and treatment administration (defined as the number of days between the first day of symptoms and the day of treatment start). Information regarding the referral channel to outpatient services was also collected (hospital-based specialist, family doctor, doctor part of the special units set up for COVID-19 at-home management, direct referral from emergency department doctor).

Outcome measures included treatment completion, side effects (patient-reported intolerance to drug altering the course of treatment, allergic reaction), and hospitalization or death due to COVID-19 progression. A composite outcome consisting of death and/or hospitalization was also created.

Information on the outcome at 28 days was captured through a standardized questionnaire. Information on outcome measures was entered into the database at this time. Data were collected using REDCap 8.11.6. (Project REDCap, USA).

2.3. Data Analysis

We analyzed data for patients who had received treatment, using available information on the occurrence of hospitalization and death. Data were analyzed with STATA 17.0 (STATACorp, College Station, TX, USA). Continuous variables were reported as medians and interquartile ranges; categorical variables were reported as absolute counts and proportions. A chi-square test was used to test for differences in categorical variables. The Kruskal–Wallis test was used to test for differences in continuous variables among the treatment groups. A survival analysis between different treatment groups was carried out using Kaplan–Meier curves and the log-rank test. The average time at risk for an event was computed as the time to the first event or day 28, whichever was earlier. Multivariable Cox regression was performed to identify independent predictors of composite outcome (28-day hospitalization and/or death related to COVID-19), calculated as hazard ratios (HR,

95%CI). Moreover, given the non-randomized assignment to the four treatment groups, a propensity score (PS) analysis using inverse probability of treatment weighting (IPTW) was performed to assess the average treatment effect (ATE) of SOT, MOL, and NRM/r in comparison to RMD. Inverse probability of treatment weighting uses weights based on the propensity score to create a synthetic sample in which the distribution of measured baseline covariates is independent of treatment assignment [24]. The following covariates were included to generate the PS: sex; age; chronic comorbidities, such as obesity, chronic kidney disease, chronic heart disease, chronic obstructive pulmonary disease, cancer, cognitive impairment, diabetes, and immunosuppression; smoking habit; vaccination status, categorized as 'not vaccinated' (none or incomplete primary schedule) or 'vaccinated' (complete primary schedule +/− booster dose); and latency between symptoms onset to antiviral administration, categorized as ≤3 or >3 days. We arbitrarily decided to enter RMD as a reference variable, since it was the group with the highest number of events (hospitalization and/or death). Standardized differences were used to compare balance in baseline covariates between the four groups before and after weighing by the inverse probability of treatment.

2.4. Ethics

The study was performed in accordance with the ethical principles of the Declaration of Helsinki and with the International Conference for Harmonization Good Clinical Practice guidelines.

3. Results

In the study period, 921 patients received early treatment for mild-to-moderate COVID-19 within the 11 ID units involved. Of these, 140 (15.2%) did not meet the inclusion criteria and were excluded (Supplementary Table S1). Of the 781 included patients (50% female, median age 66.9 years, IQR 52.3–77.9), 314 (40.2%) received SOT, 142 (18.2%) received RMD, 205 (26.3%) received MOL, and 120 (15.4%) received NRM/r. In brief, patients receiving SOT (47% female) had the lowest median age (64.7 years, IQR 50.2–77.7) and included the highest percentage of non-vaccinated (20%, 64/314) and immunocompromised people (51%, 159/314). Patients in the MOL group (42% female, median age 68.9, IQR 57.3–79.9) were the oldest, and had the highest frequency of obese people (30%, 61/205). The RMD group (59% female, median age 67.4 years, IQR 52–78.9) had the highest percentage of smokers (31%, 44/142). The NRM/r group (57.5% female, median age 66.8 years, IQR 50.3–75.6) included the highest percentage of people fully vaccinated ± a booster dose (97%, 116/120). The baseline characteristics of the study population, divided into the four treatment groups, are fully reported in Table 1.

Table 1. Baseline characteristics of the study population, divided into the four treatment groups.

	TOTAL (*n* = 781)	RMD (*n* = 142)	SOT (*n* = 314)	MOL (*n* = 205)	NRM/r (*n* = 120)	*p*-Value
Sex (*n*, %)						
– Male	394 (50.4)	59 (41.6)	166 (52.9)	118 (57.6)	51 (42.5)	
– Female	387 (49.6)	83 (58.5)	148 (47.1)	87 (42.4)	69 (57.5)	0.005
Age (median, IQR)	66.9 (52.4–77.9)	67.4 (52–78.9)	64.7 (50.2–77.8)	68.9 (57.3–79.9)	66.9 (50.3–75.6)	0.014
Vaccination (*n*, %)						
– None	108 (13.8)	17 (12)	64 (20.4)	24 (11.7)	3 (2.5)	
– One dose	14 (1.8)	2 (1.4)	8 (2.6)	3 (1.5)	1 (0.8)	
– Full schedule	144 (18.4)	24 (16.9)	67 (21.3)	46 (22.4)	7 (5.8)	
– Booster	515 (65.9)	98 (79.7)	175 (55.7)	132 (64.4)	109 (90.8)	<0.001
Time from symptoms onset to treatment (median days, IQR)	3 (2–4)	4 (2–5)	4 (3–5)	3 (2–4)	2 (2–3)	<0.001

Table 1. *Cont.*

	TOTAL (*n* = 781)	RMD (*n* = 142)	SOT (*n* = 314)	MOL (*n* = 205)	NRM/r (*n* = 120)	*p*-Value
Obese (*n*, %)	178 (22.8)	26 (18.3)	62 (19.6)	61 (29.8)	29 (24.2)	0.027
Pregnant (*n*, %)	2 (0.3)	2 (1.4)	0	0	0	–
CKD (*n*, %)	75 (9.6)	8 (5.6)	47 (15)	16 (7.8)	4 (3.3)	<0.001
CHD (*n*, %)	404 (51.7)	79 (55.6)	155 (49.4)	113 (55.1)	57 (47.5)	0.360
Cancer (*n*, %)	189 (24.2)	42 (29.6)	85 (27.1)	27 (13.2)	35 (29.2)	<0.001
COPD (*n*, %)	188 (24.1)	33 (23.2)	67 (21.3)	61 (29.8)	27 (22.5)	0.161
Cognitive impairment (*n*, %)	73 (9.4)	12 (8.5)	36 (11.5)	9 (4.4)	16 (13.3)	0.018
Stroke (*n*, %)	24 (3.1)	4 (2.8)	13 (4.1)	4 (2)	3 (2.5)	0.580
Diabetes (*n*, %)	141 (18.1)	29 (20.4)	53 (16.9)	33 (16.1)	26 (21.7)	0.501
Immunocompromised (*n*, %)	282 (36.1)	51 (35.9)	159 (50.6)	26 (12.7)	46 (38.3)	<0.001
Current or former smoker (*n*, %)	144 (18.4)	44 (31)	43 (13.7)	40 (19.5)	17 (14.2)	<0.001

Legend: RMD: remdesivir; SOT: sotrovimab; MOL: molnupiravir; NRM/r: nirmatrelvir/ritonavir; IQR: interquartile range; CKD: chronic kidney disease; CHD: chronic heart disease; COPD: chronic obstructive pulmonary disease.

Most patients were referred to the prescribing centers either by their general practitioners (*n* = 327, 42.5%) or by territorial medical units for the care of COVID-19 (*n* = 258, 33.5%); the remainder of patients were referred either by other specialists (*n* = 111, 14.4%) or by the emergency department (*n* = 45, 5.8%). In 29 cases (3.8%), the patients had direct contact with the ID specialist. No information was available for 11 patients. We found that latency between symptoms onset and treatment was significantly higher for patients treated with parenteral drugs, i.e., SOT (median 4 days, IQR 3–5) and RMD (median 4 days, IQR 2–5) compared to oral antivirals MOL (median 3 days, IQR 2–4) and NRM/r (median 3 days, IQR 2–3) (*p* = 0.001).

Outcome Data

Deaths occurred in one patient in the SOT group (0.3%) and in two patients in the RMD group (1.4%). No deaths occurred in the MOL and NRM/r groups. Eighteen patients were hospitalized due to COVID-19 progression: five (1.6%) in the SOT group, seven (4.9%) in the RMD group, four (1.9%) in the MOL group, and three (2.5%) in the NRM/r group.

Patients receiving treatment > 3 days from symptoms onset had a higher risk of meeting the composite endpoint of death or hospitalization (12/317, 3.8%) in comparison with those who started the treatment ≤ 3 days from symptoms onset (4/464, 1.3%, *p* = 0.023).

Kaplan–Meier survival curves for the composite outcome of each treatment group are reported in Figure 1.

The average time at risk for an event was 27.35 days for RMD (standard error [SE] 0.27), 27.74 days for SOT (SE 0.13), 27.67 days for MOL (SE 0.19), and 27.44 days for NRM/r (SE 0.32). Head-to-head comparison of survival curves between each treatment group showed significant differences only for RMD vs. SOT (difference in the cumulative percentage of patients with COVID-19–related hospitalization or death through day 28 was 3.3%, 95%CI −0.5–7.2%; *p*-value = 0.039). No statistical differences between other study groups were observed in the survival analysis.

Multivariable analysis performed by Cox regression showed that patients receiving SOT had a lower risk of meeting the composite outcome compared to patients in the RMD group (HR 0.14, 95%CI 0.03–0.56, *p* = 0.005), while no significant differences were evidenced between the RMD group and the MOL (HR 0.43, 95%CI 0.09–1.96, 0.273) and NRM/r groups (HR 0.51, 95%CI 0.11–2.28, 0.374). Predictors of hospitalization and/or death included a latency >3 days between symptoms onset and treatment administration (HR 1.41, 95%CI 1.07–1.85, *p* = 0.013), chronic kidney disease (HR 5.01, 95%CI 1.30–19.3, *p* = 0.019), and cancer (HR 3.11, 95%CI 1.07–9.09, *p* = 0.038), while a history of chronic heart disease resulted in a protective factor (HR 0.24, 95%CI 0.07–0.80, *p* = 0.020). Complete results of the Cox regression analysis are shown in Figure 2.

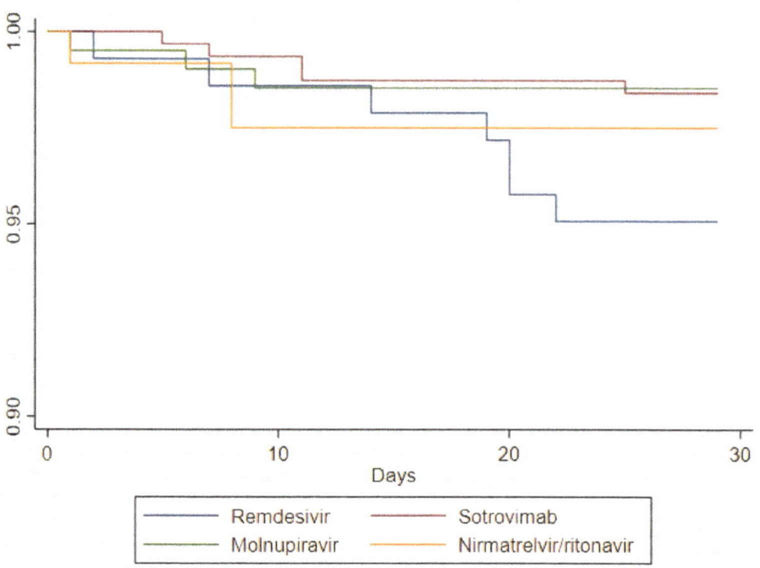

Figure 1. Kaplan–Meier survival curves of patients assigned to the four treatment groups (SOT *n* = 314; MOL *n* = 205; RMD *n* = 142; NRM/r *n* = 120). The failure event is a composite outcome of 28-day hospitalization and/or death related to COVD-19.

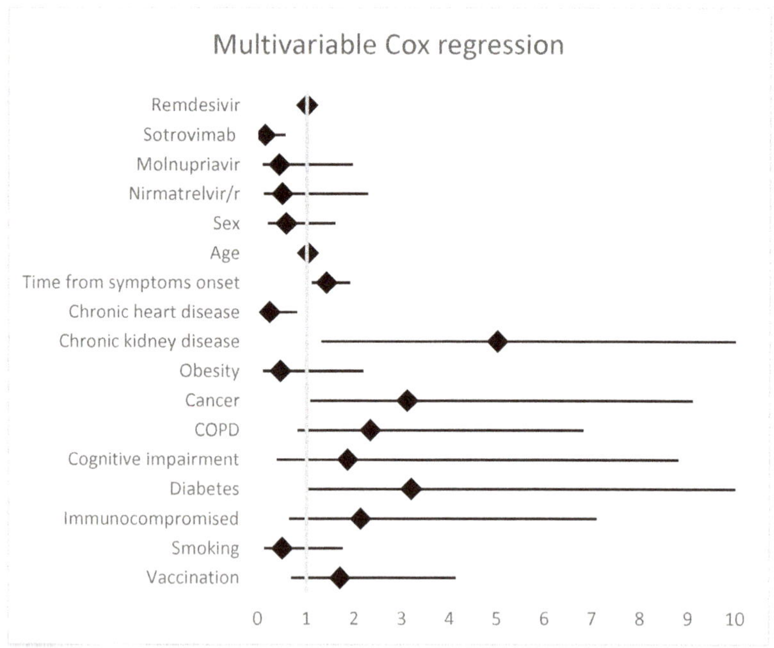

Figure 2. Multivariable Cox regression analysis. Predictors of composite outcome (28-day hospitalization and/or death related to COVD-19).

On IPTW-adjusted PS analysis, a trend in favor of SOT versus RMD was observed (ATE -0.04, 95%CI -0.07–0.002, $p = 0.063$), while no significant differences emerged when comparing the RMD group with the MOL (-0.01, 95%CI -0.07–0.04, $p = 0.659$) and NRM/r (0.00, 95%CI -0.07–0.07, $p = 0.983$) groups. Analysis of standardized differences showed good balance in baseline covariates between the four groups before and after weighting by IPTW (Supplementary Table S2).

Drug intolerance was reported by 29 patients (4%), including eight cases leading to drug discontinuation. Intolerance was reported by 5% in the MOL (10/205) and NRM/r (6/120) groups, 4% in the RMD group (5/142), and 3% in the SOT group (8/314). Discontinuation occurred only in the MOL ($n = 5$, 2.5%) and RMD groups ($n = 3$, 2.1%).

4. Discussion

Outpatient treatments for COVID-19 patients play a crucial role in the prevention of disease progression to severe forms in patients at high risk of poor outcomes [1]. However, continued evaluation of the safety and efficacy of these treatments is warranted, as new SARS-CoV-2 VOCs will emerge, and population susceptibility will change due to previous exposure and vaccine administration [13,14].

This study represents, to our knowledge, the largest multicenter report of real-life data from Italy, where the prescription of antivirals and monoclonal antibodies has been subject to strict regulation since their introduction in March 2021 [25]. Regulatory trials for all available compounds either were carried out before SARS-CoV-2 vaccine rollout or excluded vaccinated subjects [4,5,26]. Moreover, most trials did not focus on high-risk subjects, except for one trial on the use of SOT [27]. Our study population is largely representative of patients that are currently at greater risk for COVID-19, i.e., elderly patients older than 65 years old, with multiple comorbidities predisposing to severe COVID-19, albeit mostly vaccinated against COVID-19 [1,21,28,29].

All drugs showed low rates of hospitalization and/or death due to COVID-19 progression, in line with results from previous studies [16,19,30]. Multivariable analysis suggested a possible advantage in the use of SOT in comparison with RMD, while no significant differences were observed among the three antiviral agents (RMD, MOL, and NRM/r).

This study was conducted in the so-called "Omicron era": in Tuscany, the Omicron lineage B.1.1.529 was responsible for around 90% of new infections at the beginning of 2022, and reached 100% at the end of the study period in March 2022, when BA.1 and BA.2 were at 52% and 47%, respectively [31,32]. The Omicron variant has been associated with a reduced risk of hospitalization and death in the general population compared to the Delta variant, although significant variation has been observed by age [33]. Moreover, immunocompromised patients infected with the Omicron variant remain at high risk of severe outcomes, as observed in a prospective cohort of 114 solid organ transplant recipients, patients on anti-CD20 therapy, and allogenic hematopoietic stem-cell transplantation recipients, one of whom died and 23 (20%) of whom required hospital admission for a median of 11 days [34].

Chronic kidney disease and cancer were confirmed to be predictors of severe outcomes in patients with COVID-19, regardless of the use of early treatment against SARS-CoV-2. Conversely, chronic heart disease was a predictor of positive outcomes. Both severe CKD and cancer emerged as higher-risk comorbidities compared with other conditions, such as old age, chronic heart diseases, metabolic disorders, or isolated hypertension, in a multicenter cohort study carried out in Shanghai, China, during the 2022 Omicron wave [35]. Moreover, CKD patients have reduced treatment options, since the use of NRM/r and RMD is contraindicated in patients with severe renal function impairment, i.e., a glomerular filtration rate less than 30 mL/min, limiting the choice to MOL and/or mAbs.

It should be highlighted that SOT is not effective against Omicron BA.4 and BA.5, the currently dominant subvariants [7,10]. However, we decided to include SOT patients in the analysis, considering that real-life data on this compound could still be informative for the future use of other antibodies. On the other hand, no report of SARS-CoV-2 resistance to

RMD and/or oral antivirals (MOL and NRM/r) has emerged to date, and susceptibilities of BA.4 and BA.5 VoCs to the three compounds were similar to those of the ancestral SARS-CoV-2 strain [36].

All patients were prescribed the drugs within a relatively short period of time from symptoms onset, within a median 3 days for oral antivirals and 4 days for RMD and SOT, according to AIFA criteria [22]. It is worth noting that a time from symptoms onset to treatment administration longer than 3 days was a predictor of a negative outcome in our population. This finding, along with the absence of significant difference in the outcomes, supports the use of oral compounds in situations where logistics issues may delay administration of parenteral drugs [25].

Moreover, the four treatments appear to be acceptably safe in terms of adverse events, which ranged from 3 to 5% of patients, similar to those found in regulatory clinical trials and other real-life studies [16,19,30].

The main limitation of this study is its retrospective design and the non-randomized assignment to the four treatment groups. Different distributions of patients' features and comorbidities across groups are related to the different drug characteristics and/or may reflect specific attitudes of prescribers. For example, the higher frequency of chronic kidney disease in the SOT group can be explained by drug pharmacokinetics (i.e., no potential for nephrotoxicity, unlike both RMD and NRM/r). The excess of unvaccinated and immunocompromised people in the SOT group is likely to reflect a greater confidence in this compound for the frailest patients. Furthermore, NRM/r was not available until mid-February in Italy.

Furthermore, we did not collect and analyze data about COVID-19 symptoms, and we did not investigate the potential correlation between clinical manifestation and COVID-19 severity. Likewise, we could not retrieve data on the full immunization schedules of all participants: the immunization campaign in Italy has used different combinations of vaccines since its start in 2020 [37], and we cannot exclude the possibility that such variables might influence COVID-19 outcomes and drug tolerability. However, an exhaustive analysis of the role of these variables in COVID-19 patients was beyond the scope of our study. Another limitation is that we did not collect data on the time to viral clearance, nor on the presence of rebound infection, which has recently been reported after the administration of NRM/r and MOL [18,38].

5. Conclusions

In conclusion, this study represents one of the first efforts at real-life data collection on COVID-19 outpatient treatment options. We observed a low incidence of adverse events and negative outcomes with all currently used treatments, and we confirmed the paramount importance of the administration timing of early therapies against COVID-19.

Supplementary Materials: The following supporting information can be downloaded at: https://www.mdpi.com/article/10.3390/v15020438/s1, Table S1: Characteristics of the excluded patients in our cohort. Table S2: Analysis of standardized differences showed good balance in baseline covariates between the four groups before and after weighting by inverse probability of treatment weighting.

Author Contributions: Conceptualization, T.M., M.S., F.L., L.Z. and A.B. (Alessandro Bartoloni); data curation, T.M., B.R., S.M., R.P., L.P., I.R., A.T. and B.V.; formal analysis, T.M., M.S. and F.L.; investigation, T.M., M.S., B.R., R.M.A., A.B. (Anna Barbiero), F.C., G.F., E.F., M.A., S.G., G.M., S.M., R.P., C.C., A.M., L.P., I.R., F.A., A.T., M.T. (Marta Tilli), G.B., B.V., F.M., P.B., M.D.P., D.A., A.V. and D.T.; methodology, T.M., M.S., F.L. and L.Z.; project administration, T.M.; resources, G.B., A.M.P., M.T. (Mario Tumbarello), P.B., M.D.P., L.G., D.A., A.V., S.S., C.N., S.L., D.T., L.Z. and A.B. (Alessandro Bartoloni); supervision, F.L., M.T. (Mario Tumbarello), P.B., M.D.P., L.G., D.A., A.V., S.S., C.N., S.L., D.T., L.Z. and A.B. (Alessandro Bartoloni); validation, M.T. (Mario Tumbarello), S.S., C.N., S.L., D.T. and A.B. (Alessandro Bartoloni); writing—original draft, T.M., M.S. and R.M.A.; writing—review and editing, M.S., B.R., R.M.A., M.T. (Mario Tumbarello), L.Z. and A.B. (Alessandro Bartoloni). All authors have read and agreed to the published version of the manuscript.

Funding: This research received no external funding.

Institutional Review Board Statement: The study was conducted in accordance with the Declaration of Helsinki and Good Clinical Practice Standards.

Informed Consent Statement: Informed consent was waived due to the retrospective nature of the study.

Data Availability Statement: Data are available from the authors upon reasonable request.

Acknowledgments: We would like to thank all the patients and the referring GPs. We would also like to acknowledge all the physicians working in the outpatients clinics for the prescription of anti-SARS-CoV-2 early treatments. Azienda Ospedaliero-Universitaria Careggi: Nicoletta Di Lauria, Costanza Malcontenti, Costanza Fiorelli, Lucia Graziani, Andreea Miftode, Dario Nigi, Francesca Mariotti, Sara De Chiara, Alessandro Di Felice, Marco Fognani, Giuseppe Formica, and Noemi Streva. Azienda Ospedaliero-Universitaria Senese: Raffaella Corbisiero, Massimiliano Fabbiani, Lucia Migliorini, Francesca Montagnani, Francesco Pippi, Anna Sansoni, and Michele Trezzi. Azienda Ospedaliera Prato: Alessio Pampaloni, Alessandro Nerli, Giovanni Millotti, Giulia Montorzi, Elisabetta Betti, and Beatrice Anna Adri–ani. Azienda Ospedaliera Grosseto: Giorgia Angeli, Salvatore Cardaci, Tiziana Carli, Silvia Chigiotti, Leonardo Croci, Alesandro Lanari, Marco Nofri, and Francesco Strambio De Castillia. Azienda Ospedaliera Pistoia: Magdalena Viscione, Alessandra Francesca Manunta, Daniela Messeri, Claudio Fabbri, Sara Irene Bonelli, and Angela Cesarina Vivarelli. Azienda Ospedaliera Massa: Novella Cesta, Mirco Lenzi, and Giovanni Sarteschi. Azienda Ospedaliera Lucca: Sara Moneta, Giovanni Morelli, Sarah Iacopini, and Simona Fortunato. Ospedale Pediatrico Meyer: Leila Bianchi.

Conflicts of Interest: The authors declare no conflict of interest for this work. Outside of this research work, Prof. Alessandro Bartoloni reports institutional grants from MSD and ViiV and personal honoraria from GSK and Gilead.

References

1. NIH NIH COVID Treatment Guidelines. Available online: https://www.covid19treatmentguidelines.nih.gov/ (accessed on 23 January 2023).
2. Lee, T.C.; Morris, A.M.; Grover, S.A.; Murthy, S.; McDonald, E.G. Outpatient Therapies for COVID-19: How Do We Choose? *Open Forum Infect. Dis.* **2022**, *9*, ofac008. [CrossRef]
3. Bhimraj, A.; Morgan, R.L.; Shumaker, A.H.; Lavergne, V.; Cheng, V.C.; Edwards, K.M.; Gandhi, R.; Gallagher, J.; Muller, W.J.; Horo, J.C.O.; et al. Infectious Diseases Society of America Guidelines on the Treatment and Management of Patients with COVID-19. Available online: https://www.idsociety.org/globalassets/idsa/practice-guidelines/covid-19/treatment/idsa-covid-19-gl-tx-and-mgmt-v10.2.0.pdf (accessed on 25 January 2023).
4. Gupta, A.; Gonzalez-Rojas, Y.; Juarez, E.; Crespo Casal, M.; Moya, J.; Falci, D.R.; Sarkis, E.; Solis, J.; Zheng, H.; Scott, N.; et al. Early Treatment for COVID-19 with SARS-CoV-2 Neutralizing Antibody Sotrovimab. *N. Engl. J. Med.* **2021**, *385*, 1941–1950. [CrossRef]
5. Gottlieb, R.L.; Vaca, C.E.; Paredes, R.; Mera, J.; Webb, B.J.; Perez, G.; Oguchi, G.; Ryan, P.; Nielsen, B.U.; Brown, M.; et al. Early Remdesivir to Prevent Progression to Severe COVID-19 in Outpatients. *N. Engl. J. Med.* **2021**, *386*, 305–315. [CrossRef]
6. Chen, P.; Nirula, A.; Heller, B.; Gottlieb, R.L.; Boscia, J.; Morris, J.; Huhn, G.; Cardona, J.; Mocherla, B.; Stosor, V.; et al. SARS-CoV-2 Neutralizing Antibody LY-CoV555 in Outpatients with COVID-19. *N. Engl. J. Med.* **2021**, *384*, 229–237. [CrossRef]
7. Iketani, S.; Liu, L.; Guo, Y.; Liu, L.; Chan, J.F.-W.; Huang, Y.; Wang, M.; Luo, Y.; Yu, J.; Chu, H.; et al. Antibody evasion properties of SARS-CoV-2 Omicron sublineages. *Nature* **2022**, *604*, 553–556. [CrossRef]
8. Takashita, E.; Kinoshita, N.; Yamayoshi, S.; Sakai-Tagawa, Y.; Fujisaki, S.; Ito, M.; Iwatsuki-Horimoto, K.; Halfmann, P.; Watanabe, S.; Maeda, K.; et al. Efficacy of Antiviral Agents against the SARS-CoV-2 Omicron Subvariant BA.2. *N. Engl. J. Med.* **2022**, *386*, 1475–1477. [CrossRef]
9. Takashita, E.; Yamayoshi, S.; Simon, V.; van Bakel, H.; Sordillo, E.M.; Pekosz, A.; Fukushi, S.; Suzuki, T.; Maeda, K.; Halfmann, P.; et al. Efficacy of Antibodies and Antiviral Drugs against Omicron BA.2.12.1, BA.4, and BA.5 Subvariants. *N. Engl. J. Med.* **2022**, *387*, 468–470. [CrossRef]
10. Cao, Y.; Yisimayi, A.; Jian, F.; Song, W.; Xiao, T.; Wang, L.; Du, S.; Wang, J.; Li, Q.; Chen, X.; et al. BA.2.12.1, BA.4 and BA.5 escape antibodies elicited by Omicron infection. *Nature* **2022**, *608*, 593–602. [CrossRef]
11. Montgomery, H.; Hobbs, F.D.R.; Padilla, F.; Arbetter, D.; Templeton, A.; Seegobin, S.; Kim, K.; Campos, J.A.S.; Arends, R.H.; Brodek, B.H.; et al. Efficacy and safety of intramuscular administration of tixagevimab–cilgavimab for early outpatient treatment of COVID-19 (TACKLE): A phase 3, randomised, double-blind, placebo-controlled trial. *Lancet Respir. Med.* **2022**, *19*, 985–996. [CrossRef]
12. Hammond, J.; Leister-Tebbe, H.; Gardner, A.; Abreu, P.; Bao, W.; Wisemandle, W.; Baniecki, M.; Hendrick, V.M.; Damle, B.; Simón-Campos, A.; et al. Oral Nirmatrelvir for High-Risk, Nonhospitalized Adults with COVID-19. *N. Engl. J. Med.* **2022**, *386*, 1397–1408. [CrossRef]

13. Tenforde, M.W.; Self, W.H.; Adams, K.; Gaglani, M.; Ginde, A.A.; McNeal, T.; Ghamande, S.; Douin, D.J.; Talbot, H.K.; Casey, J.D.; et al. Association Between mRNA Vaccination and COVID-19 Hospitalization and Disease Severity. *JAMA* **2021**, *326*, 2043–2054. [CrossRef]

14. Johnson, A.G.; Amin, A.B.; Ali, A.R.; Hoots, B.; Cadwell, B.L.; Arora, S.; Avoundjian, T.; Awofeso, A.O.; Barnes, J.; Bayoumi, N.S.; et al. COVID-19 Incidence and Death Rates Among Unvaccinated and Fully Vaccinated Adults with and Without Booster Doses During Periods of Delta and Omicron Variant Emergence—25 U.S. Jurisdictions, April 4-December 25, 2021. *MMWR Morb. Mortal. Wkly. Rep.* **2022**, *71*, 132–138. [CrossRef]

15. COVID-19 Forecasting Team Variation in the COVID-19 infection–fatality ratio by age, time, and geography during the pre-vaccine era: A systematic analysis. *Lancet* **2022**, *399*, 1469–1488. [CrossRef]

16. Najjar-Debbiny, R.; Gronich, N.; Weber, G.; Khoury, J.; Amar, M.; Stein, N.; Goldstein, L.H.; Saliba, W. Effectiveness of Paxlovid in Reducing Severe Coronavirus Disease 2019 and Mortality in High-Risk Patients. *Clin. Infect. Dis.* **2022**, ciac443. [CrossRef]

17. Ganatra, S.; Dani, S.S.; Ahmad, J.; Kumar, A.; Shah, J.; Abraham, G.M.; McQuillen, D.P.; Wachter, R.M.; Sax, P.E. Oral Nirmatrelvir and Ritonavir in Non-hospitalized Vaccinated Patients with COVID-19. *Clin. Infect. Dis.* **2022**, ciac673. [CrossRef]

18. Malden, D.E.; Hong, V.; Lewin, B.J.; Ackerson, B.K.; Lipsitch, M.; Lewnard, J.A.; Tartof, S.Y. Hospitalization and Emergency Department Encounters for COVID-19 After Paxlovid Treatment—California, December 2021–May 2022. *MMWR Morb. Mortal. Wkly. Rep.* **2022**, *71*, 830–833. [CrossRef]

19. Wong, C.K.H.; Au, I.C.H.; Lau, K.T.K.; Lau, E.H.Y.; Cowling, B.J.; Leung, G.M. Real-world effectiveness of early molnupiravir or nirmatrelvir–ritonavir in hospitalised patients with COVID-19 without supplemental oxygen requirement on admission during Hong Kong's omicron BA.2 wave: A retrospective cohort study. *Lancet Infect. Dis.* **2022**, *22*, 1681–1693. [CrossRef]

20. Carlson, N.E.; Ginde, A.A. Real-world Use of Nirmatrelvir-Ritonavir in COVID-19 Outpatients During the Emergence of Omicron Variants BA. 2 / BA2. 12. 1. *medRxiv* **2022**. Available online: https://www.medrxiv.org/content/10.1101/2022.09.12.22279866v1 .full.pdf (accessed on 15 November 2022).

21. Butler, C.C.; Hobbs, F.D.R.; Gbinigie, O.A.; Rahman, N.M.; Hayward, G.; Richards, D.B.; Dorward, J.; Lowe, D.M.; Standing, J.F.; Khoo, S.; et al. Molnupiravir plus usual care versus usual care alone as early treatment for adults with COVID-19 at increased risk of adverse outcomes (PANORAMIC): Preliminary analysis from the United Kingdom randomised, controlled open-label, platform adaptive trial. *Lancet* **2022**, *401*, 281–293. [CrossRef]

22. AIFA Uso Degli Antivirali per COVID-19. Available online: https://www.aifa.gov.it/uso-degli-antivirali-orali-per-covid-19 (accessed on 29 November 2022).

23. Lanari, M.; Venturini, E.; Pierantoni, L.; Stera, G.; Castelli Gattinara, G.; Esposito, S.M.R.; Favilli, S.; Franzoni, E.; Fusco, E.; Lionetti, P.; et al. Eligibility criteria for pediatric patients who may benefit from anti SARS-CoV-2 monoclonal antibody therapy administration: An Italian inter-society consensus statement. *Ital. J. Pediatr.* **2022**, *48*, 7. [CrossRef]

24. Austin, P.C. An introduction to propensity score methods for reducing the effects of confounding in observational studies. *Multivar. Behav. Res.* **2011**, *46*, 399–424. [CrossRef]

25. Manciulli, T.; Lagi, F.; Barbiero, A.; Fognani, M.; Di Lauria, N.; Malcontenti, C.; Fiorelli, C.; Spinicci, M.; Ceccherini, V.; Onofrio, P.D.; et al. Implementing Early Phase Treatments for COVID-19 in Outpatient Settings: Challenges at a Tertiary Care Center in Italy and Future Outlooks. *Infect. Dis. Rep.* **2022**, *14*, 315–320. [CrossRef]

26. Jayk Bernal, A.; Gomes da Silva, M.M.; Musungaie, D.B.; Kovalchuk, E.; Gonzalez, A.; Delos Reyes, V.; Martín-Quirós, A.; Caraco, Y.; Williams-Diaz, A.; Brown, M.L.; et al. Molnupiravir for Oral Treatment of COVID-19 in Nonhospitalized Patients. *N. Engl. J. Med.* **2022**, *386*, 509–520. [CrossRef]

27. Gupta, A.; Gonzalez-Rojas, Y.; Juarez, E.; Crespo Casal, M.; Moya, J.; Rodrigues Falci, D.; Sarkis, E.; Solis, J.; Zheng, H.; Scott, N.; et al. Effect of Sotrovimab on Hospitalization or Death among High-risk Patients with Mild to Moderate COVID-19: A Randomized Clinical Trial. *JAMA* **2022**, *327*, 1236–1246. [CrossRef]

28. CDC—Centers for Disease Control and Prevention Science Brief: Evidence Used to Update the List of Underlying Medical Conditions Associated with Higher Risk for Severe COVID-19. Available online: https://www.cdc.gov/coronavirus/2019-ncov/ science/science-briefs/underlying-evidence-table.html (accessed on 1 February 2022).

29. Wang, W.; Kaelber, D.C.; Xu, R.; Berger, N.A. Breakthrough SARS-CoV-2 Infections, Hospitalizations, and Mortality in Vaccinated Patients With Cancer in the US Between December 2020 and November 2021. *JAMA Oncol.* **2022**, *8*, 1027–1034. [CrossRef]

30. Pesko, B.; Deng, A.; Chan, J.D.; Neme, S.; Dhanireddy, S.; Jain, R. Safety and Tolerability of Paxlovid (Nirmatrelvir/Ritonavir) in High-risk Patients. *Clin. Infect. Dis.* **2022**, *75*, 2049–2050. [CrossRef]

31. Isituto Superiore di Sanità (ISS). *Prevalenza e Distribuzione delle Varianti di SARS-CoV-2 di Interesse per la Sanità Pubblica in Italia*; Rapporto n. 18 del 25 marzo 2022; ISS: Rome, Italy, 2022.

32. Isituto Superiore di Sanità (ISS). *Prevalenza e Distribuzione delle Varianti di SARS-CoV-2 di Interesse per la Sanità Pubblica in Italia*; Rapporto n. 16 del 19 gennaio 2022; ISS: Rome, Italy, 2022.

33. Nyberg, T.; Ferguson, N.M.; Nash, S.G.; Webster, H.H.; Flaxman, S.; Andrews, N.; Hinsley, W.; Bernal, J.L.; Kall, M.; Bhatt, S.; et al. Comparative analysis of the risks of hospitalisation and death associated with SARS-CoV-2 omicron (B.1.1.529) and delta (B.1.617.2) variants in England: A cohort study. *Lancet* **2022**, *399*, 1303–1312. [CrossRef]

34. Malahe, S.R.K.; Hoek, R.A.S.; Dalm, V.A.S.H.; Broers, A.E.C.; den Hoed, C.M.; Manintveld, O.C.; Baan, C.C.; van Deuzen, C.M.; Papageorgiou, G.; Bax, H.I.; et al. Clinical Characteristics and Outcomes of Immunocompromised Patients With Coronavirus Disease 2019 Caused by the Omicron Variant: A Prospective, Observational Study. *Clin. Infect. Dis.* **2022**. [CrossRef]

35. Chen, X.; Wang, H.; Ai, J.; Shen, L.; Lin, K.; Yuan, G.; Sheng, X.; Jin, X.; Deng, Z.; Xu, J.; et al. Identification of CKD, bedridden history and cancer as higher-risk comorbidities and their impact on prognosis of hospitalized Omicron patients: A multi-centre cohort study. *Emerg. Microbes Infect.* **2022**, *11*, 2501–2509. [CrossRef]

36. Service, R.F. Bad news for Paxlovid? Resistance may be coming. *Science* **2022**, *377*, 138–139. [CrossRef]

37. Dellino, M.; Lamanna, B.; Vinciguerra, M.; Tafuri, S.; Stefanizzi, P.; Malvasi, A.; Di Vagno, G.; Cormio, G.; Loizzi, V.; Cazzato, G.; et al. SARS-CoV-2 Vaccines and Adverse Effects in Gynecology and Obstetrics: The First Italian Retrospective Study. *Int. J. Environ. Res. Public Health* **2022**, *19*, 13167. [CrossRef]

38. Wang, L.; Berger, N.A.; Davis, P.B.; Kaelber, D.C.; Volkow, N.D.; Xu, R. COVID-19 rebound after Paxlovid and Molnupiravir during January–June 2022. *medRxiv Prepr. Serv. Health Sci.* **2022**, 6–12. [CrossRef]

 *viruses*

Article

Mapping the Antibody Repertoires in Ferrets with Repeated Influenza A/H3 Infections: Is Original Antigenic Sin Really "Sinful"?

Tal Einav [1,*,†] , Martina Kosikova [2,†] , Peter Radvak [2] , Yuan-Chia Kuo [2] , Hyung Joon Kwon [2] and Hang Xie [2,*]

1   Basic Sciences Division and Computational Biology Program, Fred Hutchinson Cancer Research Center, Seattle, WA 98109, USA
2   Laboratory of Respiratory Viral Diseases, Division of Viral Products, Office of Vaccines Research and Review, Center for Biologics Evaluation and Research, United States Food and Drug Administration, Silver Spring, MD 20993, USA
*   Correspondence: teinav@fredhutch.org (T.E.); hang.xie@fda.hhs.gov (H.X.)
†   These authors contributed equally to this work.

Abstract: The influenza-specific antibody repertoire is continuously reshaped by infection and vaccination. The host immune response to contemporary viruses can be redirected to preferentially boost antibodies specific for viruses encountered early in life, a phenomenon called original antigenic sin (OAS) that is suggested to be responsible for diminished vaccine effectiveness after repeated seasonal vaccination. Using a new computational tool called Neutralization Landscapes, we tracked the progression of hemagglutination inhibition antibodies within ferret antisera elicited by repeated influenza A/H3 infections and deciphered the influence of prior exposures on the de novo antibody response to evolved viruses. The results indicate that a broadly neutralizing antibody signature can nevertheless be induced by repeated exposures despite OAS induction. Our study offers a new way to visualize how immune history shapes individual antibodies within a repertoire, which may help to inform future universal influenza vaccine design.

Keywords: mapping antibody repertoires; original antigenic sin; immune imprinting; repeated influenza exposures; broadly neutralizing antibody; influenza A H3N2 virus

 check for **updates**

Citation: Einav, T.; Kosikova, M.; Radvak, P.; Kuo, Y.-C.; Kwon, H.J.; Xie, H. Mapping the Antibody Repertoires in Ferrets with Repeated Influenza A/H3 Infections: Is Original Antigenic Sin Really "Sinful"? *Viruses* **2023**, *15*, 374. https://doi.org/10.3390/v15020374

Academic Editor: Jason Yiu Wing KAM

Received: 1 January 2023
Revised: 20 January 2023
Accepted: 27 January 2023
Published: 28 January 2023

1. Introduction

Rapidly evolving pathogens such as influenza frequently change their antigenicity in order to escape the host immune system, and the emergence of antigenically drifted strains necessitates the annual update of seasonal influenza vaccine components. Despite efforts to forecast which strain(s) will be most prevalent, a suboptimal or mismatched vaccine strain may occasionally be selected for vaccine production, resulting in reduced protection [1–4]. In the US, influenza vaccine effectiveness in the past decades has fluctuated significantly from 10% in the 2004–2005 season [1] to 60% in the 2010–2011 season (https://www.cdc.gov/flu/vaccines-work/effectiveness-studies.htm) (accessed on 11 April 2021) [5]. While vaccine mismatch directly accounts for this low efficacy, pre-existing host immunity also influences vaccine performance [3,6–12].

An individual's exposure history, acquired through recurrent infections and/or vaccinations, shapes their unique antibody repertoire and influences their response to newly emerging influenza viruses [6,10–19]. For example, residual antibodies from prior exposures may grant subsequent protection against viruses with similar antigenicity [15–17,20]. However, immune imprinting from viruses encountered early in life can also lead to insufficient de novo antibody response to evolved viruses—a phenomenon called original antigenic sin (OAS) [21]. While the exact mechanisms remain unknown, OAS has been associated with low antibody responses in individuals with repeated seasonal vaccination and has been hypothesized to negatively affect vaccine effectiveness in frequent vaccinees [1,9,20,22–28]. These reports provide a glimpse of the complex interplay between prior and current immunity, highlighting the influence of immune imprinting that must be addressed in the field of vaccinology. Elucidating the impact of OAS on de novo antibody responses will provide insights for future influenza vaccine development with improved performance.

In this work, we used a newly developed computational tool—Neutralization Landscapes [29]—to track the progression of the hemagglutination inhibition (HAI) responses in ferrets after repeated influenza A/H3 infections, and characterized the HAI antibody patterns induced. By mapping the HAI responses at the single-antibody scale, we demonstrated that repeated influenza A/H3 exposures, despite OAS induction, can expand the breadth of de novo HAI antibody response.

2. Materials and Methods

2.1. Viruses

The panel of H3N2 viruses used for the study included A/Philippines/2/1982 (Philippines 1982), A/Wisconsin/67/2005 (Wisconsin 2005), A/Uruguay/716/2007 (Uruguay 2007), A/Perth/16/2009 (Perth 2009), A/Victoria/361/2011 (Victoria 2011), A/Texas/50/2012 (Texas 2012), A/Switzerland/9715293/2013 (Switzerland 2013) and A/Hong Kong/4801/2014 (Hong Kong 2014), each of which has served as the prototype for the H3N2 seasonal influenza vaccine component in past decades. All H3N2 viruses were propagated in 9–10-day-old embryonated eggs, and aliquots were stored at $-80\,^{\circ}C$ until use.

2.2. Ferret Antisera

Seronegative male ferrets (Triple F Farm) at 15–16 weeks old were infected intranasally at two-week intervals with each of the four H3N2 viruses (V_1 = A/Uruguay/716/2007 or Uruguay 2007, V_2 = A/Texas/50/2012 or Texas 2012, V_3 = A/Switzerland/9715293/2013 or Switzerland 2013, and V_4 = A/Hong Kong/4801/2014 or Hong Kong 2014) [6]. After ferrets were anesthetized, approximately 10^5 focus-forming units of virus in a total of 1 mL was delivered into both nostrils per ferret at 0.5 mL per nostril [6]. Ferrets were bled via venipuncture of the cranial vena cava under anesthesia at 14 days after each infection. Sera from four ferrets in each infection scheme were collected for HAI titer determination. All procedures were carried out in accordance with a protocol approved by the Institutional Animal Care and Use Committee of the Center for Biologics Evaluation and Research, US Food and Drug Administration.

2.3. HAI Assay

Following pre-treatment with a receptor-destroying enzyme (Denka-Seiken), individual ferret sera were 2-fold serially diluted and were 1:1 (v/v) incubated with testing virus solution containing 4 hemagglutinin (HA) units per 25 µL at room temperature for 30 min before the addition of 50 µL of 0.75% guinea pig erythrocytes in the presence of 20 nM oseltamivir, as previously described [6,14]. Wells containing PBS only or virus only served as the negative and positive controls in each HAI assay performed. The endpoint HAI titer was defined as the reciprocal of the highest serum dilution that yielded a complete HA inhibition, and a titer 5 was assigned if no inhibition was observed at the starting 1:10 serum dilution. HAI geometric mean titers (GMTs) were calculated, along with the 95% confidence intervals.

3. Neutralization Landscapes and Decomposition of Ferret Antisera

Neutralization Landscapes uses monoclonal antibody data to quantify viruses cross-reactivity and enumerate the space of potential antibody inhibition profiles [29]. A landscape is a low-dimensional map where antibodies and viruses are represented as points and antibody-virus distance translates into experimentally measurable neutralization. The positions of eight H3N2 viruses were determined using neutralizing titers from six human monoclonal antibodies targeting the head of influenza HA [29]. Four antibodies (CH65 [30], 5J8 [31], C05 [32], and F045-092 [33]) target the receptor binding site, another antibody (F005-126) cross-links between two monomers within a hemagglutinin trimer [34], and the last binds to hemagglutinin's lateral patch [35]. Previous serum-based efforts using antigenic cartography suggest that these antibodies do not need to bind to the same epitope [36], and that the landscape can include H1N1- and H3N2-specific antibodies [29]. Moreover, we posit that this landscape will become more accurate as more antibodies targeting different epitopes are added. These assertions can be directly tested by assessing the landscape's ability to predict unmeasured interactions.

The 50% inhibitory concentrations ($IC_{50}s$) were determined between each antibody and virus pair, and their positions on the neutralization landscapes were fixed using two-dimensional (2D) scaling on the $\log_{10}(IC_{50})$ values, where an antibody–virus distance of d unit translates into $IC_{50} = 2^d \cdot 10^{-10}$ Molar [29]. To accommodate the HAI titers used in this study, a few cosmetic changes were made in computation. First, the spacing between gridlines was decreased by a factor of $\log_{10}(\frac{1}{2}) = 3.3$ so that 1 grid unit represents a two-fold decrease in neutralization. Second, HAI titers were converted into absolute Molar units by scanning across different conversion factors and minimizing the absolute mean error between $\log_2$(measured HAI titers) and $\log_2$(inferred HAI titers from decomposition). The optimal conversion factor translated an antibody–virus map distance of d units into an HAI titer $= 6000/2^d$. This global conversion factor had been applied to all antisera analyzed in this work.

We decomposed each antiserum by determining which set of antibody coordinates and stoichiometries best matched the measured HAI titers against the eight viruses in the panel. Decomposition proceeded by considering $n = 1, 2, 3 \dots$ antibodies until the error of the decomposition decreased below a set threshold [29]. To prevent overfitting, decomposition with an additional antibody was only accepted if it decreased the mean fold-error between measured and inferred titers by $\geq 20\%$ [29]. The relative fractions of each antibody in the mixture were allowed to vary, although each antibody must comprise $\geq 10\%$ of the mixture and the sum of all fractions must sum to 100%.

In any neutralization landscape, the abundance of each antibody in a mixture was depicted by the size of the gray circle surrounding it. For any virus lying within a gray circle, the antibody at the center of that circle was predicted to have an HAI titer ≥ 80 against it. If an antibody comprised a fraction f of a mixture, then its ability to inhibit a virus decreased by f-fold. For example, a monoclonal antibody was surrounded by a circle of radius $d_{\text{mAb}} = 6.2$ grid units ($6000/2^{d_{\text{mAb}}} = 80$), while an antibody that comprised a fraction f of the serum would be surrounded by a circle of radius $r_{\text{mAb}} - \log_2\left(\frac{1}{f}\right)$. When multiple antibodies were present in a serum, their collective inhibition or neutralization against a virus was, thus, computed using a competitive binding model,

$$IC_{50,\text{Competitive}} = \left(\textstyle\sum_j f_j / IC_{50}^{(j)}\right)^{-1},$$

where f_j represented the fractional composition of the jth antibody and $IC_{50}^{(j)}$ denoted the concentration at which this monoclonal antibody would neutralize the virus by 50% [29]. In this way, any combination of points (which determined $IC_{50}^{(j)}$ through antibody–virus map distance) at any stoichiometry (f_j) was translated into the mixture's collective HAI titer against the virus panel.

4. Results

4.1. Validation of Neutralization Landscapes

Neutralization Landscapes uses fixed virus coordinates to depict the location and magnitude of constituent neutralizing antibodies within an antiserum elicited by natural infection or vaccination [29]. A key assumption of Neutralization Landscapes is that the neutralization profiles of all dominant antibodies within a serum can be represented as individual points on the map, where the Euclidean distance d between each antibody coordinate and virus coordinate translates into an HAI titer of $6000/2^d$. This forms a basis set of individual antibody behaviors from which we can determine the minimal combination of antibodies that can replicate a serum's measurements. This process was previously validated by decomposing mixtures of 2–3 antibodies [29]; here, we extend this work to analyze ferret sera for the first time.

The resulting decompositions predict the functional behavior of the dominant antibodies within serum while neglecting the weaker or less frequent antibodies that do not affect a serum's HAI profile. Since sera may contain multiple functionally similar antibodies, we call the resulting HAI profiles "antibody signatures" as they may represent one or multiple antibodies within the response.

On the neutralization landscapes, the gray regions surrounding each antibody signature indicate an HAI titer $\geq$ 80 against any virus lying within (we refer to this virus as strongly inhibited by the mixture), and stronger antibody signatures could overwhelm the inhibition of weaker antibodies placed further out on the map.

To decompose each antiserum, we first determined which combination of coordinates and stoichiometries best matched the experimental HAI titers against the eight H3N2 viruses whose coordinates had been previously fixed using a different antibody–virus panel [37]. The results were validated in two ways. First, Figure 1A showed the decomposition of a ferret antiserum elicited by sequential influenza A/H3 infections in a neutralization landscape in which the HAI titers predicted were on average $\leq$2-fold off from the experimental measurements (Figure 1A). The same analysis was extended to include all 288 HAI measurements involved in this study, with each point representing a pair of predicted titers and corresponding experimental values in Figure 1B. It yielded a coefficient of $R^2 = 0.6$ (Figure 1B), with only 2% (6/288) of map predicted titers being $\geq$10-fold off from the experimental measurements.

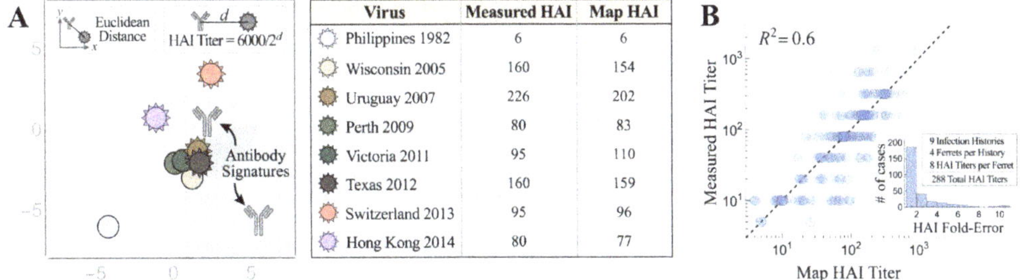

Virus	Measured HAI	Map HAI
Philippines 1982	6	6
Wisconsin 2005	160	154
Uruguay 2007	226	202
Perth 2009	80	83
Victoria 2011	95	110
Texas 2012	160	159
Switzerland 2013	95	96
Hong Kong 2014	80	77

Figure 1. Characterizing the accuracy of Neutralization Landscapes. (**A**) The decomposition of ferret antiserum elicited by sequential H3N2 infections in a neutralization landscape (left) and the resulting predicted/measured HAI titers (right). Larger antibody–virus distance on the landscape corresponds to weaker antibody inhibition, with Euclidean distance d representing an HAI titer of $6000/2^d$. The resulting antibody signatures represent the predicted inhibition profiles of the dominant antibodies within the serum. (**B**) Cumulative analysis for all ferret antisera analyzed in this work. The inset shows the distribution of fold-errors for these predictions.

Second, while we decomposed ferret antiserum using the HAI titers against all eight viruses (Figure 2A), we also performed another decomposition by using the half set of measurements to predict the HAI titers against the remaining four viruses in the panel (Figure 2B; titers against red viruses were used for direct decomposition and titers against gray viruses were inferred from the map). Despite the difficulties of triangulating the coordinates and stoichiometries of multiple antibodies using a half set of the measurements, the predicted HAI titers were on average 6.0-fold off from the experimental measurements, only slightly larger than the 2.4-fold error when the full suite of titers was used for decomposition (Figure 2C).

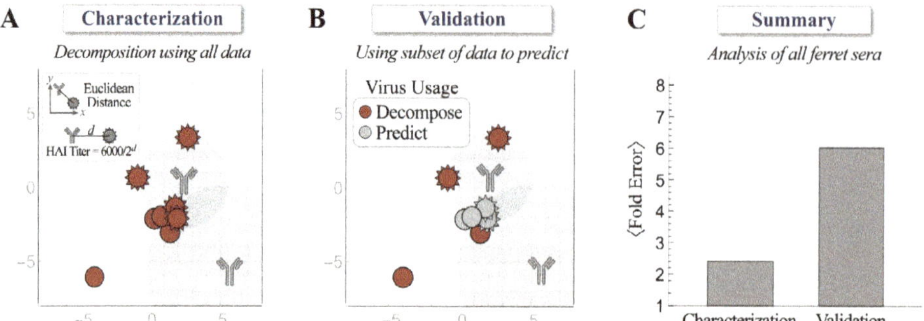

Figure 2. Validating decompositions using a subset of hemagglutination inhibition (HAI) measurements. (**A**) The decomposition of ferret antiserum using experimental HAI titers against the full set of viruses. (**B**) Decomposition of ferret antiserum using experimental HAI titers against the half set of viruses (red) to predict the titers of the remaining four viruses (gray). (**C**) The average error from both methods across all ferret antisera analyzed in this work.

Taken together, these validations demonstrated that the neutralization landscapes accurately characterize the HAI profiles of ferret antisera.

4.2. Progression of HAI Responses Following Sequential Infections

We first conducted a sequential infection experiment to demonstrate how the antibody repertoire in ferrets was shaped by recurring exposure. Figure 3A–D is the traditional way to present the neutralizing activities of HAI antibodies developed in ferrets after infection with first virus A/Uruguay/716/2007 (Uruguay 2007, denoted as V_1 throughout this work), then the second virus A/Texas/50/2012 (Texas 2012, V2) ($V_1 \rightarrow V_2$), third A/Switzerland/9715293/2013 (Switzerland 2013, V_3) ($V_1 \rightarrow V_2 \rightarrow V_3$), and fourth A/Hong Kong/4801/2014 (Hong Kong 2014, V_4) ($V_1 \rightarrow V_2 \rightarrow V_3 \rightarrow V_4$). Uruguay 2007 ($V_1$) infection elicited V_1-specific ferret HAI titers with limited cross-reactivity towards viruses that emerged before 2005 or after 2007. Following each sequential infection with V_2, V_3, and V_4, the resulting ferret antisera gradually extended the HAI cross-reactivity from V_1-specific to inhibit (with geometric mean titers (GMTs) ≥ 80) all A/H3 viruses in the panel except A/Philippines/2/1982 (Philippines 1982), which had disappeared from circulation more than three decades earlier (Figure 3B–D). Sequential infections also induced typical OAS, where ferret antisera always had lower HAI titers toward later exposed V_2, V_3 or V_4 than first encountered V_1 (Supplementary Figure S1A–C). In contrast, infection by V_2, V_3 or V_4 alone (without a priming V_1 infection) elicited higher homologous HAI GMTs (Figure 4 and Supplementary Figure S1D–F).

We then used Neutralization Landscapes to track the progression of HAI antibodies developed throughout these four infections (Figure 3E–H). Infection by V_1 showed one antibody signature that strongly inhibited V_1 and three nearby viruses—A/Wisconsin/67/2005 (Wisconsin 2005), A/Victoria/361/2011 (Victoria 2011), and Texas 2012 (note that Hong Kong 2014 also had a measured titer ≈ 80 (Figure 3A), although this is not seen on the landscape (Figure 3E)). In each subsequent infection ($V_1 \rightarrow V_2$, $V_1 \rightarrow V_2 \rightarrow V_3$, and $V_1 \rightarrow V_2 \rightarrow V_3 \rightarrow V_4$), we detected two distinct antibody signatures; one "specific" antibody (i.e., that strongly inhibited all infection strains) and another "non-specific" antibody signature (that inhibited little-to-no infection strains).

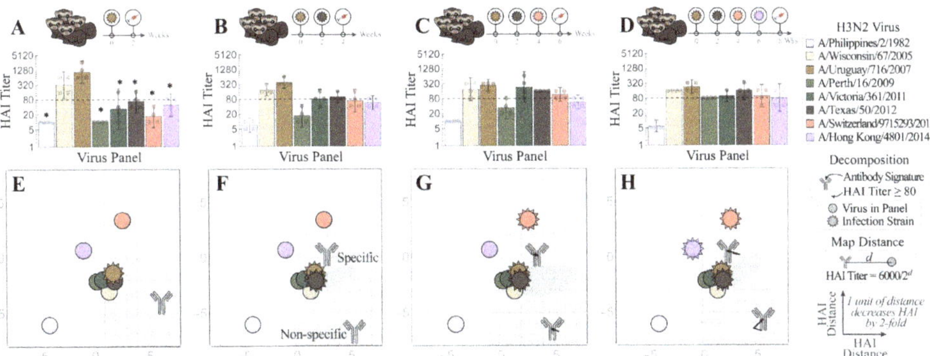

Figure 3. Tracking individual antibodies using hemagglutination inhibition (HAI) responses in ferrets during each stage of four sequential H3N2 infections. (**A–D**) Naïve ferrets were sequentially infected with V_1 = Uruguay 2007, V_2 = Texas 2012, V_3 = Switzerland 2013, and V_4 = Hong Kong 2014. HAI titers are shown for each step of the infection: (**A**) V_1, (**B**) $V_1 \rightarrow V_2$, (**C**) $V_1 \rightarrow V_2 \rightarrow V_3$, and (**D**) $V_1 \rightarrow V_2 \rightarrow V_3 \rightarrow V_4$. Individual HAI titers are presented from four ferrets (points) with geometric means (bar graphs) and 95% confidential intervals (error bars). * indicates $p < 0.05$ vs. Uruguay 2007 by Mann–Whitney test after data were log transformed. (**E–H**) These measurements were decomposed to determine the antibody signatures (the HAI profiles of the dominant antibodies) elicited after infection with (**E**) V_1 followed by (**F**) V_2, (**G**) V_3, and (**H**) V_4. Each antibody signature (gray) is predicted to have an HAI titer ≥ 80 against any virus within the gray circle, with the size of this circle proportional to the fractional composition of the antibody signature within the serum. An antibody–virus distance d denotes an HAI titer of $6000/2^d$. Virus coordinates were previously determined using a panel of monoclonal antibodies (see details in the main text). In Panels (**F–H**), infection with multiple viruses elicited both specific (HAI ≥ 80 against all infection strains) and non-specific antibody signatures. Arrows show the subtle shifts of the antibody signatures between Panels (**F**)$\rightarrow$(**G**)$\rightarrow$(**H**).

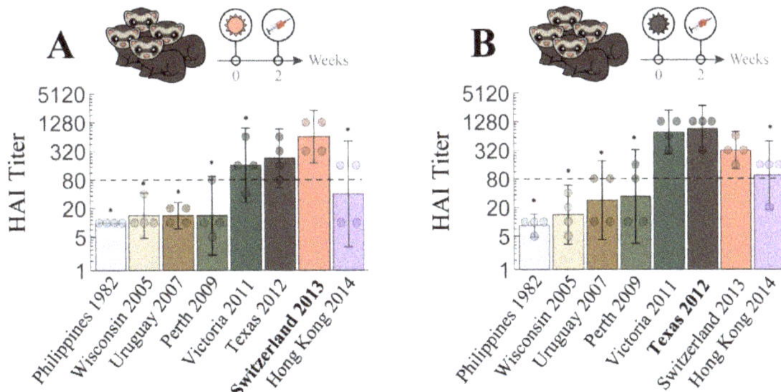

Figure 4. Cross-reactive hemagglutination inhibition (HAI) responses in ferrets with single H3N2 infection. Naïve ferrets were singly infected with either (**A**) V_3 = Switzerland 2013 or (**B**) V_2 = Texas 2012. Individual HAI titers from four ferrets (points) and geometric means (bar graphs) are shown with 95% confidential intervals (error bars). * indicates $p < 0.05$ vs. Switzerland 2013 (**A**) or vs. Texas 2012 (**B**) by Mann–Whitney test after data were log transformed.

With each additional infection, the specific antibody signature slightly moved by 1–2 units or increased its relative abundance in the mixture (shown by the size of the gray circular regions surrounding each antibody, Figure 3E–H) in a manner that kept all infection strains strongly inhibited. These antibody signatures represent the dominant inhibition profiles within each serum: weaker antibody signatures (including those elicited by earlier infections) may either be masked by these dominant profiles or the profiles of functionally similar antibodies may be combined within one antibody signature. In a sense, these landscapes provide an "Occam's razor" description of each serum using the minimum possible number of HAI profiles. Note that these landscapes do not perfectly reproduce all titers, but they are highly consistent on average (Figure 1).

Despite OAS induction, ferret antisera after the $V_1 \rightarrow V_2 \rightarrow V_3 \rightarrow V_4$ infections had all GMTs within four-fold of one another across the entire virus panel (except for the older Philippines 1982 strain), indicating extended cross-reactivity (Figure 1B–D). The individual maps of Ferrets #1–4 in this cohort also showed similar antibody patterns, i.e., that the end antisera after the $V_1 \rightarrow V_2 \rightarrow V_3 \rightarrow V_4$ infections cross-reacted with most viruses in the panel except Philippines 1982 (individual ferret traces shown in Supplementary Figures S2–S4). These progressional maps collectively suggest that a broadly neutralizing antibody signature (defined as an antibody with HAI titer ≥ 80 against multiple infection strains) can be guided into place by sequential exposures despite OAS induction.

4.3. Influence of Prior Influenza Exposures on De Novo Antibody Response

We next compared the HAI response of ferrets infected with V_4 alone with the responses elicited after one ($V_3 \rightarrow V_4$), two ($V_2 \rightarrow V_3 \rightarrow V_4$), or three prior infections ($V_1 \rightarrow V_2 \rightarrow V_3 \rightarrow V_4$), to assess how exposure history affected the de novo HAI antibody response to the latest infection by V_4. Unlike infection by V_4 alone, which elicited higher HAI titers toward itself than to most viruses in the panel (Figure 5A), ferret antisera induced after additional prior exposures had HAI GMTs toward V_4 not higher than those toward the earlier infection strains (Figure 5B–D), a typical OAS response that was also seen for every sequence of infections in Figure 3B–D.

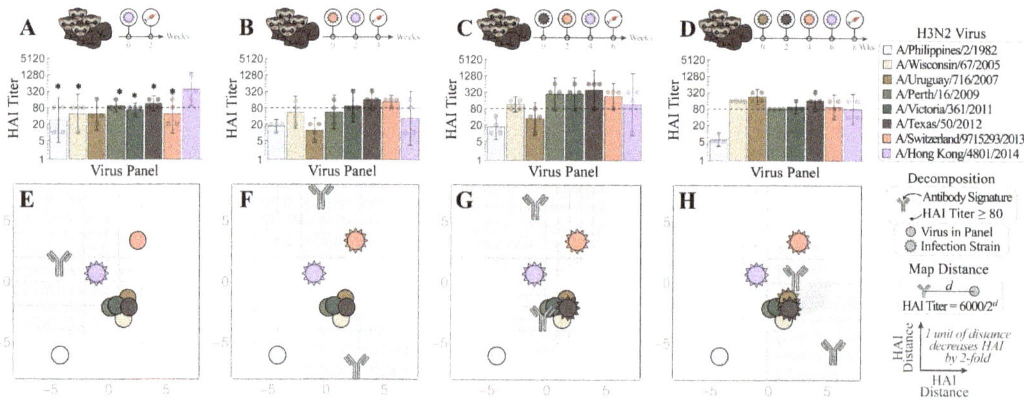

Figure 5. Mapping how exposure history shapes the ferret antibody response. Naïve ferrets were infected with (**A**) V_4 = Hong Kong 2014, or with prior exposure to (**B**) V_3 = Switzerland 2013, (**C**) V_2 = Texas 2012, and (**D**) V_1 = Uruguay 2007. All antisera were analyzed via HAI after the final infection with Hong Kong 2014. Individual HAI titers are shown from four ferrets (points) and geometric means (bar graphs) with 95% confidential intervals (error bars). * indicates $p < 0.05$ vs. Hong Kong 2014 by Mann–Whitney test after data were log transformed. (**E–H**) Each set of measurements was decomposed to determine the antibody signatures (the HAI profiles of the dominant antibodies within these sera). Each antibody signature (gray) is predicted to have an HAI titer ≥ 80 against any virus within the gray circle, with the size of this circle proportional to the fractional composition of the antibody signature within the serum. An antibody–virus distance d denotes an HAI titer of $6000/2^d$.

We then used Neutralization Landscapes to discern how prior exposures impacted subsequent antibody development at the single-antibody scale (average response shown in Figure 5E–H and individual responses shown in Figure S2). Upon infection by V_4 alone, one specific antibody signature emerged that strongly inhibited V_4 (Figure 5E). In fact, a dominant antibody signature specific for the infecting strain was observed in all individual ferrets after single infection (regardless of virus), although an additional minor antibody signature may also be seen in some ferrets (Figure S3).

Ferrets infected by two or more strains consistently showed a polyclonal response, with a specific antibody signature targeting the infection strains and a non-specific antibody signature targeting other viruses in the panel (Figure 3F–H and Figure 5F–H). While multiple infections often resulted in a single predicted antibody signature that strongly inhibits all infection strains (Figure 5F,H), one triple infection resulted in a broad response mediated by two distinct antibody signatures (Figure 5G: Switzerland 2013 and Hong Kong 2014 were strongly inhibited by the top antibody signature, while Texas 2012 was strongly inhibited by the bottom signature). Multiple infections always resulted in OAS, since any antibody signature that strongly inhibited a prior infection strain exhibited weaker inhibition against V_4 (i.e., V_4 lies further from the center of the gray antibody circles than the earlier infection strains, Figure 5F–H). Nevertheless, following the $V_1 \rightarrow V_2 \rightarrow V_3 \rightarrow V_4$ infections, ferrets developed antibodies that inhibited not only all four infection strains but also the other H3 viruses in the panel, except Philippines 1982.

Taken together, these results suggest that prior exposures affect the antibody response, but that a broadly neutralizing antibody signature can be induced by repeated exposures, even in the presence of OAS.

5. Discussion

It is estimated that most humans are infected with influenza by the age of 3 and continue to be reinfected by antigenically drifted strains every 5–10 years [38,39]. Given the variability in infection histories and the stochastic processes involved in each specific infection, it is exceedingly difficult to determine the composition of pre-existing immunity and how it affects an individual's antibody repertoire. Traditionally, antigenic cartography recomputes the antigenic positions of viruses based on the inhibition capacity of sera used in each study [3,6,14,40,41]. This does not take into account the relationships between viruses inferred from previous studies and complicates efforts to map the antibody repertoire within individual sera. In contrast, the newly developed Neutralization Landscapes uses fixed virus coordinates (determined by a panel of well-defined monoclonal antibodies) to characterize the polyclonal antibodies of the sera [29].

This framework utilizes serological assays to peer into the antibody response to assess the number and inhibition profiles of the antibodies within. Such questions cannot be directly addressed from serum measurements, yet they are crucial to resolve when polyclonal serum is dominated by a single antibody signature (that can be highly susceptible to virus escape mutants) versus multiple antibodies working together. Moreover, this framework can assess the breadth and potency of the antibodies elicited by each virus exposure, providing a vantage to study how preexisting immunity shapes the subsequent antibody response.

These landscapes emphasize that knowing the cross-reactivity relationships between viruses (i.e., the coordinates of each strain on the landscape) and a ferret's infection history is insufficient to fully predict the antibody response. Landscapes breaks this degeneracy by using HAI titers to search through the space of antibody phenotypes and describes the antibody signatures within sera.

In this study, we combined Neutralization Landscapes with a ferret reinfection model to dissect the collective HAI antibody responses after repeated infections, and deciphered the influence of prior exposures on de novo antibodies. Using four recent H3N2 vaccine strains—Uruguay 2007 (V_1) from the 2008 to 2010 seasons, Texas 2012 (V_2) from 2013 to 2015, Switzerland 2013 (V_3) from 2015 to 2016, and Hong Kong 2014 (V_4) from 2016 to

2018—we tracked ferret HAI antibody footprints after each step of the sequential infections $V_1{\rightarrow}V_2{\rightarrow}V_3{\rightarrow}V_4$ and mapped de novo HAI responses under four scenarios of prior exposures (V_4 alone, $V_3{\rightarrow}V_4$, $V_2{\rightarrow}V_3{\rightarrow}V_4$, and $V_1{\rightarrow}V_2{\rightarrow}V_3{\rightarrow}V_4$). After two or more infections, we found that the ferret antibody repertoire often contained at least one "specific" antibody signature that strongly inhibited all infection strains, and at least one "non-specific" antibody signature that weakly interacted with other H3 viruses in the panel. Along each step of the exposures $V_1{\rightarrow}V_2{\rightarrow}V_3{\rightarrow}V_4$, both the specific and non-specific signatures tended to move closer to the infection strains. Eventually, a single cross-reactive antibody signature emerged that potently inhibited all four infection strains and the rest of the virus panel except Philippines 1982 (Figure 3H). These results are consistent with the hypothesis that with each additional infection, antibodies are refocused on conserved epitopes or structural regions across different infection strains [42–44].

In a classical immune response to the same antigen, the reaction from the primary exposure is greatly magnified in subsequent encounters. OAS occurs upon exposure to antigenically related viruses, where the response is skewed more heavily towards the earlier infection strains than to the latest strain. While this OAS phenomenon has been suggested to decrease vaccine effectiveness [9,18,27,45–47], seasonal vaccination can persistently extend the number of strains that the human antibody repertoire potently inhibits even when antibodies against earlier viruses are back-boosted [40]. In this study, we observed OAS with each subsequent infection, regardless of the total number of exposures (Figure 3B–D and Figure 5B–D). Despite the ubiquity of OAS, the cross-reactivity of ferret antisera increased with each additional infection. Previously, we also demonstrated that repeated A/H3 infections enhanced antibody avidity toward both early and later exposed viruses, resulting in extended antibody cross-reactivity, despite the induction of OAS costs de novo antibodies specific for contemporary viruses [6]. A recent study has also reported that ferrets sequentially immunized with antigenically different recombinant H3 HAs develop broadly neutralizing antibody responses and are more resistant to antigenically distinct viruses [48]. In humans, seasonal vaccination can also persistently extend cross-reactive antibody landscapes, especially with antigenically advanced vaccine strains, although antibodies against early exposed viruses are back-boosted by seasonal vaccination as well [40]. It is reported that both OAS and non-OAS antibodies originate from clonally related B cells and target the same general regions of HA, although OAS antibodies bind with low affinities [49]. Understanding the mechanisms that drive clonal selection of non-OAS antibodies with broad cross-reactivity is crucial for next generation universal influenza vaccine development.

In this study we also noticed that the inhibition profiles of ferret antibodies following $V_1{\rightarrow}V_2$ or $V_3{\rightarrow}V_4$ infection were different, despite both schemes showing OAS. Antibodies derived from $V_1{\rightarrow}V_2$ infection strongly inhibited both V_1 and V_2 (Figure 3B,F), whereas antibodies generated after $V_3{\rightarrow}V_4$ infection produced a strong HAI response only against V_3 and a weak inhibition against V_4 (Figure 5B,F). Since the first infection by V_1 or V_3 resulted in high homologous neutralizing titers (HAI $\geq$ 320) in all eight ferrets, we hypothesize that the different outcomes for the subsequent infection are due to the antigenic distance between the first and second infecting strains. In the neutralization landscapes, V_1 (Uruguay 2007) and V_2 (Texas 2012) are antigenically similar (within 0.6 antigenic units) and, hence, resemble primary and secondary infections by nearly identical viruses, while V_3 (Switzerland 2013) and V_4 (Hong Kong 2014) are separated by 4.6 antigenic units and are analogous to infection by two distinct strains [29]. Further experiments are warranted to verify whether similar antibody responses correlate with virus antigenic distances.

Of note, the Neutralization Landscapes in this study used fixed virus coordinates that were pre-determined by human monoclonal antibodies, which may fundamentally differ from those built on ferret antisera [29,37]. For example, Uruguay 2007 (V_1) and Texas 2012 (V_2) are considered antigenically distinct by ferret antisera raised from a single infection (Figures 3A and 4B), whereas these two strains are considered antigenically similar and lie close together on the neutralization landscapes (Supplementary Figure S5) [3].

These antigenic differences between human and ferret antibody-based characterizations may arise because of differences between the HAI and neutralization assays, or because humans have complex immune histories that imprint an antibody repertoire, whereas reference ferret antisera are raised in naïve ferrets exposed to a single virus strain [3]. Moreover, some inconsistencies between the landscape titers and the measurements are expected so to avoid overfitting the intrinsic noise of the HAI assay or the heterogeneous responses between individual ferrets. Nevertheless, the full suite of HAI titers presented on the neutralization landscapes shows an average two-fold error to the experimental measurements (Figures 1B and 2C), demonstrating that the antigenic relationships among the majority of the viruses in the panel are the same across humans and ferrets.

In summary, by tracking the changes in the inhibition profile of ferret antisera induced by repeated influenza A/H3 infections, we demonstrated that a broadly neutralizing antibody could be guided along the map after a series of infections. We further show that prior immune history can heavily influence the ferret antibody repertoire. In ferrets exposed to two or more viruses, a broadly neutralizing antibody signature that potently inhibited all infection strains was, nevertheless, produced at the expense of de novo HAI antibodies (Figure 3H). While our current work was focused on HA head-specific antibodies, it does not consider antibodies directed towards the HA stem and neuraminidase that have also been shown to exhibit OAS and may influence the dynamics of this system [50,51]. Hence, complementing this framework with binding or neutralization landscapes would help to resolve the head- versus stem-directed antibody response. Ongoing work will refine these antibody trajectories across multiple infections and multiple regions of an influenza virus, which will help to develop strategies that further expand the broadly neutralizing antibody pool via vaccination to protect against emerging influenza strains.

Supplementary Materials: The following supporting information can be downloaded at: https://www.mdpi.com/article/10.3390/v15020374/s1, Figure S1: Hemagglutination inhibition (HAI) responses in ferrets with sequential H3N2 infections; Figure S2: Tracing individual ferret responses through four sequential infections; Figure S3: Tracing individual ferret responses with four different prior infection histories; Figure S4: Antibody repertoires from single infections of ferrets; Figure S5: Comparing neutralization and HAI virus landscapes.

Author Contributions: H.X. conceived and designed the ferret study. M.K., P.R., H.X. and Y.-C.K. conducted the ferret infection experiments. M.K., H.X., P.R. and H.J.K. performed the HAI assays. T.E. conceived and designed Neutralization Landscapes. T.E., M.K. and H.X. analyzed the HAI data. H.X., T.E. and M.K. wrote and finalized the paper. All authors have read and agreed to the published version of the manuscript.

Funding: This work was supported in part by the intramural research fund of the Center for Biologics Evaluation and Research, US Food and Drug Administration (to H.X.). Tal Einav is a Damon Runyon Fellow supported by the Damon Runyon Cancer Research Foundation (DRQ 01-20).

Institutional Review Board Statement: Not Applicable.

Informed Consent Statement: Not Applicable.

Data Availability Statement: Raw HAI datasets generated in the study are available on request from the corresponding author.

Conflicts of Interest: The authors declare no conflict of interest.

Disclaimer: The findings and conclusions in this article have not been formally disseminated by US Food and Drug Administration and should not be construed to represent any Agency determination or policy.

References

1. Belongia, E.A.; Kieke, B.A.; Donahue, J.G.; Greenlee, R.T.; Balish, A.; Foust, A.; Lindstrom, S.; Shay, D.K. Effectiveness of inactivated influenza vaccines varied substantially with antigenic match from the 2004–2005 season to the 2006–2007 season. *J. Infect. Dis.* **2009**, *199*, 159–167. [CrossRef] [PubMed]

2. Skowronski, D.M.; Janjua, N.Z.; De, S.G.; Sabaiduc, S.; Eshaghi, A.; Dickinson, J.A.; Fonseca, K.; Winter, A.L.; Gubbay, J.B.; Krajden, M.; et al. Low 2012-13 influenza vaccine effectiveness associated with mutation in the egg-adapted H3N2 vaccine strain not antigenic drift in circulating viruses. *PLoS ONE* **2014**, *9*, e92153. [CrossRef] [PubMed]

3. Xie, H.; Wan, X.F.; Ye, Z.; Plant, E.P.; Zhao, Y.; Xu, Y.; Li, X.; Finch, C.; Zhao, N.; Kawano, T.; et al. H3N2 Mismatch of 2014-15 Northern Hemisphere Influenza Vaccines and Head-to-head Comparison between Human and Ferret Antisera derived Antigenic Maps. *Sci. Rep.* **2015**, *5*, 15279. [CrossRef]

4. Flannery, B.; Clippard, J.; Zimmerman, R.K.; Nowalk, M.P.; Jackson, M.L.; Jackson, L.A.; Monto, A.S.; Petrie, J.G.; McLean, H.Q.; Belongia, E.A.; et al. Early estimates of seasonal influenza vaccine effectiveness—United States, January 2015. *Morb. Mortal. Wkly. Rep.* **2015**, *64*, 10–15.

5. Treanor, J.J.; Talbot, H.K.; Ohmit, S.E.; Coleman, L.A.; Thompson, M.G.; Cheng, P.Y.; Petrie, J.G.; Lofthus, G.; Meece, J.K.; Williams, J.V.; et al. Effectiveness of seasonal influenza vaccines in the United States during a season with circulation of all three vaccine strains. *Clin. Infect. Dis.* **2012**, *55*, 951–959. [CrossRef] [PubMed]

6. Kosikova, M.; Li, L.; Radvak, P.; Ye, Z.; Wan, X.F.; Xie, H. Imprinting of Repeated Influenza A/H3 Exposures on Antibody Quantity and Antibody Quality: Implications for Seasonal Vaccine Strain Selection and Vaccine Performance. *Clin. Infect. Dis.* **2018**, *67*, 1523–1532. [CrossRef]

7. Treanor, J. What Happens Next Depends on What Happened First. *Clin. Infect. Dis.* **2018**, *67*, 1533–1534. [CrossRef]

8. Henry, C.; Palm, A.E.; Krammer, F.; Wilson, P.C. From Original Antigenic Sin to the Universal Influenza Virus Vaccine. *Trends Immunol.* **2018**, *39*, 70–79. [CrossRef]

9. Monto, A.S.; Malosh, R.E.; Petrie, J.G.; Martin, E.T. The Doctrine of Original Antigenic Sin: Separating Good From Evil. *J. Infect. Dis.* **2017**, *215*, 1782–1788. [CrossRef]

10. Liu, F.; Tzeng, W.P.; Horner, L.; Kamal, R.P.; Tatum, H.R.; Blanchard, E.G.; Xu, X.; York, I.; Tumpey, T.M.; Katz, J.M.; et al. Influence of Immune Priming and Egg Adaptation in the Vaccine on Antibody Responses to Circulating A(H1N1)pdm09 Viruses After Influenza Vaccination in Adults. *J. Infect. Dis.* **2018**, *218*, 1571–1581. [CrossRef]

11. Andrews, S.F.; Kaur, K.; Pauli, N.T.; Huang, M.; Huang, Y.; Wilson, P.C. High preexisting serological antibody levels correlate with diversification of the influenza vaccine response. *J. Virol.* **2015**, *89*, 3308–3317. [CrossRef] [PubMed]

12. Andrews, S.F.; Huang, Y.; Kaur, K.; Popova, L.I.; Ho, I.Y.; Pauli, N.T.; Henry Dunand, C.J.; Taylor, W.M.; Lim, S.; Huang, M.; et al. Immune history profoundly affects broadly protective B cell responses to influenza. *Sci. Transl. Med.* **2015**, *7*, 316ra192. [CrossRef] [PubMed]

13. Li, Y.; Myers, J.L.; Bostick, D.L.; Sullivan, C.B.; Madara, J.; Linderman, S.L.; Liu, Q.; Carter, D.M.; Wrammert, J.; Esposito, S.; et al. Immune history shapes specificity of pandemic H1N1 influenza antibody responses. *J. Exp. Med.* **2013**, *210*, 1493–1500. [CrossRef] [PubMed]

14. Xie, H.; Li, L.; Ye, Z.; Li, X.; Plant, E.P.; Zoueva, O.; Zhao, Y.; Jing, X.; Lin, Z.; Kawano, T.; et al. Differential Effects of Prior Influenza Exposures on H3N2 Cross-reactivity of Human Postvaccination Sera. *Clin. Infect. Dis.* **2017**, *65*, 259–267. [CrossRef] [PubMed]

15. Xie, H.; Jing, X.; Li, X.; Lin, Z.; Plant, E.; Zoueva, O.; Yang, H.; Ye, Z. Immunogenicity and cross-reactivity of 2009-2010 inactivated seasonal influenza vaccine in US adults and elderly. *PLoS ONE* **2011**, *6*, e16650. [CrossRef] [PubMed]

16. Xie, H.; Li, X.; Gao, J.; Lin, Z.; Jing, X.; Plant, E.; Zoueva, O.; Eichelberger, M.C.; Ye, Z. Revisiting the 1976 "swine flu" vaccine clinical trials: Cross-reactive hemagglutinin and neuraminidase antibodies and their role in protection against the 2009 H1N1 pandemic virus in mice. *Clin. Infect. Dis.* **2011**, *53*, 1179–1187. [CrossRef] [PubMed]

17. Linderman, S.L.; Chambers, B.S.; Zost, S.J.; Parkhouse, K.; Li, Y.; Herrmann, C.; Ellebedy, A.H.; Carter, D.M.; Andrews, S.F.; Zheng, N.Y.; et al. Potential antigenic explanation for atypical H1N1 infections among middle-aged adults during the 2013–2014 influenza season. *Proc. Natl. Acad. Sci. USA* **2014**, *111*, 15798–15803. [CrossRef]

18. Choi, Y.S.; Baek, Y.H.; Kang, W.; Nam, S.J.; Lee, J.; You, S.; Chang, D.Y.; Youn, J.C.; Choi, Y.K.; Shin, E.C. Reduced antibody responses to the pandemic (H1N1) 2009 vaccine after recent seasonal influenza vaccination. *Clin. Vaccine Immunol.* **2011**, *18*, 1519–1523. [CrossRef]

19. Sasaki, S.; He, X.S.; Holmes, T.H.; Dekker, C.L.; Kemble, G.W.; Arvin, A.M.; Greenberg, H.B. Influence of prior influenza vaccination on antibody and B-cell responses. *PLoS ONE* **2008**, *3*, e2975. [CrossRef]

20. Zhang, A.; Stacey, H.D.; Mullarkey, C.E.; Miller, M.S. Original Antigenic Sin: How First Exposure Shapes Lifelong Anti-Influenza Virus Immune Responses. *J. Immunol.* **2019**, *202*, 335–340. [CrossRef] [PubMed]

21. Francis, T., Jr. On the Doctrine of Original Antigenic Sin. *Proc. Am. Philos. Soc.* **1960**, *104*, 572–578.

22. McLean, H.Q.; Thompson, M.G.; Sundaram, M.E.; Meece, J.K.; McClure, D.L.; Friedrich, T.C.; Belongia, E.A. Impact of repeated vaccination on vaccine effectiveness against influenza A(H3N2) and B during 8 seasons. *Clin. Infect. Dis.* **2014**, *59*, 1375–1385. [CrossRef] [PubMed]

23. Ohmit, S.E.; Petrie, J.G.; Malosh, R.E.; Fry, A.M.; Thompson, M.G.; Monto, A.S. Influenza vaccine effectiveness in households with children during the 2012–2013 season: Assessments of prior vaccination and serologic susceptibility. *J. Infect. Dis.* **2015**, *211*, 1519–1528. [CrossRef] [PubMed]

24. Ohmit, S.E.; Thompson, M.G.; Petrie, J.G.; Thaker, S.N.; Jackson, M.L.; Belongia, E.A.; Zimmerman, R.K.; Gaglani, M.; Lamerato, L.; Spencer, S.M.; et al. Influenza vaccine effectiveness in the 2011–2012 season: Protection against each circulating virus and the effect of prior vaccination on estimates. *Clin. Infect. Dis.* **2014**, *58*, 319–327. [CrossRef]

25. Saito, N.; Komori, K.; Suzuki, M.; Morimoto, K.; Kishikawa, T.; Yasaka, T.; Ariyoshi, K. Negative impact of prior influenza vaccination on current influenza vaccination among people infected and not infected in prior season: A test-negative case-control study in Japan. *Vaccine* **2017**, *35*, 687–693. [CrossRef] [PubMed]

26. Rondy, M.; Launay, O.; Castilla, J.; Costanzo, S.; Puig-Barbera, J.; Gefenaite, G.; Larrauri, A.; Rizzo, C.; Pitigoi, D.; Syrjanen, R.K.; et al. Repeated seasonal influenza vaccination among elderly in Europe: Effects on laboratory confirmed hospitalised influenza. *Vaccine* **2017**, *35*, 4298–4306. [CrossRef]

27. Lewnard, J.A.; Cobey, S. Immune History and Influenza Vaccine Effectiveness. *Vaccines* **2018**, *6*, 28. [CrossRef] [PubMed]

28. Skowronski, D.M.; Chambers, C.; De Serres, G.; Sabaiduc, S.; Winter, A.L.; Dickinson, J.A.; Gubbay, J.B.; Fonseca, K.; Drews, S.J.; Charest, H.; et al. Serial Vaccination and the Antigenic Distance Hypothesis: Effects on Influenza Vaccine Effectiveness During A(H3N2) Epidemics in Canada, 2010–2011 to 2014-2015. *J. Infect. Dis.* **2017**, *215*, 1059–1099. [CrossRef] [PubMed]

29. Einav, T.; Creanga, A.; Andrews, S.F.; McDermott, A.B.; Kanekiyo, M. Harnessing Low Dimensionality to Visualize the Antibody-Virus Landscape for Influenza. *Nat. Comput. Sci.* **2022**. [CrossRef]

30. Whittle, J.R.; Zhang, R.; Khurana, S.; King, L.R.; Manischewitz, J.; Golding, H.; Dormitzer, P.R.; Haynes, B.F.; Walter, E.B.; Moody, M.A.; et al. Broadly neutralizing human antibody that recognizes the receptor-binding pocket of influenza virus hemagglutinin. *Proc. Natl. Acad. Sci. USA* **2011**, *108*, 14216–14221. [CrossRef]

31. Krause, J.C.; Tsibane, T.; Tumpey, T.M.; Huffman, C.J.; Basler, C.F.; Crowe, J.E., Jr. A broadly neutralizing human monoclonal antibody that recognizes a conserved, novel epitope on the globular head of the influenza H1N1 virus hemagglutinin. *J. Virol.* **2011**, *85*, 10905–10908. [CrossRef] [PubMed]

32. Ekiert, D.C.; Kashyap, A.K.; Steel, J.; Rubrum, A.; Bhabha, G.; Khayat, R.; Lee, J.H.; Dillon, M.A.; O'Neil, R.E.; Faynboym, A.M.; et al. Cross-neutralization of influenza A viruses mediated by a single antibody loop. *Nature* **2012**, *489*, 526–532. [CrossRef] [PubMed]

33. Ohshima, N.; Iba, Y.; Kubota-Koketsu, R.; Asano, Y.; Okuno, Y.; Kurosawa, Y. Naturally occurring antibodies in humans can neutralize a variety of influenza virus strains, including H3, H1, H2, and H5. *J. Virol.* **2011**, *85*, 11048–11057. [CrossRef] [PubMed]

34. Iba, Y.; Fujii, Y.; Ohshima, N.; Sumida, T.; Kubota-Koketsu, R.; Ikeda, M.; Wakiyama, M.; Shirouzu, M.; Okada, J.; Okuno, Y.; et al. Conserved neutralizing epitope at globular head of hemagglutinin in H3N2 influenza viruses. *J. Virol.* **2014**, *88*, 7130–7144. [CrossRef] [PubMed]

35. Kanekiyo, M.; Joyce, M.G.; Gillespie, R.A.; Gallagher, J.R.; Andrews, S.F.; Yassine, H.M.; Wheatley, A.K.; Fisher, B.E.; Ambrozak, D.R.; Creanga, A.; et al. Mosaic nanoparticle display of diverse influenza virus hemagglutinins elicits broad B cell responses. *Nat. Immunol.* **2019**, *20*, 362–372. [CrossRef]

36. Smith, D.J.; Lapedes, A.S.; de Jong, J.C.; Bestebroer, T.M.; Rimmelzwaan, G.F.; Osterhaus, A.D.; Fouchier, R.A. Mapping the antigenic and genetic evolution of influenza virus. *Science* **2004**, *305*, 371–376. [CrossRef]

37. Creanga, A.; Gillespie, R.A.; Fisher, B.E.; Andrews, S.F.; Lederhofer, J.; Yap, C.; Hatch, L.; Stephens, T.; Tsybovsky, Y.; Crank, M.C.; et al. A comprehensive influenza reporter virus panel for high-throughput deep profiling of neutralizing antibodies. *Nat. Commun.* **2021**, *12*, 1722. [CrossRef]

38. Bodewes, R.; de Mutsert, G.; van der Klis, F.R.; Ventresca, M.; Wilks, S.; Smith, D.J.; Koopmans, M.; Fouchier, R.A.; Osterhaus, A.D.; Rimmelzwaan, G.F. Prevalence of antibodies against seasonal influenza A and B viruses in children in Netherlands. *Clin. Vaccine Immunol.* **2011**, *18*, 469–476. [CrossRef]

39. Kucharski, A.J.; Lessler, J.; Read, J.M.; Zhu, H.; Jiang, C.Q.; Guan, Y.; Cummings, D.A.; Riley, S. Estimating the Life Course of Influenza A(H3N2) Antibody Responses from Cross-Sectional Data. *PLoS Biol.* **2015**, *13*, e1002082. [CrossRef] [PubMed]

40. Fonville, J.M.; Wilks, S.H.; James, S.L.; Fox, A.; Ventresca, M.; Aban, M.; Xue, L.; Jones, T.C.; Le, N.M.; Pham, Q.T.; et al. Antibody landscapes after influenza virus infection or vaccination. *Science* **2014**, *346*, 996–1000. [CrossRef] [PubMed]

41. Fonville, J.M.; Fraaij, P.L.; de, M.G.; Wilks, S.H.; van, B.R.; Fouchier, R.A.; Rimmelzwaan, G.F. Antigenic Maps of Influenza A(H3N2) Produced With Human Antisera Obtained after Primary Infection. *J. Infect. Dis.* **2015**, *213*, 31–38. [CrossRef] [PubMed]

42. Sahini, L.; Tempczyk-Russell, A.; Agarwal, R. Large-scale sequence analysis of hemagglutinin of influenza A virus identifies conserved regions suitable for targeting an anti-viral response. *PLoS ONE* **2010**, *5*, e9268. [CrossRef]

43. Wong, T.M.; Allen, J.D.; Bebin-Blackwell, A.G.; Carter, D.M.; Alefantis, T.; DiNapoli, J.; Kleanthous, H.; Ross, T.M. Computationally Optimized Broadly Reactive Hemagglutinin Elicits Hemagglutination Inhibition Antibodies against a Panel of H3N2 Influenza Virus Cocirculating Variants. *J. Virol.* **2017**, *91*, e01581-17. [CrossRef] [PubMed]

44. Wu, K.W.; Chien, C.Y.; Li, S.W.; King, C.C.; Chang, C.H. Highly conserved influenza A virus epitope sequences as candidates of H3N2 flu vaccine targets. *Genomics* **2012**, *100*, 102–109. [CrossRef] [PubMed]

45. Skowronski, D.M.; Chambers, C.; Sabaiduc, S.; De Serres, G.; Winter, A.L.; Dickinson, J.A.; Gubbay, J.B.; Drews, S.J.; Martineau, C.; Charest, H.; et al. Beyond Antigenic Match: Possible Agent-Host and Immuno-epidemiological Influences on Influenza Vaccine Effectiveness During the 2015-2016 Season in Canada. *J. Infect. Dis.* **2017**, *216*, 1487–1500. [CrossRef] [PubMed]

46. Smith, D.J.; Forrest, S.; Ackley, D.H.; Perelson, A.S. Variable efficacy of repeated annual influenza vaccination. *Proc. Natl. Acad. Sci. USA* **1999**, *96*, 14001–14006. [CrossRef]
47. Yewdell, J.W.; Santos, J.J.S. Original Antigenic Sin: How Original? How Sinful? *Cold Spring Harb. Perspect. Med.* **2020**, *11*, a038786. [CrossRef] [PubMed]
48. Chiba, S.; Hatta, M.; Pattinson, D.; Yasuhara, A.; Neumann, G.; Kawaoka, Y. Ferret model to mimic the sequential exposure of humans to historical H3N2 influenza viruses. *Vaccine* **2022**, *41*, 590–597. [CrossRef]
49. Linderman, S.L.; Hensley, S.E. Antibodies with 'Original Antigenic Sin' Properties Are Valuable Components of Secondary Immune Responses to Influenza Viruses. *PLoS Pathog.* **2016**, *12*, e1005806. [CrossRef]
50. Arevalo, C.P.; Le Sage, V.; Bolton, M.J.; Eilola, T.; Jones, J.E.; Kormuth, K.A.; Nturibi, E.; Balmaseda, A.; Gordon, A.; Lakdawala, S.S.; et al. Original antigenic sin priming of influenza virus hemagglutinin stalk antibodies. *Proc. Natl. Acad. Sci. USA* **2020**, *117*, 17221–17227. [CrossRef]
51. Rajendran, M.; Nachbagauer, R.; Ermler, M.E.; Bunduc, P.; Amanat, F.; Izikson, R.; Cox, M.; Palese, P.; Eichelberger, M.; Krammer, F. Analysis of Anti-Influenza Virus Neuraminidase Antibodies in Children, Adults, and the Elderly by ELISA and Enzyme Inhibition: Evidence for Original Antigenic Sin. *mBio* **2017**, *8*, e02281-16. [CrossRef]

Article

Assessment of Immunogenic and Antigenic Properties of Recombinant Nucleocapsid Proteins of Five SARS-CoV-2 Variants in a Mouse Model

Alexandra Rak [1], Nikolay Gorbunov [2], Valeria Kostevich [2], Alexey Sokolov [2], Polina Prokopenko [1], Larisa Rudenko [1] and Irina Isakova-Sivak [1,*]

[1] Department of Virology, Institute of Experimental Medicine, Saint Petersburg 197022, Russia
[2] Department of Molecular Genetics, Institute of Experimental Medicine, Saint Petersburg 197022, Russia
*   Correspondence: isakova.sivak@iemspb.ru

Abstract: COVID-19 cases caused by new variants of highly mutable SARS-CoV-2 continue to be identified worldwide. Effective control of the spread of new variants can be achieved through targeting of conserved viral epitopes. In this regard, the SARS-CoV-2 nucleocapsid (N) protein, which is much more conserved than the evolutionarily influenced spike protein (S), is a suitable antigen. The recombinant N protein can be considered not only as a screening antigen but also as a basis for the development of next-generation COVID-19 vaccines, but little is known about induction of antibodies against the N protein via different SARS-CoV-2 variants. In addition, it is important to understand how antibodies produced against the antigen of one variant can react with the N proteins of other variants. Here, we used recombinant N proteins from five SARS-CoV-2 strains to investigate their immunogenicity and antigenicity in a mouse model and to obtain and characterize a panel of hybridoma-derived monoclonal anti-N antibodies. We also analyzed the variable epitopes of the N protein that are potentially involved in differential recognition of antiviral antibodies. These results will further deepen our knowledge of the cross-reactivity of the humoral immune response in COVID-19.

Keywords: SARS-CoV-2; nucleocapsid phosphoprotein; monoclonal antibody; immunogenicity; recombinant protein; cross-reactivity; epitopes

Citation: Rak, A.; Gorbunov, N.; Kostevich, V.; Sokolov, A.; Prokopenko, P.; Rudenko, L.; Isakova-Sivak, I. Assessment of Immunogenic and Antigenic Properties of Recombinant Nucleocapsid Proteins of Five SARS-CoV-2 Variants in a Mouse Model. *Viruses* **2023**, *15*, 230. https://doi.org/10.3390/v15010230

Academic Editor: Jason Yiu Wing KAM

Received: 4 January 2023
Revised: 10 January 2023
Accepted: 11 January 2023
Published: 13 January 2023

1. Introduction

During almost three years of the global pandemic caused by SARS-CoV-2, the virus genome has evolved significantly, generating multiple lineages and variants that have spread readily around the world [1–3]. Moreover, such variability in the viral antigenic properties has resulted in a significant decrease in COVID-19-vaccine effectiveness against genetically evolved SARS-CoV-2 variants and caused widespread vaccine breakthrough infections [4,5]. This is mainly because the vast majority of COVID-19 vaccines elicit Spike-specific antibodies, but the Spike viral protein is highly variable and easily mutates to escape population immunity [6,7]. In contrast to the Spike protein, the viral nucleocapsid (N) protein is highly conserved among all SARS-CoV-2 variants and has 90% similarity to the SARS-CoV-1 N protein [8]. High levels of N-specific IgG antibodies are induced upon COVID-19, since N is the most abundantly expressed SARS-CoV-2 protein [9]. Furthermore, the N protein is one of the major targets for virus-specific T-cell responses [10], which makes this antigen a promising target for COVID-19 vaccine development [11]. Indeed, it was shown that a Spike-based vaccine supplemented with the N protein conferred acute protection in both the lung and the brain after a challenge, while a Spike-based vaccine alone provided acute protection only in the lung [12].

N is a flexible and multivalent RNA-binding protein that contains three dynamic disordered regions: the N-terminal domain (NTD), the linker and the C-terminal domain

(CTD). These regions undergo liquid–liquid phase separation when mixed with RNA [9]. A central β-hairpin of the N molecule contains a serine/arginine-rich region (residues 176–209) that serves as a regulatory element [13]. A recent study found that the N protein can be readily identified on the surfaces of SARS-CoV-2-infected and surrounding cells, where it is bound via electrostatic high-affinity binding to heparan sulfate and heparin [14]. This surface localization of the N protein makes it accessible to anti-N antibodies, which activate Fc-receptor-expressing innate immune cells. Furthermore, N is able to bind, with high affinity, to 11 human chemokines, including CXCL12β: a leukocyte chemotaxis factor, inhibited by N, from SARS-CoV-2, SARS-CoV-1 and MERS-CoV [14].

The N protein is an important antigen for development of COVID-19 diagnostics. First of all, this antigen can be directly detected in biological fluids for the purpose of diagnosis in the early stages of infection. The specificity and relatively high sensitivity of this direct analysis has been shown in fluorescence immunochromatographic (FIC) assays [15] and ELISA tests [16,17]. Moreover, strong positive correlation was observed between elevated plasma N-antigens and odds of pulmonary damage severity, resulting in worsened clinical outcomes [17,18]; therefore, N-level measurement upon hospital admission may improve risk stratification through identification of patients with implicit odd of severe diseases [18]. An N-antigen-based assay may be performed in a simple self-test format, although its sensitivity and diagnostic accuracy would be lower compared to those of an RT-PCR assay [19]. On the other hand, the N protein may be used as a basis for ELISAs that detect antiviral antibodies and are similar in specificity and sensitivity to those based on S-protein fragments. Full-length and truncated forms are suitable for development of test systems; the latter showed greater sensitivity in analysis of mouse, rabbit and human sera and seems to be a better serological marker for evaluating SARS-CoV-2 immunogenicity [20]. It was shown that antibodies against S and N proteins in COVID-19 convalescents were both detectable for up to 200 days after a positive SARS-CoV-2 RT-PCR test [21], but demonstrated markedly different trends in signal intensity: anti-N antibodies were characterized with lower persistence [22].

Despite the highly conservative nature of the N protein, it still undergoes slow evolutionary changes, which can potentially affect its tertiary structure [23]. As a consequence, the sensitivity and specificity of N-based COVID-19 diagnostic tools may be compromised. Previously, we observed cross-reactivity of anti-N antibodies, raised to the ancestral SARS-CoV-2 virus, through demonstration of a strong positive correlation in the magnitudes of anti-N (B.1) antibodies and antibodies specific to various variants of concern (VOCs) [24]. However, little is known about the immunogenicity and antigenicity of the N proteins of these VOCs or whether slight antigenic differences can affect the performances of N-based diagnostic tools. Here, we assessed the immunogenicities of the recombinant N proteins of five SARS-CoV-2 strains belonging to different lineages, as well as cross-reactivity of induced anti-N antibodies. In addition, we generated and characterized several monoclonal antibodies (mAbs) raised to the N protein of the B.1 virus, with different epitope specificities. Our results revealed the varied recognition repertoire of antiviral antibodies generated in response to immunization with the N proteins of different VOCs and cross-reactivity of anti-N (B.1) mAbs.

2. Materials and Methods

2.1. Cells, Viruses and Proteins

African green monkey kidney Vero E6 cells were obtained from the American Type Culture Collection (ATCC) and maintained in DMEM supplemented with 10% fetal bovine serum (FBS) and 1× antibiotic–antimycotic (AA) (all from Capricorn Scientific, Ebsdorfergrund, Germany). Three SARS-CoV-2 viruses were obtained from the Smorodintsev Research Institute of Influenza (Saint Petersburg, Russia). HCoV-19/Russia/StPetersburg 3524/2020 (B.1 Lineage, Wuhan), hCoV-19/Russia/SPE-RII-32759S/2021 (B.1.617.2 Lineage, Delta) and hCoV-19/Russia/SPE-RII-6243V1/2021 (B.1.1.529 Lineage, Omicron). These viruses were grown on Vero E6 cells using DMEM supplemented with 2% FBS, 10 mM of HEPES and $1\times$ AA (all from Capricorn Scientific, Ebsdorfergrund, Germany) at 37 °C and 5% CO_2. After full cytopathic effect was reached, the virus-containing media was harvested, clarified via low-speed centrifugation and stored at -70 °C in aliquots. All experiments with live SARS-CoV-2 were performed in a biosafety-level-3 laboratory.

Recombinant N proteins were expressed in *Escherichia coli* cells as previously described [24], using full-length sequences of the following SARS-CoV-2 strains:

- hCoV-19/Russia/StPetersburg-3524/2020 (B.1 Lineage, Wuhan);
- hCoV-19/Russia/SPE-RII-27029S/2021 (B.1.351 Lineage, Beta);
- hCoV-19/Japan/TY7-503/2021 (P.1 Lineage, Gamma);
- hCoV-19/Russia/SPE-RII-32759S/2021 (B.1.617.2 Lineage, Delta);
- hCoV-19/Russia/SPE-RII-6243V1/2021 (B.1.1.529 Lineage, Omicron).

2.2. Mice

All experiments were performed in compliance with relevant laws and institutional guidelines and approved by the local Ethical Committee of Institute of Experimental Medicine (protocol No.1/22, dated 18 February 2022). Female CBA and BALB/c mice that weighed 18 to 20 g were purchased from the Stolbovaya breeding nursery (Moscow region, Russia). The CBA mice were immunized intraperitoneally three times with 20 µg of the recombinant N protein in an AlumVax adjuvant (1:1 v/v) (OZ Biosciences, San Diego, CA, USA) at 14-day intervals. Blood samples were collected 14 days after the final immunization, and sera were stored at -20 °C.

2.3. Assessment of Virus-Specific Antibodies in Mouse Serum Samples

Serum IgG antibodies specific to N proteins were measured with an ELISA. Briefly, high-binding 96-well plates (Thermo Fisher Scientific, Waltham, MA, USA) were coated with purified recombinant N proteins, 100 ng per well, in a carbonate–bicarbonate buffer (pH 7.4) overnight at 4 °C. Then, the plates were blocked with 1% BSA in PBS (pH 7.4) for 40 min at 37 °C and washed 3 times with PBS-T (PBS with 0.1% Tween 20). Serum samples were diluted 5-fold in PBS-T (1:500 to 1:121,500) and added to wells, followed by incubation for 1 h at 37 °C. Each sample was tested in duplicate. After washing, an HRP-conjugated goat antimouse IgG secondary antibody (Bio-Rad, Hercules, CA, USA) was added to each well and incubated at 37 °C for 1 h. Then, the plates were finally washed and developed with 1-Step TMB Substrate Solution (HEMA, Moscow, Russia) for 15 min. After the reaction with 1 M of H_2SO_4 was stopped, the resulting absorbance was measured at a wavelength of 450 nm (OD_{450}) using an xMark Microplate Spectrophotometer (Bio-Rad, Hercules, CA, USA). The area under the OD_{450} curve (AUC) values were calculated as a trapezoidal square for each serum sample and expressed in arbitrary units.

2.4. Monoclonal Antibodies

Production of mAbs was obtained through methods of hybridoma technology [25]. BALB/c mice were immunized with 10 µg of the N protein (B.1), emulsified with a complete Freund's adjuvant, in plantar aponeurosis of the hind limbs. After 4 weeks, the animals were immunized subcutaneously with 10 µg of the N protein (B.1) mixed with an incomplete adjuvant. On the 30th day after the second immunization, the animals were boosted intravenously with 5 µg of the N protein (B.1) in saline. The lymphocytes of the inguinal and abdominal lymph nodes were isolated 4 days later and mixed with Sp 2/0 myeloma cells at a ratio of 2:1. Cell fusion was performed in prewarmed 50% 1500 kDa polyethylene glycol (PEG) for 1.5 min, followed by dropwise addition of an equal volume of RPMI-1640 medium. After hybridization, cells were pelleted and cocultured with peritoneal macrophages (at a 10:1 ratio) in 96-well culture plates. To select hybridomas, we used a RPMI-1640 medium that contained 10% FBS, 10^{-4} M of hypoxanthine, 4×10^{-7} M of aminopterin and 1.6×10^{-5} M of thymidine. Primary screening of clones was performed via ELISAs of culture media from hybridomas that used the recombinant N protein (B.1) as a coating antigen. Then, the ability of the mAbs to recognize the N proteins of different

SARS-CoV-2 strains was evaluated using the same ELISAs. Hybridomas that produced specific antibodies against N proteins were then subcloned, and individual clones were expanded in 175 cm^2 flasks for intraperitoneal injection into animals (2 million cells per mouse). Isotype determination was performed using an ISO-2 kit (Sigma, St. Louis, MO, USA) according to the instructions of the manufacturer. Antibodies were purified from the ascitic fluid of mice, which was collected 12–18 days after intraperitoneal injection of hybridoma cells through protein-A-affinity chromatography with the MabSelect sorbent (GE HealthCare, Chicago, IL, USA), using the manufacturer's protocol.

2.5. SDS-PAGE and Western blot

Sodium dodecyl sulfate–polyacrylamide gel electrophoresis (SDS-PAGE) was used to check the structure and purity of obtained anti-N mAbs, while the ability of anti-N mAbs to detect linear epitopes of N proteins of five SARS-CoV-2 strains was assessed via western blotting. Purified mAbs and recombinant N proteins were resolved in reduced conditions on a 10% polyacrylamide gel at 120 V for 1 h before being stained with colloidal Coomassie G-250 solution (Bio-Rad, Hercules, USA) for 1 h at room temperature or semi-dry transferred to 0.45 μm nitrocellulose membranes for 2 h at 100 V. Blots were blocked overnight at 4 °C with 5% skimmed milk in PBS-T and then treated with anti-N mAbs diluted 1:1000 in block buffer for 1 h at 37 °C. Then, goat anti-mouse HRP-conjugated secondary antibody (Bio-Rad, Hercules, CA, USA) diluted 1:3000 in blocking solution was added to the triple-washed blots for 1 h at 37 °C. After three washes with PBS-T, the blots were developed with 0.05% solution of diaminobenzidine (Sigma, St. Louis, MO, USA) in PBS containing 1% hydrogen peroxide. Finally, the membranes were washed with water and the images were captured using Gel Doc EZ Gel Documentation System (Bio-Rad, Hercules, CA, USA).

2.6. Cell ELISA and Immunocytochemistry

A cell-based ELISA was used to check the specificities of purified mAbs to native viral antigens. A 14C2 monoclonal antibody (Abcam, Cambridge, UK) that binds the M2e protein of the influenza virus was used as a negative control. For the assay, 96-well plates were seeded with 4×10^4 Vero E6 cells per well the day before virus inoculation. Cell monolayers were rinsed twice with PBS prior to inoculation with 50 μL of SARS-CoV-2 diluted in DMEM supplemented with 2% FBS, AA and 10 mM of HEPES to reach a multiplicity of infection (MOI) of 0.01. After adsorption at 37 °C for 1 h, the cells were overlaid with 100 μL of the same culture medium and incubated at 37 °C for 1 or 3 days for B.1 or P.1/B.1.1.529 viruses, respectively. After incubation, the culture medium was carefully removed and the cells were fixed with 2% formalin in PBS at 4 °C overnight. Then, the fixative solution was removed and the plates were washed with PBS-T and blocked with 3% skim milk in PBS-T for 1 h at 37 °C. Then, 50 μL of 3-fold dilutions of mAbs in PBS-T (from 15 to 0.02 μg/mL) were added to the wells, and the plates were incubated for 1 h at 37 °C. After being washed with PBS-T, the plates were treated with a 1:3000 solution of goat antimouse immunoglobulins conjugated with horseradish peroxidase (Bio-Rad, Hercules, CA, USA) for 1 h at 37 °C. Finally, the plates were thoroughly washed and stained with 1-Step TMB Substrate Solution (HEMA, Moscow, Russia). The resulting OD was measured using a BioRad Model 680 microplate reader (Bio-Rad, Hercules, CA, USA) at a wavelength of 450 nm.

Immunocytochemical analysis was performed in a similar way. Vero E6 cells were seeded on 6-well tissue-culture plates at a dose of 1×10^6 cells per well. After 2 days of incubation with B.1, P.1 or B.1.1.529 viruses, these cells were fixed and sequentially treated with 5 μg/mL of mAbs and goat antimouse HRP-conjugated immunoglobulins. Then the plates were stained with 3-amino-9-ethylcarbazole (Sigma, St. Louis, MO, USA) in the presence of 1% H_2O_2, according to the manufacturer's protocol.

2.7. Statistical Analysis

Data were analyzed using the statistical tool of GraphPad Prism 6.0 Software. Compliance with normal distribution was checked with the Shapiro–Wilk test. The Wilcoxon matched-pairs test was used to compare serum IgG antibody responses to different N antigens. One-way ANOVA with the Tukey post hoc test was used to examine the significance of differences between several study groups. The significance level was set at $p < 0.05$.

3. Results

3.1. Assessment of Immunogenicity of N Proteins

To investigate the immunogenicity and antigenicity of the N proteins of the ancestral SARS-CoV-2 strain and different VOCs, we immunized mice with recombinant N proteins of five SARS-CoV-2 strains [24] and measured the magnitudes of N-specific IgG responses, and the ability of these antibodies to cross-react with other N antigens, using an in-house ELISA protocol. Testing serum samples with homologous N antigens revealed significant differences in the levels of induced anti-N antibodies in different immunization groups; the highest IgG levels were noted in mice immunized with the ancestral B.1 (Wuhan) antigen compared to VOCs B.1.351 (Beta), P.1 (Gamma), B.1.617.2 (Delta) and B.1.1.529 (Omicron) (Figure 1). The strongest differences in immunogenic properties were noted between the N protein of the ancestral B.1 (Wuhan) virus and the N proteins of the P.1, B.1.617.2 and B.1.1.529 VOCs (Figure 1b, $p < 0.0001$). Since all of the N proteins were expressed and administered to mice in an identical manner, the variable levels of the anti-N antibodies were most likely due to the differences in protein sequences and the content of the linear and/or spatial B-cell epitopes located in the N proteins.

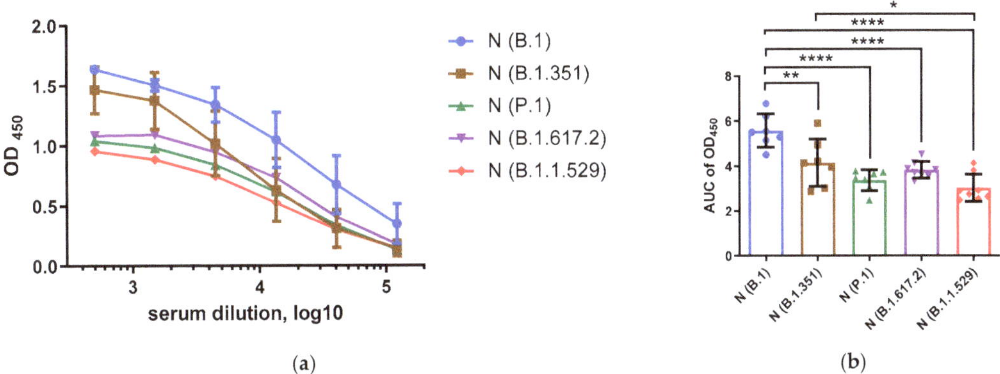

Figure 1. Assessment of homologous anti-N antibody levels in serum samples of mice immunized with B.1 (blue), B.1.351(brown), P.1 (green), B.1.617.2 (magenta) and B.1.1.529 (red). Hereafter, the color designations of the N proteins of different SARS-CoV-2 VOCs are the same. The OD_{450} values are shown in (**a**), and the area under the OD_{450} curve (AUC) values are shown in (**b**). These data were compared using ANOVA with the Tukey post hoc test. * $p < 0.05$, ** $p < 0.01$, **** $p < 0.0001$.

To further characterize the antigenic properties of the N proteins of the five SARS-CoV-2 variants, we compared the magnitudes of IgG responses, in mice immunized with each N protein, against all five recombinant N antigens: B.1 (Wuhan), B.1.351 (Beta), P.1 (Gamma), B.1.617.2 (Delta) and B.1.1.529 (Omicron). We used the area under the OD_{450} curve (AUC) values as a measure of intensity of antibody binding to the N antigen, then normalized the AUC values of each heterologous antigen to the AUC data for homologous antigens. This artificial parameter allowed assessment of cross-reactivity of mouse immune sera against various N antigens, regardless of the magnitude of the response. Interestingly, the antibody that was raised to the N (B.1) protein bound to the heterologous N antigens to a lesser extent, suggesting accumulation of escape mutations in this protein with virus evolution (Figure 2A). Similarly, the antibodies induced via the N (B.1.351) protein were less likely to be recognized by the N proteins of SARS-CoV-2 variants that circulated at later times (Figure 2B). In general, the intensity of antibody binding from the heterologous N antigens was reduced compared to that of the homologous protein (Figure 2), although several exceptions were noted: the N (B.1) protein bound to the IgG raised to the N (B.1.351) protein better than did the homologous antigen (Figure 2B); the N (B.1.617.2) antigen recognized a higher proportion of antibodies raised to the N (P.1) protein (Figure 2C); and the N (B.1.351) antigen revealed higher levels of IgG raised to the N (B.1.1.529) protein (Figure 2E). These data indicate that different N proteins can generate different subsets of N-specific antibodies; this should be taken into account as a limitation in development of N-based vaccines and serology assays. In particular, our data from the N (P.1), N (B.1.617.2) and N (B.1.1.529) immunization groups suggest that N-specific antibodies raised to infection with evolutionarily diverse SARS-CoV-2 will be poorly recognized by the N antigen from the ancestral B.1 strain, which is currently present in the majority of N-based serological tests (Figure 2C,D,E).

3.2. B- and T-cell N-Protein Epitopes, including Substitutions

To understand which differences in the amino acid sequences of N proteins can be associated with diverse recognition of N-specific antibodies raised to various N antigens, we aligned the N-protein sequences of five SARS-CoV-2 strains, mapped their unique substitutions on the N proteins on the structural scheme (Figure 3) and listed all substitutions in Table S1. Furthermore, using the Immune Epitope Database (IEDB) resources, we identified linear B-cell (Table S2) and T-cell (Table S3) epitopes that contained these mutations.

The epitope coverage of SARS-CoV-2 proteins is well-studied: 15,905 viral epitopes have been deposited in IEDB, including 1159 N-protein epitopes, which have been described in 96 studies. In total, 791 of these N-protein epitopes have been identified as B-cell-mediated immune-response targets. The majority of B-cell epitopes of the N protein identified to date have been deposited at the IEDB based on the studies of Hotop et al. [26], Mishra et al. [27], Heffron et al. [28], Gregory et al. [29] and Schwarz et al. [30]. A total of 174 N-protein epitopes were confirmed in T-cell assays and are class I epitopes described as having protective potential in humans by Tarke et al. [31] and Heide et al. [32] and in transgenic mice by Zhuang et al. [33].

Noteworthily, there was no linear accumulation of mutations in the N protein over time, since each VOC has its own set of mutations compared to the ancestral B.1 (Wuhan) variant (Table S1), suggesting that all studied SARS-CoV-2 lineages evolved independently. Nevertheless, a number of mutations seem to correlate with the immunogenicity of the recombinant N protein (Figure 1). Thus, the highest numbers of mutations were found in the antigens of the B.1.1.529 and B.1.617.2 variants (Figure 3), which affected a high proportion of the linear B-cell epitopes established for the B.1 N protein (Table S1). A closer analysis of the cross-reactivity of the N-specific antibodies induced via each recombinant N protein (Figure 2) and the amino acid differences between the particular N immunogens and each N protein used as an antigen, carried out in ELISA (Table S1), suggested that the residues at positions 13, 80, 203 and 204 had the greatest influence on N-protein antigenicity. Interestingly, the deletion of the three amino acid residues at positions 31–33 did not result in significant impairment of the antigenic properties of the N protein; sera from mice immunized with the P.1 (Beta) and B.1.617.2 (Delta) N proteins recognized the B.1.1.529 (Omicron) N antigen at the same level as the homologous antigens, and vice versa (Figure 2). Notably, the substitutions at positions 13, 203 and 204 were noted for several VOCs, and given that they can significantly change the antigenicity and/or immunogenicity of the N protein, it can be assumed that they are major escape mutations that drive the evolution of SARS-CoV-2.

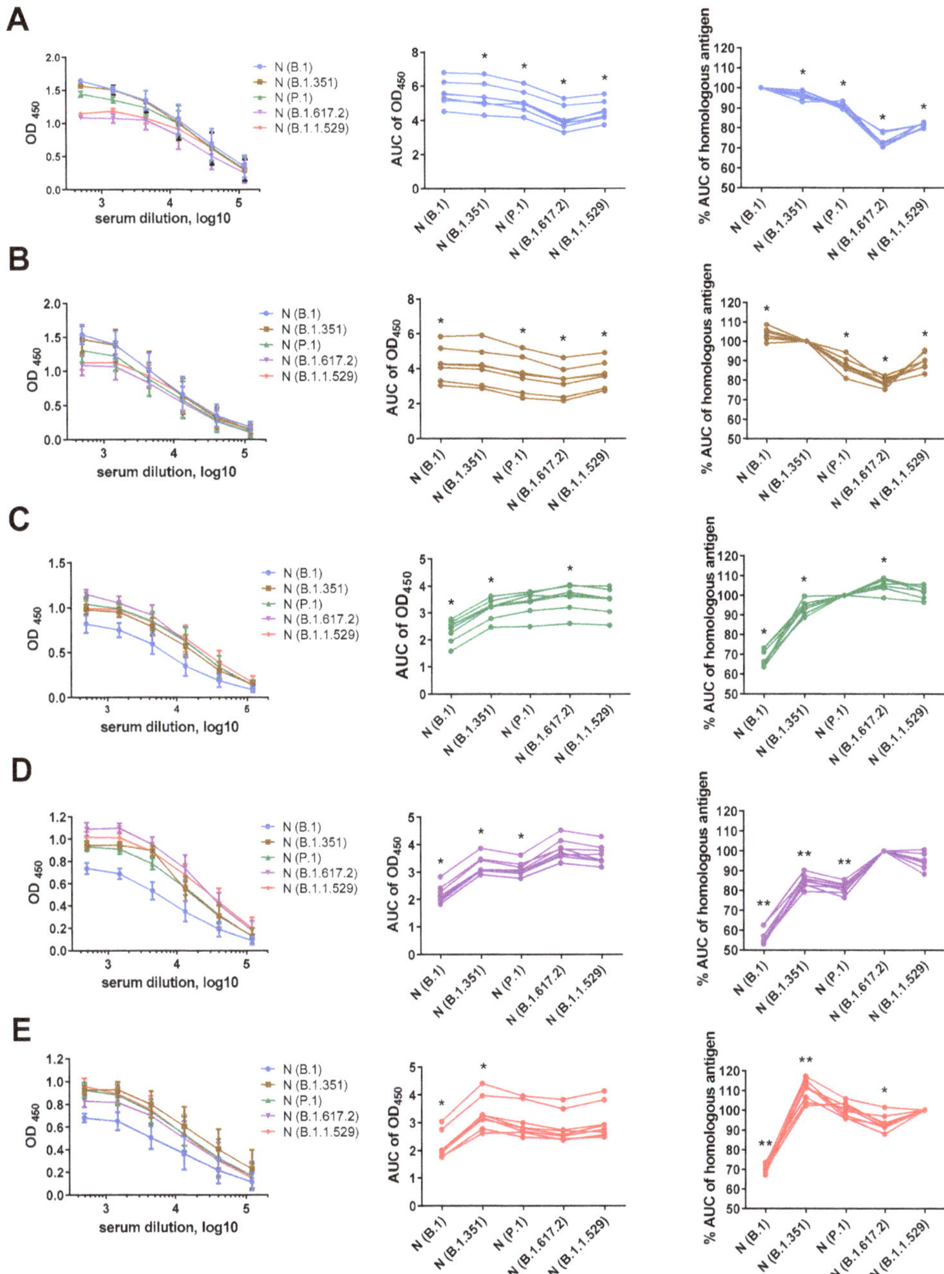

Figure 2. N-based ELISAs of serum samples of mice immunized with B.1 (**A**, blue), B.1.351 (**B**, brown), P.1 (**C**, green), B.1.617.2 (**D**, magenta) and B.1.1.529 (**E**, red). The left panel shows the mean OD_{450} values in ELISAs with the indicated N antigen. The area under the OD_{450} curve (AUC) values for each serum sample tested against five N antigens are shown in the middle panel. The right panel shows the AUC values normalized to the homologous antigen. These data were compared using the Wilcoxon matched-pairs test. * $p < 0.05$, ** $p < 0.01$.

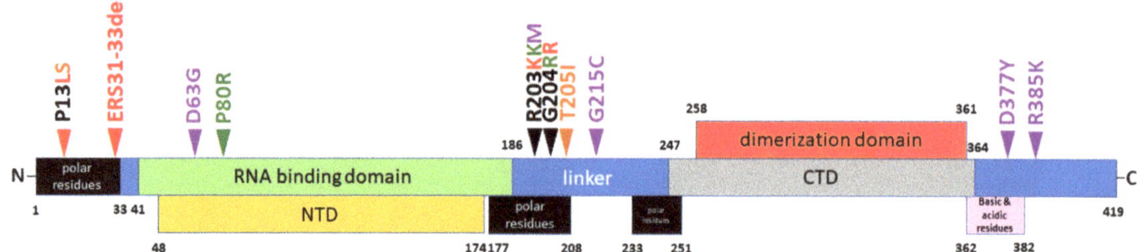

Figure 3. Scheme of N-protein domains and variable amino acid residues characterizing five different variants of the SARS-CoV-2 virus used in this work. The locations of the functional domains of the N protein are given according to [9]. The substitutions present in the N-protein sequences of the B.1.351 (brown), P.1 (green), B.1.617.2 (magenta) and B.1.1.529 (red) strains are indicated with arrows.

3.3. Generation and Characterization of Anti-N Monoclonal Antibodies

To further characterize the antigenic properties of the N proteins of different VOCs, we generated four mAbs (NCL2, NCL5, NCL7 and NCL10) through immunization of BALB/c mice with the N protein (B.1), using standard hybridoma technology. Purified anti-N antibodies were isotyped as follows: NCL2 and NCL10 were of the IgG1 isotype, while NCL5 and NCL7 were of the IgG2a isotype. We characterized these mAbs using the SDS-PAGE (Figure 4a) and Western blot (Figure 4b) analyses. The electrophoretic mobility of the mAbs corresponded to the expected molecular weights (~160 kDa and ~55 + 25 kDa in nonreducing and reducing conditions, respectively). The SDS-PAGE analysis of the purified mAbs demonstrated the absence of significant amounts of contaminating components (Figure 4a). For study of the specificity of the newly generated mAbs, those mAbs were used in the WB analysis to detect recombinant N proteins; there, an additional recombinant N protein of seasonal coronavirus OC43 was used, along with five SARS-CoV-2 antigens (Figure 4b). Interestingly, NCL2, NCL5 and NCL7 recognized all of the SARS-CoV-2 N antigens but not the OC43 N protein, whereas the NCL10 antibody did not recognize the N protein of the B.1.1.529 strain but was able to bind to the N protein of seasonal coronavirus OC43 (Figure 4b). These data suggest that the NCL10 antibody recognized B-cell epitopes with mutations specific to the Omicron variant only, most probably the region with the deletion of residues 31–33 (Figure 3). In addition, mutation P13L could be responsible for the altered recognition of the NCL10 antibody. However, the N-terminal region of the N protein differs significantly between the OC43 and SARS-CoV-2 viruses; therefore, it remains to be elucidated exactly which epitope is targeted by the NCL10 mAb.

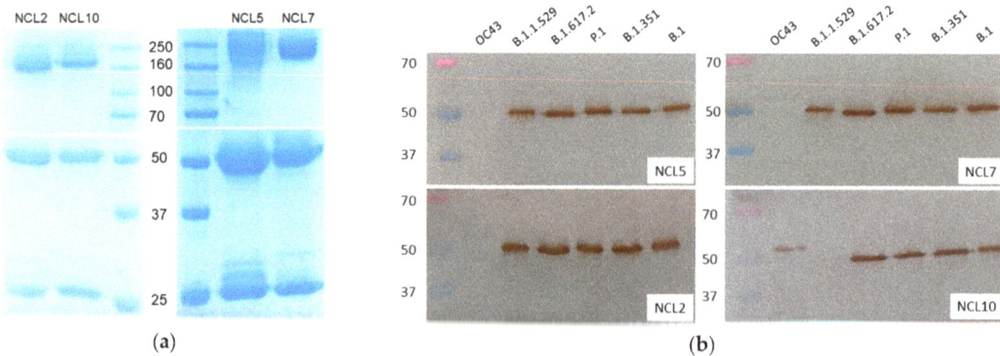

(a) **(b)**

Figure 4. SDS-PAGE (**a**) and Western blot (**b**) analyses of the purified anti-N antibodies. The upper panel of the SDS-PAGE shows the results of the analysis in nonreducing conditions, and the lower panel shows the results of the analysis in the presence of 1% β-mercaptoethanol.

Detection of native N antigens, which are present in virus-infected cells, using the cell ELISA approach recapitulated the results of the Western blot analysis: while NCL2/5/7 mAbs universally recognized the B.1, P.1 and B.1.1.529 virus-infected cells, the NCL10 antibody was unable to bind to the B.1.1.529 SARS-CoV-2-infected cells (Figure 5a). Moreover, the OD_{450} values, which were determined based on the concentration of the resulting virus–antibody complexes, were different between NCL10 and the three remaining mAbs when identical antibody concentration was used, suggesting that the NCL10 antibody binds with lower intensity to N antigens, probably due to the lower affinity of this binding. An immunocytochemical analysis further demonstrated that NCL10 mAb does not recognize the N protein of the B.1.1.529 VOC (Figure 5b). Future studies with generations of escape mutants after passaging SARS-CoV-2 variants in the presence of generated mAbs will allow the precise epitope mapping thereof.

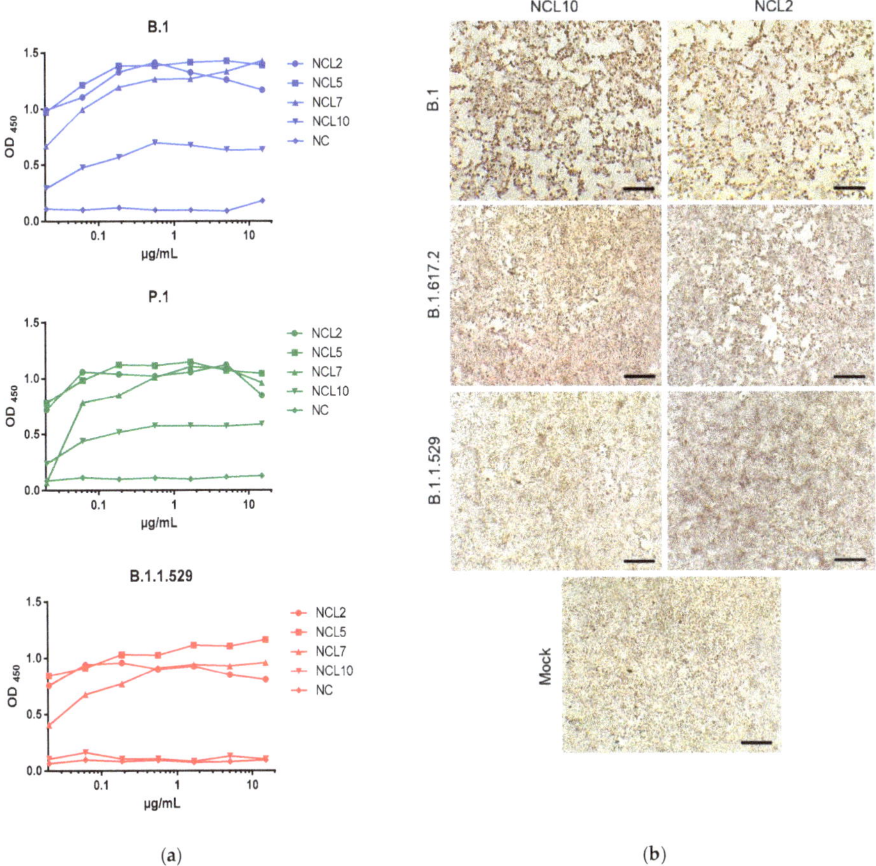

(a) (b)

Figure 5. Analysis of specificity of N-specific mAbs in a Vero E6 cell model. (**a**) Dependence of absorbance intensity (OD_{450}) on the concentration (µg/mL) of anti-N mAbs or 14C2 mAbs (NC, anti-influenza M2e antibodies). Vero E6 cells were infected with the B.1, B.1.617.2 or B.1.1.529 SARS-CoV-2 variant, followed by cell ELISA with indicated mAbs. (**b**) Immunocytochemical analysis of SARS-CoV-2 N protein in infected Vero E6 cells after 48 h of incubation. The different patterns of cytopathic effects caused by three different strains of SARS-CoV-2 are visualized. The positive red staining was developed as a result of treatment with the NCL2 and NCL10 (except for the B.1.1.529 detection) mAbs. The mock comprised the untreated cells. Scale bars: 100 µm.

4. Discussion

The impact of N-protein mutations on the reliability of data obtained with approved test systems has not yet been sufficiently studied [31]. In addition, development of N-based COVID-19 vaccines also requires studies of immunogenicity and the possibility of generations of anti-N antibodies with a putative autoimmune effect [32]. Previously, we established an in-house ELISA protocol based on the recombinant N proteins of different SARS-CoV-2 VOCs to evaluate the cross-reactivity of N-specific antibodies in COVID-19 convalescents [24]. A strong positive correlation in the magnitude of anti-N (B.1) antibodies and those of antibodies specific to four other VOCs in COVID-19-recovered patients was found, suggesting that N-binding antibodies are highly cross-reactive and the most immunogenic epitopes within this protein are not under selective pressure. Here, we assessed the antigenicities and immunogenicities of N proteins of different VOCs in a mouse model through comparison of the magnitudes and the cross-reactivities of the anti-N antibodies generated in response to immunization with the recombinant N proteins. The strongest impact on the magnitude of homologous antibody responses was revealed for the N proteins of the P.1, B.1.617.2 and B.1.1.529 strains (Figures 1 and 2). The amino acid sequences of these antigens are known to contain a large number of mutations compared to that of the ancestral B.1 strain (Figure 3), and the linear epitopes that carry them have been identified as targets for antiviral antibodies through multiple B-cell assays (Table S2). Thus, we hypothesize that differences in the specificities of anti-N antibodies produced in response to immunization with similar antigens may be due to evolutionarily determined variability in immunogenic epitopes, leading to emergence of escape mutations. T-cell epitopes of N proteins that contain variable regions may also serve as triggers of T-cell-mediated immune response, according to the results of the activation and binding analyses (Table S3). Most of these epitopes were shown to bind to recombinant human MHC molecules of the HLA-A*01:01, HLA-A*02:01 and HLA-B*07:02 alleles. Although we did not assess T-cell responses in the mouse model, the presence of mutations within the established T-cell epitopes of various HLA alleles may indicate variability of N-protein-induced, T-cell-mediated immune response in humans infected with different SARS-CoV-2 strains.

During the first two years of the pandemic (2019–2020), the main detectable mutations in the SARS-CoV-2 genome were D614G (81.5%) in the S protein and a combination of R203K/G204R (37%) in the N protein [33]. Later, the adaptive nature of the R203K/G204R mutations in the N proteins of the P.1 and B.1.1.529 variants was shown in silico, and large-scale phylogenetic analysis indicated that the R203K/G204R was associated with the elevated transmissibility and infectivity of the B.1.1.7 SARS-CoV-2 variant. A positive correlation has been found between COVID-19 severity and frequency of 203K/204R [34]. In addition, the R203K/G204R combination appeared to increase nucleocapsid phosphorylation resistance via inhibition of the GSK-3 kinase, resulting in increased virus replication. Similar occasional alanine substitutions also occurred at positions 203 and 204 and led to the same outcomes, indicating an evolutionary trend towards ablation of the ancestral RG motif to increase SARS-CoV-2 infectivity [35]. The virus-like particles that bore the RG203/204KR mutations in the N protein stimulated an augmented humoral immune response and enhanced neutralization in immunized mouse sera [36]. COVID-19 induced with viruses with R203K and G204R resulted in inferior clinical outcomes [37]. The 203/204 mutations occurred in the phosphorylated "RS-motif" [38], which is localized within the serine-rich region of 181–213: a target for phosphorylation that allows recruitment of the host RNA helicase DDX1, thus facilitating template readthrough and synthesis of longer subgenomic mRNA [38].

The results of a docking analysis indicated that the site for binding of targeted antiviral agents is localized in the region of 66–134 in the RNA-binding domain of the N protein [39]. Therefore, mutations in this region, such as P80R in the P.1 SARS-CoV-2 N protein, may influence the sensitivity of the virus to these drugs.

P13L/S substitution unique to the N proteins of the B.1.1529 and B.1.351 strains is located in the CD8 immunodominant region that is targeted by cells that exhibit central and effector memory phenotypes [40]. Furthermore, peptides that contain this substitution have been identified as positive in an HLA-binding analysis of MHC class II epitopes [41] and as targets for antiviral antibodies in COVID-19 convalescents [42]. R203M mutation is localized into peptides for which specific binding of anti-SARS-COV-2 antibodies was detectable [26]. Peptides that contained residue at 205 elicited interferon-γ production in convalescent donors [43], while peptides with residues at 215 and 377 showed specific Ab binding and IFN releasing as a result of T-cell stimulation [27,41,44]. Epitopes that included amino acid residues at 31–33, which are absent in the B.1.1.529 variant, exhibited low-intensity binding of Abs but stimulated IFNγ release in COVID-19 convalescents [26,44]. Peptides that included residue of the N protein at 63 provided strong IFNγ release, as was demonstrated with cell sorting and ELISPOT assays in COVID-19 convalescents and peptide-vaccinated people [45]. P80R mutation is localized in peptides that provide binding, T-cell activation and IFNγ release after in vivo administration or restimulation [44,46,47].

Thus, mutations P13L, ERS31-33 del, D63G, R203K/M, G204R and D377Y, unique to the P.1, B.1.617.2 and B.1.1.529 strains, can be considered variations that appeared in the viral genome as a result of attempts to evade host immune response. Published data on peptides that contain variations that uniquely distinguish the N proteins of different SARS-CoV-2 strains from each other, as well as our results on the varying specificities of anti-N antibodies produced in response to immunization with N proteins of different VOCs, allowed us to consider these substitutions as escape mutations that arose in the viral genome as a result of immune pressure.

It should be noted that our study was limited to assessment of N-based humoral immune responses but not T-cell immunity. As the mouse major histocompatibility complex (MHC) is known to be remarkably different from the human leukocyte antigen (HLA) system [48], our animal model is not applicable for assessment of T-cell-mediated immune responses triggered via N-protein immunization. To study T-cell-involving reactions with maximal approximation to the human organism, use of HLA transgenic humanized mouse models is required [49,50].

To additionally test N-protein antigenicity and obtain a useful tool for detection of viral antigens, we also generated, in this work, a number of mAbs against the N protein of the B.1 strain and examined their specificities towards different VOCs. Previously, several mAbs against the N proteins of different VOCs have been obtained and described. Hodge et al. compared the properties of flexible and rigid anti-N antibodies and showed the utility of the latter for the highly sensitive ELISA detection of the N protein NTD [51]. In another study, a panel of 41 mAbs against the N protein of the B.1 strain was obtained in order to develop latex-based lateral flow immunoassays (LFIAs). These test systems allowed detection of as low as 8 pg of a purified protein or 625 $TCID_{50}$/mL of a virus, and they cross-reacted with the P.1 and B.1.617.2 variants [52]. The use of N-specific detection by highly sensitive nanobodies C2 and E2, which are specific to B.1, P.1 and B.1.617.2, was demonstrated by Isaacs et al. [53]. Molecular modeling of the interaction of anti-N mAbs specific to 501Y.V1-V3, obtained by Yamaoka et al. [54], revealed binding with exterior protein surface epitopes, and immunochromatographic test systems, combined with silver amplification technology, were developed. Lee et al. produced mAbs against conserved N-protein peptides and developed a laboratory-confirmed sandwich ELISA as a rapid biosensor, allowing detection of as low as 4×10^3 $TCID_{50}$/reaction for SARS-CoV-2 or 1 ng/mL of the recombinant N protein [55]. An original immunochromatographic test-strip method based on the 14 anti-N mAbs was proposed as a tool for detection of the B.1 and B.1.617.2 variants [56].

Our obtained mAbs, named NCL2, NCL5 and NCL7, appear to be suitable for detection of N proteins in all five SARS-CoV-2 variants, whereas the NCL10 mAb was unable to recognize the N protein of the B.1.1.529 strain (Figures 4b and 5). Notably, the latter mAb was also able to recognize the recombinant N protein of seasonal coronavirus OC43; although this property does not allow use of NCL10 for diagnostic purposes, it indicates the presence of similar epitopes in the molecules of the two evolutionarily diverse viruses, and this mAb may be applicable for studying the structures and evolution of SARS-CoV-2 proteins. Our further studies will be devoted to development of test systems based on the obtained mAbs and defining the affinity of their interaction with viral antigens.

Overall, our results demonstrate the slow-evolving nature of the SARS-CoV-2 N protein, which affects the specificity profile of anti-N antibodies and should be considered as a limitation in development of N-based vaccines and test systems.

Supplementary Materials: The following supporting information can be downloaded at: https://www.mdpi.com/article/10.3390/v15010230/s1, Table S1: Amino acid substitutions between the N protein used for immunization and the N proteins of different SARS-CoV-2 strains used as antigens in ELISA; Table S2: B-cell epitopes of SARS-CoV-2 N protein of B.1 (Wuhan) strain deposited in the Immune Epitope Database which contain variable amino acid residues; Table S3: List of variable T-cell epitopes of N protein of SARS-CoV-2 with confirmed binding to HLA molecules.

Author Contributions: Conceptualization, I.I.-S. and A.R.; methodology, A.R., N.G., P.P. and I.I.-S.; software, I.I.-S. and A.R.; investigation, A.R., N.G., V.K., A.S., P.P. and I.I.-S.; data curation, I.I.-S. and L.R.; writing—original draft preparation, A.R. and I.I.-S.; writing—review and editing, L.R.; supervision, L.R.; project administration, I.I.-S.; funding acquisition, I.I.-S. All authors have read and agreed to the published version of the manuscript.

Funding: This research was funded by the Russian Science Foundation, grant number 21-75-30003.

Institutional Review Board Statement: This study was approved by the Ethics Committee of the Institute of Experimental Medicine (protocol No.1/22, dated 18 February 2022).

Informed Consent Statement: Not applicable.

Data Availability Statement: The data presented in this study are available on request from the corresponding authors.

Conflicts of Interest: The authors declare no conflict of interest.

References

1. Evolutionary Insight into the Emergence of SARS-CoV-2 Variants of Concern | Nature Medicine. Available online: https://www.nature.com/articles/s41591-022-01892-2 (accessed on 16 December 2022).
2. Chakraborty, C.; Sharma, A.R.; Bhattacharya, M.; Agoramoorthy, G.; Lee, S.-S. Evolution, Mode of Transmission, and Mutational Landscape of Newly Emerging SARS-CoV-2 Variants. *Mbio* **2021**, *12*, e01140-21. [CrossRef] [PubMed]
3. Bano, I.; Sharif, M.; Alam, S. Genetic drift in the genome of SARS COV-2 and its global health concern. *J. Med. Virol.* **2022**, *94*, 88–98. [CrossRef]
4. Hacisuleyman, E.; Hale, C.; Saito, Y.; Blachere, N.E.; Bergh, M.; Conlon, E.G.; Schaefer-Babajew, D.J.; DaSilva, J.; Muecksch, F.; Gaebler, C.; et al. Vaccine Breakthrough Infections with SARS-CoV-2 Variants. *N. Engl. J. Med.* **2021**, *384*, 2212–2218. [CrossRef] [PubMed]
5. Andrews, N.; Stowe, J.; Kirsebom, F.; Toffa, S.; Rickeard, T.; Gallagher, E.; Gower, C.; Kall, M.; Groves, N.; O'Connell, A.-M.; et al. COVID-19 Vaccine Effectiveness against the Omicron (B.1.1.529) Variant. *N. Engl. J. Med.* **2022**, *386*, 1532–1546. [CrossRef] [PubMed]
6. Guruprasad, L. HumanSARS CoV-2 spike protein mutations. *Proteins* **2021**, *89*, 569–576. [CrossRef] [PubMed]
7. Harvey, W.T.; Carabelli, A.M.; Jackson, B.; Gupta, R.K.; Thomson, E.C.; Harrison, E.M.; Ludden, C.; Reeve, R.; Rambaut, A.; COVID-19 Genomics UK (COG-UK) Consortium; et al. SARS-CoV-2 variants, spike mutations and immune escape. *Nat. Rev. Microbiol.* **2021**, *19*, 409–424. [CrossRef]
8. Bai, Z.; Cao, Y.; Liu, W.; Li, J. The SARS-CoV-2 Nucleocapsid Protein and Its Role in Viral Structure, Biological Functions, and a Potential Target for Drug or Vaccine Mitigation. *Viruses* **2021**, *13*, 1115. [CrossRef]
9. Cubuk, J.; Alston, J.J.; Incicco, J.J.; Singh, S.; Stuchell-Brereton, M.D.; Ward, M.D.; Zimmerman, M.I.; Vithani, N.; Griffith, D.; Wagoner, J.A.; et al. The SARS-CoV-2 nucleocapsid protein is dynamic, disordered, and phase separates with RNA. *Nat. Commun.* **2021**, *12*, 1936. [CrossRef]
10. Grifoni, A.; Weiskopf, D.; Ramirez, S.I.; Mateus, J.; Dan, J.M.; Moderbacher, C.R.; Rawlings, S.A.; Sutherland, A.; Premkumar, L.; Jadi, R.S.; et al. Targets of T Cell Responses to SARS-CoV-2 Coronavirus in Humans with COVID-19 Disease and Unexposed Individuals. *Cell* **2020**, *181*, 1489–1501.e15. [CrossRef]
11. Dutta, N.K.; Mazumdar, K.; Gordy, J.T. The Nucleocapsid Protein of SARS–CoV-2: A Target for Vaccine Development. *J. Virol.* **2020**, *94*, e00647-20. [CrossRef]
12. Dangi, T.; Class, J.; Palacio, N.; Richner, J.M.; Penaloza MacMaster, P. Combining spike- and nucleocapsid-based vaccines improves distal control of SARS-CoV-2. *Cell Rep.* **2021**, *36*, 109664. [CrossRef]

13. Redzic, J.S.; Lee, E.; Born, A.; Issaian, A.; Henen, M.A.; Nichols, P.J.; Blue, A.; Hansen, K.C.; D'Alessandro, A.; Vögeli, B.; et al. The inherent dynamics and interaction sites of the SARS-CoV-2 nucleocapsid N-terminal region. *J. Mol. Biol.* **2021**, *433*, 1671081. [CrossRef]

14. López-Muñoz, A.D.; Kosik, I.; Holly, J.; Yewdell, J.W. Cell surface SARS-CoV-2 nucleocapsid protein modulates innate and adaptive immunity. *Sci. Adv.* **2022**, *8*, eabp9770. [CrossRef]

15. Diao, B.; Wen, K.; Zhang, J.; Chen, J.; Han, C.; Chen, Y.; Wang, S.; Deng, G.; Zhou, H.; Wu, Y. Accuracy of a nucleocapsid protein antigen rapid test in the diagnosis of SARS-CoV-2 infection. *Clin. Microbiol. Infect.* **2021**, *27*, 289.e1–289.e4. [CrossRef]

16. Li, X.; Xiong, M.; Deng, Q.; Guo, X.; Li, Y. The utility of SARS-CoV-2 nucleocapsid protein in laboratory diagnosis. *J. Clin. Lab. Anal.* **2022**, *36*, e24534. [CrossRef]

17. ACTIV-3/TICO Study Group; Rogers, A.J.; Wentworth, D.; Phillips, A.; Shaw-Saliba, K.; Dewar, R.L.; Aggarwal, N.R.; Babiker, A.G.; Chang, W.; Dharan, N.J.; et al. The Association of Baseline Plasma SARS-CoV-2 Nucleocapsid Antigen Level and Outcomes in Patients Hospitalized With COVID-19. *Ann. Intern. Med.* **2022**, *175*, 1401–1410. [CrossRef]

18. Wick, K.D.; Leligdowicz, A.; Willmore, A.; Carrillo, S.A.; Ghale, R.; Jauregui, A.; Chak, S.S.; Nguyen, V.; Lee, D.; Jones, C.; et al. Plasma SARS-CoV-2 nucleocapsid antigen levels are associated with progression to severe disease in hospitalized COVID-19. *Crit. Care* **2022**, *26*, 278. [CrossRef]

19. Sukumaran, A.; Suvekbala, V.; Krishnan, R.A.; Thomas, R.E.; Raj, A.; Thomas, T.; Abhijith, B.L.; Jose, J.; Paul, J.K.; Vasudevan, D.M. Diagnostic Accuracy of SARS-CoV-2 Nucleocapsid Antigen Self-Test in Comparison to Reverse Transcriptase–Polymerase Chain Reaction. *J. Appl. Lab. Med.* **2022**, *7*, 871–880. [CrossRef]

20. Yue, L.; Cao, H.; Xie, T.; Long, R.; Li, H.; Yang, T.; Yan, M.; Xie, Z. N-terminally truncated nucleocapsid protein of SARS-CoV-2 as a better serological marker than whole nucleocapsid protein in evaluating the immunogenicity of inactivated SARS-CoV-2. *J. Med. Virol.* **2021**, *93*, 1732–1738. [CrossRef]

21. Poore, B.; Nerenz, R.D.; Brodis, D.; Brown, C.I.; Cervinski, M.A.; Hubbard, J.A. A comparison of SARS-CoV-2 nucleocapsid and spike antibody detection using three commercially available automated immunoassays. *Clin. Biochem.* **2021**, *95*, 77–80. [CrossRef]

22. Choudhary, H.R.; Parai, D.; Dash, G.C.; Peter, A.; Sahoo, S.K.; Pattnaik, M.; Rout, U.K.; Nanda, R.R.; Pati, S.; Bhattacharya, D. IgG antibody response against nucleocapsid and spike protein post-SARS-CoV-2 infection. *Infection* **2021**, *49*, 1045–1048. [CrossRef] [PubMed]

23. Rahman, M.S.; Islam, M.R.; Alam, A.S.M.R.U.; Islam, I.; Hoque, M.N.; Akter, S.; Rahaman, M.; Sultana, M.; Hossain, M.A. Evolutionary dynamics of SARS-CoV-2 nucleocapsid protein and its consequences. *J. Med. Virol.* **2021**, *93*, 2177–2195. [CrossRef] [PubMed]

24. Rak, A.; Donina, S.; Zabrodskaya, Y.; Rudenko, L.; Isakova-Sivak, I. Cross-Reactivity of SARS-CoV-2 Nucleocapsid-Binding Antibodies and Its Implication for COVID-19 Serology Tests. *Viruses* **2022**, *14*, 2041. [CrossRef] [PubMed]

25. Zhang, C. Hybridoma Technology for the Generation of Monoclonal Antibodies. In *Antibody Methods and Protocols*; Proetzel, G., Ebersbach, H., Eds.; Methods in Molecular Biology; Humana Press: Totowa, NJ, USA, 2012; Volume 901, pp. 117–135. ISBN 978-1-61779-930-3.

26. Hotop, S.-K.; Reimering, S.; Shekhar, A.; Asgari, E.; Beutling, U.; Dahlke, C.; Fathi, A.; Khan, F.; Lütgehetmann, M.; Ballmann, R.; et al. Peptide microarrays coupled to machine learning reveal individual epitopes from human antibody responses with neutralizing capabilities against SARS-CoV-2. *Emerg. Microbes Infect.* **2022**, *11*, 1037–1048. [CrossRef] [PubMed]

27. Mishra, N.; Huang, X.; Joshi, S.; Guo, C.; Ng, J.; Thakkar, R.; Wu, Y.; Dong, X.; Li, Q.; Pinapati, R.S.; et al. Immunoreactive peptide maps of SARS-CoV-2. *Commun. Biol.* **2021**, *4*, 225. [CrossRef] [PubMed]

28. Heffron, A.S.; McIlwain, S.J.; Amjadi, M.F.; Baker, D.A.; Khullar, S.; Sethi, A.K.; Palmenberg, A.C.; Shelef, M.A.; O'Connor, D.H.; Ong, I.M. The Landscape of Antibody Binding in SARS-CoV-2 Infection. *PLoS Biol.* **2021**, *19*, e3001265. [CrossRef]

29. Gregory, D.J.; Vannier, A.; Duey, A.H.; Roady, T.J.; Dzeng, R.K.; Pavlovic, M.N.; Chapin, M.H.; Mukherjee, S.; Wilmot, H.; Chronos, N.; et al. Repertoires of SARS-CoV-2 epitopes targeted by antibodies vary according to severity of COVID-19. *Virulence* **2022**, *13*, 890–902. [CrossRef]

30. Schwarz, T.; Heiss, K.; Mahendran, Y.; Casilag, F.; Kurth, F.; Sander, L.E.; Wendtner, C.-M.; Hoechstetter, M.A.; Müller, M.A.; Sekul, R.; et al. SARS-CoV-2 Proteome-Wide Analysis Revealed Significant Epitope Signatures in COVID-19 Patients. *Front. Immunol.* **2021**, *12*, 629185. [CrossRef]

31. Thakur, S.; Sasi, S.; Pillai, S.G.; Nag, A.; Shukla, D.; Singhal, R.; Phalke, S.; Velu, G.S.K. SARS-CoV-2 Mutations and Their Impact on Diagnostics, Therapeutics and Vaccines. *Front. Med.* **2022**, *9*, 815389. [CrossRef]

32. Matyushkina, D.; Shokina, V.; Tikhonova, P.; Manuvera, V.; Shirokov, D.; Kharlampieva, D.; Lazarev, V.; Varizhuk, A.; Vedekhina, T.; Pavlenko, A.; et al. Autoimmune Effect of Antibodies against the SARS-CoV-2 Nucleoprotein. *Viruses* **2022**, *14*, 1141. [CrossRef]

33. Troyano-Hernáez, P.; Reinosa, R.; Holguín, Á. Evolution of SARS-CoV-2 Envelope, Membrane, Nucleocapsid, and Spike Structural Proteins from the Beginning of the Pandemic to September 2020: A Global and Regional Approach by Epidemiological Week. *Viruses* **2021**, *13*, 243. [CrossRef]

34. Wu, H.; Xing, N.; Meng, K.; Fu, B.; Xue, W.; Dong, P.; Tang, W.; Xiao, Y.; Liu, G.; Luo, H.; et al. Nucleocapsid mutations R203K/G204R increase the infectivity, fitness, and virulence of SARS-CoV-2. *Cell Host Microbe* **2021**, *29*, 1788–1801.e6. [CrossRef]

35. Johnson, B.A.; Zhou, Y.; Lokugamage, K.G.; Vu, M.N.; Bopp, N.; Crocquet-Valdes, P.A.; Kalveram, B.; Schindewolf, C.; Liu, Y.; Scharton, D.; et al. Nucleocapsid mutations in SARS-CoV-2 augment replication and pathogenesis. *PLOS Pathog.* **2022**, *18*, e1010627. [CrossRef]

36. Raheja, H.; Das, S.; Banerjee, A.; Dikshaya, P.; Deepika, C.; Mukhopadhyay, D.; Ramachandra, S.G.; Das, S. RG203KR Mutations in SARS-CoV-2 Nucleocapsid: Assessing the Impact Using a Virus-Like Particle Model System. *Microbiol. Spectr.* **2022**, *10*, e00781-22. [CrossRef]

37. Nagy, Á.; Pongor, S.; Győrffy, B. Different mutations in SARS-CoV-2 associate with severe and mild outcome. *Int. J. Antimicrob. Agents* **2020**, *57*, 106272. [CrossRef]

38. Peng, T.-Y.; Lee, K.-R.; Tarn, W.-Y. Phosphorylation of the arginine/serine dipeptide-rich motif of the severe acute respiratory syndrome coronavirus nucleocapsid protein modulates its multimerization, translation inhibitory activity and cellular localization: Phosphorylation of SARS CoV-N Protein RS Motif. *FEBS J.* **2008**, *275*, 4152–4163. [CrossRef]

39. Tatar, G.; Ozyurt, E.; Turhan, K. Computational drug repurposing study of the RNA binding domain of SARS-CoV-2 nucleocapsid protein with antiviral agents. *Biotechnol. Prog.* **2021**, *37*, e3110. [CrossRef]

40. Peng, Y.; Mentzer, A.J.; Liu, G.; Yao, X.; Yin, Z.; Dong, D.; Dejnirattisai, W.; Rostron, T.; Supasa, P.; Liu, C.; et al. Broad and strong memory CD4+ and CD8+ T cells induced by SARS-CoV-2 in UK convalescent individuals following COVID-19. *Nat. Immunol.* **2020**, *21*, 1336–1345. [CrossRef]

41. Tarke, A.; Sidney, J.; Kidd, C.K.; Dan, J.M.; Ramirez, S.I.; Yu, E.D.; Mateus, J.; da Silva Antunes, R.; Moore, E.; Rubiro, P.; et al. Comprehensive analysis of T cell immunodominance and immunoprevalence of SARS-CoV-2 epitopes in COVID-19 cases. *Cell Rep. Med.* **2021**, *2*, 100204. [CrossRef]

42. Holenya, P.; Lange, P.J.; Reimer, U.; Woltersdorf, W.; Panterodt, T.; Glas, M.; Wasner, M.; Eckey, M.; Drosch, M.; Hollidt, J.-M.; et al. Peptide microarray-based analysis of antibody responses to SARS-CoV-2 identifies unique epitopes with potential for diagnostic test development. *Eur. J. Immunol.* **2021**, *51*, 1839–1849. [CrossRef]

43. Keller, M.D.; Harris, K.M.; Jensen-Wachspress, M.A.; Kankate, V.V.; Lang, H.; Lazarski, C.A.; Durkee-Shock, J.; Lee, P.-H.; Chaudhry, K.; Webber, K.; et al. SARS-CoV-2–specific T cells are rapidly expanded for therapeutic use and target conserved regions of the membrane protein. *Blood* **2020**, *136*, 2905–2917. [CrossRef]

44. Heide, J.; Schulte, S.; Kohsar, M.; Brehm, T.T.; Herrmann, M.; Karsten, H.; Marget, M.; Peine, S.; Johansson, A.M.; Sette, A.; et al. Broadly directed SARS-CoV-2-specific CD4+ T cell response includes frequently detected peptide specificities within the membrane and nucleoprotein in patients with acute and resolved COVID-19. *PLOS Pathog.* **2021**, *17*, e1009842. [CrossRef] [PubMed]

45. Heitmann, J.S.; Bilich, T.; Tandler, C.; Nelde, A.; Maringer, Y.; Marconato, M.; Reusch, J.; Jäger, S.; Denk, M.; Richter, M.; et al. A COVID-19 peptide vaccine for the induction of SARS-CoV-2 T cell immunity. *Nature* **2022**, *601*, 617–622. [CrossRef] [PubMed]

46. Zhuang, Z.; Lai, X.; Sun, J.; Chen, Z.; Zhang, Z.; Dai, J.; Liu, D.; Li, Y.; Li, F.; Wang, Y.; et al. Mapping and role of T cell response in SARS-CoV-2–infected mice. *J. Exp. Med.* **2021**, *218*, e20202187. [CrossRef] [PubMed]

47. Minervina, A.A.; Pogorelyy, M.V.; Kirk, A.M.; Crawford, J.C.; Allen, E.K.; Chou, C.-H.; Mettelman, R.C.; Allison, K.J.; Lin, C.-Y.; Brice, D.C.; et al. SARS-CoV-2 antigen exposure history shapes phenotypes and specificity of memory CD8+ T cells. *Nat. Immunol.* **2022**, *23*, 781–790. [CrossRef]

48. Akkina, R. New generation humanized mice for virus research: Comparative aspects and future prospects. *Virology* **2013**, *435*, 14–28. [CrossRef]

49. Zhu, M.; Niu, B.; Liu, L.; Yang, H.; Qin, B.; Peng, X.; Chen, L.; Liu, Y.; Wang, C.; Ren, X.; et al. Development of a humanized HLA-A30 transgenic mouse model. *Anim. Model. Exp. Med.* **2022**, *5*, 350–361. [CrossRef]

50. Brumeanu, T.-D.; Vir, P.; Karim, A.F.; Kar, S.; Benetiene, D.; Lok, M.; Greenhouse, J.; Putmon-Taylor, T.; Kitajewski, C.; Chung, K.K.; et al. Human-Immune-System (HIS) humanized mouse model (DRAGA: HLA-A2.HLA-DR4.Rag1KO.IL-2RγcKO.NOD) for COVID-19. *Hum. Vaccines Immunother.* **2022**, *18*, 2048622. [CrossRef]

51. Hodge, C.D.; Rosenberg, D.J.; Grob, P.; Wilamowski, M.; Joachimiak, A.; Hura, G.L.; Hammel, M. Rigid monoclonal antibodies improve detection of SARS-CoV-2 nucleocapsid protein. *Mabs* **2021**, *13*, 1905978. [CrossRef]

52. Lu, R.-M.; Ko, S.-H.; Chen, W.-Y.; Chang, Y.-L.; Lin, H.-T.; Wu, H.-C. Monoclonal Antibodies against Nucleocapsid Protein of SARS-CoV-2 Variants for Detection of COVID-19. *Int. J. Mol. Sci.* **2021**, *22*, 12412. [CrossRef]

53. Isaacs, A.; Amarilla, A.A.; Aguado, J.; Modhiran, N.; Albornoz, E.A.; Baradar, A.A.; McMillan, C.L.D.; Choo, J.J.Y.; Idris, A.; Supramaniam, A.; et al. Nucleocapsid Specific Diagnostics for the Detection of Divergent SARS-CoV-2 Variants. *Front. Immunol.* **2022**, *13*, 926262. [CrossRef] [PubMed]

54. Yamaoka, Y.; Miyakawa, K.; Jeremiah, S.S.; Funabashi, R.; Okudela, K.; Kikuchi, S.; Katada, J.; Wada, A.; Takei, T.; Nishi, M.; et al. Highly specific monoclonal antibodies and epitope identification against SARS-CoV-2 nucleocapsid protein for antigen detection tests. *Cell Rep. Med.* **2021**, *2*, 100311. [CrossRef] [PubMed]

55. Lee, J.-H.; Jung, Y.; Lee, S.-K.; Kim, J.; Lee, C.-S.; Kim, S.; Lee, J.-S.; Kim, N.-H.; Kim, H.-G. Rapid Biosensor of SARS-CoV-2 Using Specific Monoclonal Antibodies Recognizing Conserved Nucleocapsid Protein Epitopes. *Viruses* **2022**, *14*, 255. [CrossRef] [PubMed]

56. Xie, C.; Ding, H.; Ding, J.; Xue, Y.; Lu, S.; Lv, H. Preparation of highly specific monoclonal antibodies against SARS-CoV-2 nucleocapsid protein and the preliminary development of antigen detection test strips. *J. Med. Virol.* **2022**, *94*, 1633–1640. [CrossRef]

 *viruses*

Article

Assessing the Pre-Vaccination Anti-SARS-CoV-2 IgG Seroprevalence among Residents and Staff in Nursing Home in Niigata, Japan, November 2020

Keita Wagatsuma [1,2,*], Sayaka Yoshioka [1], Satoru Yamazaki [3], Ryosuke Sato [3], Wint Wint Phyu [1], Irina Chon [1], Yoshiki Takahashi [3], Hisami Watanabe [1] and Reiko Saito [1]

[1] Division of International Health (Public Health), Graduate School of Medical and Dental Sciences, Niigata University, Niigata 951-8510, Japan
[2] Japan Society for the Promotion of Science, Tokyo 102-0083, Japan
[3] Niigata City Public Health and Sanitation Center, Niigata 950-0914, Japan
* Correspondence: waga@med.niigata-u.ac.jp; Tel.: +81-25-227-2129

Abstract: An outbreak of coronavirus disease 2019 (COVID-19) occurred in a nursing home in Niigata, Japan, November 2020, with an attack rate of 32.0% (63/197). The present study was aimed at assessing the pre-vaccination seroprevalence almost half a year after the COVID-19 outbreak in residents and staff in the facility, along with an assessment of the performance of the enzyme-linked immunosorbent assay (ELISA) and the chemiluminescent immunoassay (CLIA), regarding test seropositivity and seronegativity in detecting immunoglobulin G (IgG) anti-severe acute respiratory syndrome 2 (SARS-CoV-2) antibodies (anti-nucleocapsid (N) and spike (S) proteins). A total of 101 people (30 reverse transcription PCR (RT-PCR)-positive and 71 RT-PCR-negative at the time of the outbreak in November 2020) were tested for anti-IgG antibody titers in April 2021, and the seroprevalence was approximately 40.0–60.0% for residents and 10.0–20.0% for staff, which was almost consistent with the RT-PCR test results that were implemented during the outbreak. The seropositivity for anti-S antibodies showed 90.0% and was almost identical to the RT-PCR positives even after approximately six months of infections, suggesting that the anti-S antibody titer test is reliable for a close assessment of the infection history. Meanwhile, seropositivity for anti-N antibodies was relatively low, at 66.7%. There was one staff member and one resident that were RT-PCR-negative but seropositive for both anti-S and anti-N antibody, indicating overlooked infections despite periodical RT-PCR testing at the time of the outbreak. Our study indicated the impact of transmission of SARS-CoV-2 in a vulnerable elderly nursing home in the pre-vaccination period and the value of a serological study to supplement RT-PCR results retrospectively.

Keywords: SARS-CoV-2; nursing home; epidemiology; seroprevalence; pre-vaccination

 check for updates

Citation: Wagatsuma, K.; Yoshioka, S.; Yamazaki, S.; Sato, R.; Phyu, W.W.; Chon, I.; Takahashi, Y.; Watanabe, H.; Saito, R. Assessing the Pre-Vaccination Anti-SARS-CoV-2 IgG Seroprevalence among Residents and Staff in Nursing Home in Niigata, Japan, November 2020. *Viruses* 2022, 14, 2581. https://doi.org/10.3390/v14112581

Academic Editor: Jason Yiu Wing KAM

Received: 1 November 2022
Accepted: 20 November 2022
Published: 21 November 2022

Publisher's Note: MDPI stays neutral with regard to jurisdictional claims in published maps and institutional affiliations.

1. Introduction

Since the first report of the novel severe acute respiratory syndrome coronavirus 2 (SARS-CoV-2) in Wuhan, China, at the end of 2019, the coronavirus disease 2019 (COVID-19) pandemic has become a threat to public health on a global scale [1]. In Japan, the COVID-19 pandemic started on 16 January 2020, with the first confirmed case being a returnee from Wuhan, China, and the number of infections increasing exponentially from January to April 2020 and gradually spreading to prefectures in Japan, including the Niigata Prefecture. In this context, multiple transmission clusters associated with the "three Cs"—closed spaces with poor ventilation, crowded places with many people nearby, and close-contact settings where many people gather in close quarters—have been identified in nurseries, nursing homes, hospitals, care facilities, schools, and other locations to date [2–5].

Since most infected individuals remain mild or asymptomatic, it is widely accepted that the volume of unreported cases of COVID-19 is substantial. Several previous studies have described how the proportion of asymptomatic infections reached more than 20.0% in the elderly population in 2020 [6,7]. Indeed, despite many mild cases, some may progress to severe diseases, resulting in an estimated infection fatality ratio of as high as approximately 6.4% in people over 70 years old in 2020 [8–11]. In this context, nursing home residents are a highly vulnerable population to the spread of SARS-CoV-2 and have accounted for a significant proportion of the virus-induced disease burden during the ongoing pandemic worldwide [12]. Furthermore, because of Japan's super-aged society, community super spreading occurred from the resident community of older adults, and the transmission was sustained among people in that age group [5]. Although detection of SARS-CoV-2 viral ribonucleic acid (RNA) by real-time reverse transcription PCR (RT-PCR) is generally considered the gold standard for diagnosis of COVID-19, the high proportion of asymptomatic individuals in COVID-19, as noted above, may also underestimate the incidence and prevalence of the disease [13]. Therefore, the surveillance of anti-SARS-CoV-2 immunoglobulin G (IgG) serum antibody titer is one of the useful methods for precise determination of the number of affected individuals in the target population of a community, in addition to detection of the viral genome by RT-PCR [13]. Given this high proportion of unreported infections and the poor prognosis for elderly patients, it is important to conduct serological surveys to gain a complete picture of the COVID-19 disease dynamics and burden in target groups.

In this present study, we aimed to assess seroprevalence almost half a year after the COVID-19 outbreak that occurred in November 2020 in a nursing home in Niigata, Japan, by measuring IgG anti-SARS-CoV-2 antibodies to anti-nucleocapsid (N) and anti-spike (S) proteins among pre-vaccination residents and staff utilizing the enzyme-linked immunosorbent assay (ELISA) and the chemiluminescent immunoassay (CLIA). Seroprevalence numbers will not only provide a measure of the cumulative incidence of SARS-CoV-2 infections but also provide additional insight into the usefulness of comparing anti-N and anti-S antibodies following infection with the virus.

2. Materials and Methods

2.1. Study Design and Participants

An outbreak of COVID-19 occurred in an elderly nursing home with 97 residents and 81 staff in Niigata, Japan, in November 2020. We conducted a cross-sectional sero-epidemiological study in April 2021 to measure anti-SARS-CoV-2 IgG antibodies (i.e., anti-N and anti-S proteins) for the remaining elderly residents and staff approximately six months after the outbreak. After written informed consent was obtained, blood (serum) was collected from the forearm vein using a winged needle (21G) and EDTA-2Na/F-treated vacuum blood collection tubes (5 mL), and serum samples were centrifuged before measurement and stored in a freezer at −20 °C until testing. Epidemiological data such as age (years), sex (i.e., male or female), and occupation of staff (i.e., doctor, nurse, caregiver, or clerk) were collected. Additional individual characteristics, such as comorbidities and anthropometric measurements, were not collected. All individuals were sampled before the COVID-19 messenger RNA (mRNA) vaccination.

2.2. Measurement of Quantitative Antibody Levels in Serum

The anti-N and anti-S SARS-CoV-2 antibodies were measured by the ELISA method from DENKA (Tokyo, Japan) and the commercial CLIA method from Abbott (Chicago, IL, USA). Specifically, the two immunoassays included are as follows: (i) The ABBOTT SARS-CoV-2 IgG assay (Abbott, Chicago, IL, USA), which is a CMIA for the qualitative detection of IgG antibodies that target the anti-N and anti-S antibodies [14–16]. The positive cut-off index was ≥1.4 (S/N ratio) for anti-N antibodies and >50.0 AU/mL for anti-S antibodies for the Abbott. These tests were performed using the high-throughput ARCHITECT i2000SR. (ii) The DENKA SARS-CoV-2 IgG assay (DENKA, Tokyo, Japan), which is an ELISA method, is similarly used for the detection of IgG antibodies against N and S antigens using a prototype indirect enzyme immunoassay (DK20-COV4E) [17]. Each well of a 96-well microplate was coated with recombinant SARS-CoV-2 S and N proteins. Serum specimens diluted at a 1:200 ratio with dilution buffer were added to each well. After one hours of incubation at room temperature, the wells were washed three times with washing buffer. Horseradish peroxidase-conjugated goat anti-human IgG antibodies were added to each well, and the plate was incubated at room temperature for 1 h. After five washes, the substrate was added to each well, and the plate was incubated at room temperature. Reactions were stopped by the addition of reaction stopper. Finally, optical density (OD) 450 and OD 630 were measured with a Sunrise™ plate reader (Tecan, Männedorf, Switzerland). Antibody titers were calculated in units of binding antibody unit (BAU)/mL with calibrators assigned to the first World Health Organization (WHO) international standard for anti-SARS-CoV-2 immunoglobulin (National Institute for Biological Standards and Control (NIBSC) code 20/136) [18,19]. The positive cut-off index was $\geq$30.0 index BAU/mL for anti-N antibodies and $\geq$50.0 BAU/mL for anti-S antibodies for the DENKA. All tests were performed and interpreted according to the manufacturer's instructions for each immunoassay, respectively, in a biosafety level 2 (BSL-2) capacity laboratory.

2.3. Statistical Analysis

Data were described as the median [interquartile range (IQR)] for continuous variables and frequency (%) for categorical variables. Test seropositivity and seronegativity of the DENKA and Abbott methods for anti-N and anti-S antibodies sampled in April 2021 were calculated based on the results of the RT-PCR that was implemented at the time of the outbreak in November 2020 [20]. A Cohen's kappa statistic (κ) was estimated for each of the anti-N and anti-S antibodies to assess the level of interrater concordance between the two assays of the DENKA and Abbott methods beyond chance. The κ coefficient value was classified as slight (0.00 to 0.20), fair (0.21 to 0.40), moderate (0.41 to 0.60), substantial (0.61 to 0.80), and almost perfect (0.81 to 1.00) according to Landis and Koch criteria [21]. Spearman's rank-order correlation coefficient (ρ) was used to investigate the linear associations between anti-N IgG antibodies and anti-S IgG antibodies for each method of DENKA and Abbott. Statistical significance was set at $p < 0.05$, using a two-tailed test. All analyses were performed using EZR version 1.27 [22].

2.4. Ethical Consideration

This study was approved by the Niigata University Ethical Committee (approval number 2020–0429) and followed the Declaration of Helsinki (as revised in 2013). Participation in the study was voluntary, and a written informed consent was obtained from each participant.

3. Results

In November 2020, a COVID-19 outbreak occurred in an elderly nursing home in Niigata, Japan, with a total of 178 persons, 97 residents, and 81 staff. None of the elderly or staff received COVID-19 vaccines at the time of the outbreak because the COVID-19 mRNA vaccination program in Japan started in February 2021. Active epidemiological investigations initiated on 16 November 2020, when this nursing home informed the Niigata City Public Health and Sanitation Center, Niigata, Japan, that several residents were symptomatic with rapid antigen diagnostic tests positive for SARS-CoV-2. RT-PCR testing for the residents and staff was carried out by the local Public Health and Sanitation Center at the time of the outbreak to identify cases. As a result, a total of 63 cases (attack rate of 32.0%, 63/197), i.e., 56 residents (attack rate of 57.7%, 56/97), and 7 staff (attack rate of 8.6%, 7/81), were identified by RT-PCR. The first round of RT-PCR was performed for all residents and staff on the same and the following days, i.e., Novemebr 16 and 17, when this facility reported the outbreak to the authority. Then RT-PCR was repeated for all residents and staff almost weekly up to December 16, 2020, until no more positive RT-PCR results were identified. All of the residents and staff who were positive with RT-PCR were hospitalized due to the Japanese government's policy for quarantine purposes during 2020. There were no direct deaths from COVID-19 among residents or staff.

To understand the temporal dynamics of the outbreak, epidemic curves were constructed for 58 of the 63 (92.1%) initial cases (Figure 1A), for whom the date of illness onset was known. The outbreak started on 10 November 2020, when one resident (the index case) developed a fever and two additional residents developed symptoms on the following day. One staff member became symptomatic three days after the onset of the index elderly resident. The epidemic peaked on 14 and 15 November 2020, and ended in two weeks, suggesting the infection may have spread from a single source of exposure in a short time period. This facility is a two-story building, and the initial cases were on the second floor. The attack rate of the elderly on the second floor was higher (85.4%; 41/48) than on the first floor (46.8%; 15/32), suggesting COVID-19 spread quickly in the closed settings and nearly all residents on the second floor were infected. Notably, all five asymptomatic RT-PCR positives were elderly residents and, but no were staff included. As a source of infection, there is a possibility that an asymptomatic staff member that was not detected by RT-PCR may have introduced the virus to this facility. Given that the initial elderly cases were staying in the facility long before the onset of there outbreak, there is no chance of them introducing infections, except for through daily contact with the staff. Although the local Public Health and Sanitation Center obtained demographic information for RT-PCR positive cases, such as age, staying floor for residents, or occupation for staff, no such information was available for RT-PCR negatives.

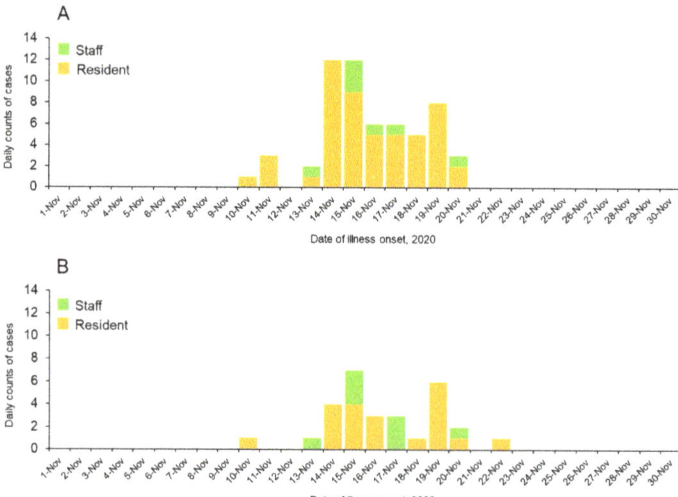

Figure 1. Epidemic histogram of SARS-CoV-2 outbreak in nursing care home in Niigata, Japan, November 2020. (**A**) Epidemic histogram of reverse transcription PCR (RT-PCR) positives at the time of the outbreak (*n* = 58). (**B**) Epidemic histogram of RT-PCR positives who participated in this study (*n* = 29). Daily counts of confirmed cases by RT-PCR tests are described as a function of the day of illness onset. Note that five persons whose date of illness onset by active epidemiological investigations was unknown were excluded in (**A**). Yellow and green bars correspond to staff and residents, respectively.

In the present study, we conducted a cross-sectional sero-epidemiological study in April 2021 for 103 persons (i.e., 41 residents and 62 staff) in a nursing home in Niigata, Japan, approximately six months after the outbreak occurred in November 2020 (Table 1). It should be noted that only 57.8% (103/178) of the initial population and 47.6% (30/63) of RT-PCR positives participated in this study owing to reasons such as declining to participate, leaving or retirement of staff, or deaths of elderly residents that were not related to COVID-19 during the five months after the outbreak. Of these study populations, 9.7% (6/62) of the staff and 58.5% (24/41) of the residents were RT-PCR positive.

Table 1. Epidemiological characteristics of the study population (*n* = 103).

Characteristic	Staff (*n* = 62, 60.1%)		Resident (*n* = 41, 39.9%)	
RT-PCR Test Result	Positive (*n* = 6, 9.7%)	Negative (*n* = 56, 90.3%)	Positive (*n* = 24, 58.5%)	Negative (*n* = 17, 41.5%)
Age (years), median (IQR)	34.0 (30.0–44.0)	50.0 (38.8–56.0)	90.0 (86.0–93.0)	93.0 (86.0–97.0)
Sex, *n* (%)				
Male	2 (33.3)	8 (14.3)	0 (0.0)	4 (23.5)
Female	4 (66.7)	48 (85.7)	24 (100.0)	13 (76.5)
Occupation, *n* (%)				
Doctor	0 (0.0)	4 (7.1)	NA	NA
Nurse	0 (0.0)	14 (25.0)	NA	NA
Caregiver	6 (100.0)	31 (55.4)	NA	NA
Clerk	0 (0.0)	7 (12.5)	NA	NA

Abbreviations: RT-PCR, reverse transcription PCR; SD, standard deviation; IQR, interquartile range; NA, not available. Notes: Data are displayed as median (interquartile range [IQR]) or *n* (%).

The median age was 49.0 years (IQR: 35.3–55.8) for staff and 90.3 years (IQR: 86.0–94.0) for residents, and most participants were females (≥66.7%) (Table 1). Majority of staff members were caregivers (60.0%, 37/62), followed by nurses (25.0%, 14/62), clerks (11.0%, 11/62), and doctors (4.0%, 4/62). Among staff, most cases were negative for RT-PCR results (90.3%, 56/63), whereas among residents, the positive and negative groups were each approximately half of the cases (58.5% for positive and 41.5% for negative), which was compatible with the initial investigation conducted by the local Public Health and Sanitation Center that COVID-19 infection rates were higher in elderly residents than in staff members. Note that among the staff, only the caregivers were RT-PCR positive, and no other occupations (doctor, nurse, and clerk) were positive. The epidemic curve of the 29 RT-PCR positives who participated in this study (Figure 1B) is almost like the original outbreak of RT-PCR positives (Figure 1A).

For a total of 101 patients, excluding the two who were unable to have serum samples collected, IgG antibody titers were measured (Figure 2). For the staff, the positivity rate of anti-N antibody titers was 9.7% (6/62) for DENKA and 9.7% (6/62) for Abbott (Figure 2A), while those of anti-S antibody were 9.7% (6/62) for DENKA and 12.9% (8/62) for Abbott (Figure 2B). For the residents, the positivity of anti-N antibodies was 48.7% (19/39) for DENKA and 41.0% (16/39) for Abbott (Figure 2C), while those of anti-S antibodies were 56.4% (22/39) for DENKA and 56.4% (22/39) for Abbott (Figure 2D). Overall, these results showed a higher prevalence of positive antibody titers in residents than in staff for both anti-S and anti-N antibodies, consistent with a higher number of infections in residents based on RT-PCR testing during the outbreak.

To quantify the diagnostic performance of the IgG antibody titer assay conducted in this study, the seropositivity and seronegativity of the DENKA and Abbott methods were calculated based on RT-PCR results (Table 2, Figure S1). For DENKA, the seropositivity and seronegativity of anti-N antibodies were 66.7% (20/30) and 93.0% (66/71), respectively, while the seropositivity and seronegativity of anti-S antibodies were 90.0% (27/30) and 97.2% (69/71), respectively. For Abbott, the seropositivity and seronegativity for anti-N antibodies were 66.7% (20/30) and 97.2% (69/71) respectively, while the seropositivity and seronegativity for anti-S antibody titers were 90.0% (27/30) and 95.8% (68/71), respectively. These results show that the seropositivity and seronegativity were almost identical between the DENKA and Abbott methods for anti-N and anti-S antibodies, respectively; however, the seropositivity of anti-N antibodies was relatively lower than that of anti-S antibodies at approximately six months after the outbreak. When divided by residents or staff, the seropositivity and seronegativity for anti-N and anti-S antibodies showed similar tendencies between DENKA and Abbott and were in agreement with the overall results (Figure S1).

Notably, there were two RT-PCR-negative but positive anti-S and anti-N antibody titers in one staff member and one resident without any symptoms, suggesting that a small number of asymptomatic infections were missed. Specifically, the staff was a caregiver in this nursing home. This staff's anti-S and anti-N antibodies were 331.6 AU/mL and 3.55 Index (S/N ratio) for Abbott, and 56.4 BAU/mL and 30.0 BAU/mL for DENKA, respectively, indicating that this staff was seropositive. Meanwhile, the resident showed anti-S and anti-N antibodies that were positive (568.4 AU/mL) and 2.23 index (S/N ratio) for Abbott, and positive 150.9 BAU/mL and 79.6 BAU/mL for DENKA, respectively. Comparing the match between the DENKA and Abbott methods, Cohen's kappa statistic showed a level of 0.75 (95% confidence interval [CI], 0.59 to 0.90) and 0.87 (95% CI, 0.76 to 0.98) for anti-N and anti-S antibodies, respectively, demonstrating substantial high concordance. In addition, when also divided by residents or staff, the statistics for anti-N and anti-S antibodies showed similar estimates between the two assays (Table S1). When assessing the association between IgG titers for anti-S and anti-N antibodies, a significant linear correlation was observed for both methods: 0.64 ($p < 0.01$) for DENKA and 0.66 ($p < 0.01$) for Abbott. This indicates that patients with higher anti-S IgG showed higher rates of anti-N antibody positivity, but in turn, patients with lower anti-S IgG tended to show negative anti-N antibodies (Figure S2).

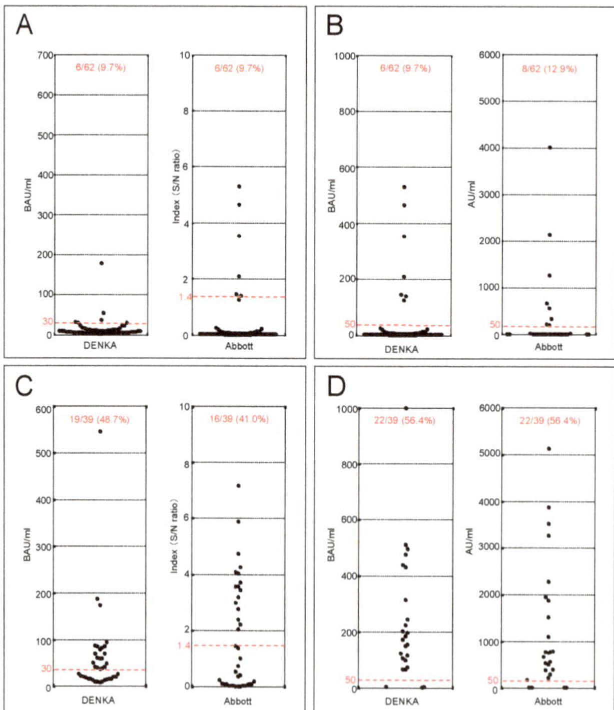

Figure 2. IgG antibody responses of SARS-CoV-2 in nursing care home in Niigata, Japan, April 2021 (n = 101). (**A**) Responses of severe acute respiratory syndrome 2 (SARS-CoV-2) in staff's anti-nucleocapsid (N) IgG antibodies by DENKA (Tokyo, Japan) and Abbott (Chicago, IL, USA). (**B**) Responses of SARS-CoV-2 in staff's anti-spike (S) IgG antibodies by DENKA and Abbott. (**C**) Responses of SARS-CoV-2 in resident's anti-N IgG antibodies by DENKA and Abbott. (**D**) Responses of SARS-CoV-2 in resident's anti-S IgG antibodies by DENKA and Abbott. The percentage of positive cases (%) is described and the red dotted line indicates the threshold for positivity (i.e., positive cutoff index).

Table 2. Test seropositivity and seronegativity of the DENKA and Abbott methods for anti-N and anti-S IgG antibodies among patients with COVID-19 at the nursing home in Niigata, Japan in April 2021 ($n = 101$).

DENKA (Tokyo, Japan)	Anti-N IgG Antibody	Anti-S IgG Antibody
Seropositivity (%)	66.7 (20/30)	90.0 (27/30)
Seronegativity (%)	93.0 (66/71)	97.2 (69/71)
Abbott (Chicago, IL, USA)	**Anti-N IgG Antibody**	**Anti-S IgG Antibody**
Seropositivity (%)	66.7 (20/30)	90.0 (27/30)
Seronegativity (%)	97.2 (69/71)	95.8 (68/71)

Abbreviations: N, nucleocapsid; S, spike; IgG, immunoglobulin G. Notes: Seropositivity and seronegativity of the DENKA and Abbott methods were calculated using the results of the antibody titer as reference.

4. Discussion

In the present study, we measured anti-SARS-CoV-2 IgG antibodies targeting anti-N and anti-S proteins by two laboratory-based immunoassay testing methods (i.e., DENKA and Abbott) using serum specimens collected from unvaccinated residents and staff of a nursing home in Niigata, Japan. Two main findings were obtained: first, virus transmission spreads within an enclosed environment, and the sero-prevalence in April 2021 among the residents and staff was approximately 40.0–60.0% and 10.0–20.0%, respectively, which was in close agreement with the initial RT-PCR test positive results (57.7% RT-PCR positive for residents and 8.6% for staff) at the time of the outbreak in November 2020, indicating that the infection rates were almost seven times higher among residents. Secondly, seropositivity for anti-S antibodies showed high concordance with RT-PCR positive results even after approximately six months of infections (90.0% for both DENKA and Abbott). The seronegativity for anti-S antibodies also showed high concordance with RT-PCR negatives (97.2% for DENKA and 95.8% for Abbott). Meanwhile, seropositivity for anti-N antibodies remained low at 66.7% for both DENKA and Abbott. It should also be noted that there were two RT-PCR-negative results that had positive anti-S and anti-N antibody titers (one staff and one resident).

In this nursing home, the initial RT-PCR-based infection of elderly residents (57.7%) was almost seven times higher than that of staff (8.6%), indicating a higher risk of infection among the elderly residents, and serological results supported the similar findings. In this study, there were no direct deaths from COVID-19. One impressive paper from January–May 2020 in Japan by Iritani et al. reported that the number and size of clusters in elderly care homes were independently associated with higher mortality rates in all 47 prefectures in Japan, underlining the importance of infection control in such facilities to avoid pressure on local healthcare [23]. Indeed, a report from Nagasaki City, Nagasaki, Japan, showed that although the number of age-specific occurrences per population was not as high for residents of elderly care facilities, deaths were overwhelmingly higher in these facilities (incidence rate of 48.1 per 100,000 person-year) and approximately twice as high as for community-dwelling older adults [24]. In addition, consistent with our present results, a large cohort study of approximately 1500 residents and more than 3000 staff in 201 long-term care facilities in the United Kingdom (UK) between June 2020 and May 2021, measuring anti-N antibodies, found that seropositivity in the last 11 months was 34.6% for residents and 26.1% for staff, suggesting a higher rate of infection among residents [25]. It is imperative to protect care home residents who are vulnerable to COVID-19 infections from potential sources of SARS-CoV-2 through rapid screening and response measures to minimize the scale of outbreaks once the infection is introduced [26].

Interestingly, only caregivers were infected among the staff (doctor, nurse, caregiver, and clerk) in this nursing home. Indeed, one notable paper on healthcare workers in United States of America (USA) hospitals and nursing homes in July–August 2020 showed that seroprevalence also varied by occupation [27]. More specifically, for example, nurses (4.2%) and receptionists/medical assistants (4.1%) were more likely to be seropositive than physicians (2.2%) for hospitals, while in nursing homes, nursing assistants (19.9%) and social workers/case managers/counselors (21.7%) were more likely to be seropositive than occupational/physical/speech therapists (9.8%). Although their studies were not able to explicitly assess the reasons for these differences (i.e., heterogenicity) in seropositivity between occupations, it was observed that occupations with more direct contact with older people tended to be the most frequently infected. Taken together, these findings suggested that strict infection control measures should be implemented, as well as education for the group of healthcare workers who have frequent contact with the elderly, because healthcare workers can initiate and spread the infections quickly among vulnerable groups.

The present study showed that the results of the SARS-CoV-2 IgG antibody test against anti-S antibodies from DENKA and Abbott generally matched and were almost in agreement with the RT-PCR-positive results, even approximately six months after natural infection. Meanwhile, the seropositivity of the anti-N antibody was relatively lower than that of the anti-S antibody. Recent literature suggests that IgG antibodies to the N protein decrease over time, while responses to the S protein are more stable over a longer period [28–31]. Besides, antibodies against the S protein is reportedly more specific than antibodies against the N protein due to lower cross-reactivity with other seasonal coronaviruses [32]. An impressive study investigated seropositivity patterns at different intervals (i.e., 2, 6, and 13 months) after an outbreak in the Lithuanian private sector in April 2020, when approximately one third of employees (94 out of 300) tested positive via RT-PCR [20]. This study showed that six months after the outbreak, 95.0% of 59 previously infected individuals had virus-specific anti-S antibodies, irrespective of the severity of infection, suggesting that specific antibodies persisted for longer than 6 months in the majority of cases, consistent with our results.

In this present study, there were two individuals, a member of staff and a resident, who were RT-PCR negative but had positive anti-S and anti-N antibody titers. There is a possibility that this asymptomatic seropositive staff can be the source of infection, but other staff could be the source since 19 staff members declined to participate in this study, left the facility, or retired from it. Indeed, one extensive systematic review, including 34 studies in 2020, demonstrated a large unexplained false-negative of RT-PCR and suggested this was due to missed cases of asymptomatic infections (tau-squared = 1.39) [33,34]. Therefore, in addition to repeated RT-PCR testing, it is useful to conduct surveys with additional serological testing in cohorts of individuals to supplement RT-PCR results and capture asymptomatic cases missed by PCR, as in this study [34]. Those additional studies may give important information to clarify what the infection source was and what kind of infection controls should be taken to prevent future spread.

The findings in this report are subject to at least seven technical limitations. First, there is limited epidemiological information on the presumed cause of infection in the index case (i.e., the first resident who became ill on 10 November 2020) does not allow a detailed description of the transmission chains. Besides, there is incomplete epidemiologic information about potential visitors to associate with the case and limited information about interactions outside the nursing home that may have contributed to the initial virus entry. Altogether, this present study enrolled a small number of persons (i.e., 103) and was limited to describing their epidemiological characteristics and was unable to more objectively explore in depth the potential drivers associated with the exposure and transmission dynamics in this nursing home. Second, we were unable to perform viral genome analysis (e.g., whole genome sequencing), which, when combined with the lack of detailed epidemiologic information, makes it impossible to fully characterize the transmission patterns in this nursing home [35,36]. Three, it was not possible to collect epidemiological information on symptoms for each case, so the associations between RT-PCR testing results and symptomatic/asymptomatic symptoms could not be scrutinized. Importantly, previous research has suggested that infection during the pre-symptomatic period or from asymptomatic individuals may have been potential drivers in infection transmission within the facility, suggesting that they are likely to have contributed to transmission [37–40]. In particular, a study in a similar contextual setting to ours from Belgium suggests that approximately 14.0% and 50.0% of pre-vaccination seropositive staff and residents, respectively, did not report previous COVID-19 symptoms [41]. Fifth, the IgG antibody titer assay in this present study did not assess specific neutralizing antibody levels against SARS-CoV-2 owing to technical challenges, which may have underestimated exposure in the study population [42,43]. Future research will focus on more detailed quantifications of specific neutralizing antibody titer assays to explicitly conclude seroprevalence. Sixth, the present study only collected specific epidemiological data on the subjects (i.e., age, sex, and occupation for staff), which made it difficult to explicitly scrutinize the association between serum antibody titers and other crucial factors (e.g., comorbidities and anthropometric measurements). Though not available for this study, these factors could help disentangle the directionality of exact transmission and may be helpful for future studies. Finally, seroprevalence was estimated at a single time point approximately six months after natural infection, and no serology data were obtained soon after the outbreak, which limits generalizability. Ultimately, there remains room to examine the shift of antibody titers at multiple time points in the future.

Notwithstanding these limitations, the present study consistently demonstrated that the point pre-vaccination seroprevalence among the residents was higher compared to staff members in this outbreak in a nursing home in Niigata, Japan. Besides, the diagnostic performance in pre-vaccination residents and staff of a nursing home showed a relatively high match with RT-PCR results after approximately six months, partially highlighting that the anti-S IgG antibody tests may be useful as a diagnostic tool to scrutinize the possibility of COVID-19 infection. This present study highlights the value of serological analysis to understand the extent of SARS-CoV-2 circulation in this high-risk setting (e.g., long-term nursing homes), which also demonstrates the importance of repeatedly performing RT-PCR screening as an epidemic control measure for infectious diseases. Further studies are needed at the same institution to determine whether the post-vaccination anti-SARS-CoV-2 IgG antibodies, including neutralizing antibodies, are protective against the re-infection and, if so, the duration of protection among residents and staff.

Supplementary Materials: The following supporting information can be downloaded at: https://www.mdpi.com/article/10.3390/v14112581/s1, Figure S1: Seroprevalence in RT-PCR-positive and RT-PCR-negative participants (n = 101); Figure S2: Bivariate relationship between anti-S IgG antibody and anti-N IgG antibody for the DENKA and Abbott methods (n = 101); Table S1: Concordance assessment between the DENKA and Abbott methods for anti-N and anti-S IgG antibodies by resident or staff utilizing Cohen's kappa statistic (n = 101).

Author Contributions: Conceptualization, K.W., S.Y. (Sayaka Yoshioka), H.W. and R.S. (Reiko Saito); Methodology, K.W. and S.Y. (Sayaka Yoshioka); Software, K.W.; Validation, S.Y. (Sayaka Yoshioka); Formal Analysis, K.W.; Investigation, K.W., S.Y. (Sayaka Yoshioka), S.Y. (Satoru Yamazaki), R.S. (Ryosuke Sato), W.W.P., I.C., Y.T., H.W. and R.S. (Reiko Saito); Resources, K.W., S.Y. (Sayaka Yoshioka), S.Y. (Satoru Yamazaki), W.W.P., I.C., Y.T., H.W. and R.S. (Reiko Saito); Data Curation, S.Y. (Sayaka Yoshioka); Writing—Original Draft Preparation, K.W.; Writing—Review and Editing, S.Y. (Sayaka Yoshioka) and R.S. (Reiko Saito); Visualization, K.W.; Supervision, H.W. and R.S. (Reiko Saito); Project Administration, R.S. (Reiko Saito); Funding Acquisition, K.W. and R.S. (Reiko Saito). All authors have read and agreed to the published version of the manuscript.

Funding: K.W. received funding from the Grants-in-Aid for Scientific Research (KAKENHI) by the Japan Society for the Promotion of Science (JSPS) (22J23183); the Community Medical Research Grant by the Niigata City Medical Association (GC03220213); and the Tsukada Medical Research Grant (grant number not available). R.S. (Reiko Saito) received funding from the Japan Initiative for Global Research Network on Infectious Diseases (J-GRID) by the Japan Agency for Medical Research and Development (AMED) (21wm0125005h0002); the KAKENHI by the JSPS (21K10414); the Health and Labor Sciences Research Grants, Ministry of Health, Labor, and Welfare, Japan (H30-Shinkougyousei-Shitei-004); and the Niigata Prefecture Coronavirus Infectious Disease Control Research and Human Resources Development Support Fund (grant number not available). The funder of the study had no role in the study's design, data collection, data analysis, data interpretation, or writing of the paper.

Institutional Review Board Statement: The study was conducted in accordance with the Declaration of Helsinki and approved by the Ethics Committee of Niigata University (protocol code 2020–0429).

Informed Consent Statement: Informed consent was obtained from all subjects involved in the study.

Data Availability Statement: Anonymized datasets generated during the study are available on request from the corresponding author.

Acknowledgments: We would like to thank the staff of all participating medical institutions and the staff of the Division of International Health (Public Health), Graduate School of Medical and Dental Sciences, Niigata University, Niigata, Japan. We are deeply thankful to the staff of the Niigata City Public Health and Sanitation Center for their field epidemiological investigations. We thank Yukio Matsuda, Naito Shinichi, and Nobuko Asazuma for the epidemiological investigation, as well as all the doctors, nurses, and staff members at the study site. We also thank Makoto Naito, Hirohisa Yoshizawa, Hidehiro Hasegawa, Makoto Yamazaki, and Kensuke Ozawa for technical support.

Conflicts of Interest: The authors declare no conflict of interest.

References

1. Heymann, D.L.; Shindo, N. COVID-19: What is next for public health? *Lancet* **2020**, *395*, 542–545. [CrossRef]
2. Furuse, Y.; Sando, E.; Tsuchiya, N.; Miyahara, R.; Yasuda, I.; Ko, Y.K.; Saito, M.; Morimoto, K.; Imamura, T.; Shobugawa, Y.; et al. Clusters of Coronavirus Disease in Communities, Japan, January-April 2020. *Emerg. Infect. Dis.* **2020**, *26*, 2176–2179. [CrossRef]
3. Wagatsuma, K.; Sato, R.; Yamazaki, S.; Iwaya, M.; Takahashi, Y.; Nojima, A.; Oseki, M.; Abe, T.; Phyu, W.W.; Tamura, T.; et al. Genomic Epidemiology Reveals Multiple Introductions of Severe Acute Respiratory Syndrome Coronavirus 2 in Niigata City, Japan, Between February and May 2020. *Front. Microbiol.* **2021**, *12*, 749149. [CrossRef] [PubMed]
4. Wagatsuma, K.; Phyu, W.W.; Osada, H.; Tang, J.W.; Saito, R. Geographic Correlation between the Number of COVID-19 Cases and the Number of Overseas Travelers in Japan, Jan-Feb, 2020. *Jpn. J. Infect. Dis.* **2021**, *74*, 157–160. [CrossRef] [PubMed]
5. Furuse, Y.; Tsuchiya, N.; Miyahara, R.; Yasuda, I.; Sando, E.; Ko, Y.K.; Imamura, T.; Morimoto, K.; Imamura, T.; Shobugawa, Y.; et al. COVID-19 case-clusters and transmission chains in the communities in Japan. *J. Infect.* **2022**, *84*, 248288. [CrossRef] [PubMed]
6. Nishiura, H.; Kobayashi, T.; Miyama, T.; Suzuki, A.; Jung, S.M.; Hayashi, K.; Kinoshita, R.; Yang, Y.; Yuan, B.; Akhmetzhanov, A.R.; et al. Estimation of the asymptomatic ratio of novel coronavirus infections (COVID-19). *Int. J. Infect. Dis.* **2020**, *94*, 154–155. [CrossRef]
7. Mizumoto, K.; Kagaya, K.; Zarebski, A.; Chowell, G. Estimating the asymptomatic proportion of coronavirus disease 2019 (COVID-19) cases on board the Diamond Princess cruise ship, Yokohama, Japan, 2020. *Eurosurveillance* **2020**, *25*, 2000180. [CrossRef]
8. Russell, T.W.; Hellewell, J.; Jarvis, C.I.; van Zandvoort, K.; Abbott, S.; Ratnayake, R.; Cmmid Covid-Working, G.; Flasche, S.; Eggo, R.M.; Edmunds, W.J.; et al. Estimating the infection and case fatality ratio for coronavirus disease (COVID-19) using age-adjusted data from the outbreak on the Diamond Princess cruise ship, February 2020. *Eurosurveillance* **2020**, *25*, 2000256. [CrossRef] [PubMed]
9. Munayco, C.; Chowell, G.; Tariq, A.; Undurraga, E.A.; Mizumoto, K. Risk of death by age and gender from CoVID-19 in Peru, March-May, 2020. *Aging* **2020**, *12*, 13869–13881. [CrossRef]
10. Undurraga, E.A.; Chowell, G.; Mizumoto, K. COVID-19 case fatality risk by age and gender in a high testing setting in Latin America: Chile, March-August 2020. *Infect. Dis. Poverty* **2021**, *10*, 11. [CrossRef] [PubMed]
11. Yanez, N.D.; Weiss, N.S.; Romand, J.A.; Treggiari, M.M. COVID-19 mortality risk for older men and women. *BMC Public Health* **2020**, *20*, 1742. [CrossRef] [PubMed]
12. Grabowski, D.C.; Mor, V. Nursing Home Care in Crisis in the Wake of COVID-19. *JAMA* **2020**, *324*, 23–24. [CrossRef]
13. Speletas, M.; Kyritsi, M.A.; Vontas, A.; Theodoridou, A.; Chrysanthidis, T.; Hatzianastasiou, S.; Petinaki, E.; Hadjichristodoulou, C. Evaluation of Two Chemiluminescent and Three ELISA Immunoassays for the Detection of SARS-CoV-2 IgG Antibodies: Implications for Disease Diagnosis and Patients' Management. *Front. Immunol.* **2020**, *11*, 609242. [CrossRef] [PubMed]
14. Manthei, D.M.; Whalen, J.F.; Schroeder, L.F.; Sinay, A.M.; Li, S.H.; Valdez, R.; Giacherio, D.A.; Gherasim, C. Differences in Performance Characteristics Among Four High-Throughput Assays for the Detection of Antibodies Against SARS-CoV-2 Using a Common Set of Patient Samples. *Am. J. Clin. Pathol.* **2021**, *155*, 267–279. [CrossRef] [PubMed]
15. Bryan, A.; Pepper, G.; Wener, M.H.; Fink, S.L.; Morishima, C.; Chaudhary, A.; Jerome, K.R.; Mathias, P.C.; Greninger, A.L. Performance Characteristics of the Abbott Architect SARS-CoV-2 IgG Assay and Seroprevalence in Boise, Idaho. *J. Clin. Microbiol.* **2020**, *58*, e00941-20. [CrossRef] [PubMed]
16. Maine, G.N.; Lao, K.M.; Krishnan, S.M.; Afolayan-Oloye, O.; Fatemi, S.; Kumar, S.; VanHorn, L.; Hurand, A.; Sykes, E.; Sun, Q. Longitudinal characterization of the IgM and IgG humoral response in symptomatic COVID-19 patients using the Abbott Architect. *J. Clin. Virol.* **2020**, *133*, 104663. [CrossRef]
17. Kozawa, K.; Miura, H.; Kawamura, Y.; Higashimoto, Y.; Ihira, M.; Yoshikawa, T. Unremarkable antibody responses against various infectious agents after inoculation with the BNT162b2 COVID-19 vaccine. *J. Med. Virol.* **2022**, *94*, 4583–4585. [CrossRef] [PubMed]

18. World Health Organization. WHO/BS.2020.2403 Establishment of the WHO International Standard and Reference Panel for anti-SARS-CoV-2 Antibody. Available online: https://www.who.int/publications/m/item/WHO-BS-2020.2403 (accessed on 15 August 2022).

19. Kristiansen, P.A.; Page, M.; Bernasconi, V.; Mattiuzzo, G.; Dull, P.; Makar, K.; Plotkin, S.; Knezevic, I. WHO International Standard for anti-SARS-CoV-2 immunoglobulin. *Lancet* **2021**, *397*, 1347–1348. [CrossRef]

20. Kučinskaitė-Kodzė, I.; Simanavičius, M.; Šimaitis, A.; Žvirblienė, A. Persistence of SARS-CoV-2-Specific Antibodies for 13 Months after Infection. *Viruses* **2021**, *13*, 2313. [CrossRef] [PubMed]

21. Landis, J.R.; Koch, G.G. The measurement of observer agreement for categorical data. *Biometrics* **1977**, *33*, 159–174. [CrossRef]

22. Kanda, Y. Investigation of the freely available easy-to-use software 'EZR' for medical statistics. *Bone Marrow Transplant.* **2013**, *48*, 452–458. [CrossRef]

23. Iritani, O.; Okuno, T.; Hama, D.; Kane, A.; Kodera, K.; Morigaki, K.; Terai, T.; Maeno, N.; Morimoto, S. Clusters of COVID-19 in long-term care hospitals and facilities in Japan from 16 January to 9 May 2020. *Geriatr. Gerontol. Int.* **2020**, *20*, 715–719. [CrossRef]

24. Shimizu, K.; Maeda, H.; Sando, E.; Fujita, A.; Tashiro, M.; Tanaka, T.; Izumikawa, K.; Motomura, K.; Morimoto, K. Epidemiology of SARS-CoV-2 infection in nursing facilities and the impact of their clusters in a Japanese core city. *J. Infect. Chemother.* **2022**, *28*, 955–961. [CrossRef]

25. Krutikov, M.; Palmer, T.; Tut, G.; Fuller, C.; Azmi, B.; Giddings, R.; Shrotri, M.; Kaur, N.; Sylla, P.; Lancaster, T.; et al. Prevalence and duration of detectable SARS-CoV-2 nucleocapsid antibodies in staff and residents of long-term care facilities over the first year of the pandemic (VIVALDI study): Prospective cohort study in England. *Lancet Healthy Longev.* **2022**, *3*, e13–e21. [CrossRef]

26. Bernadou, A.; Bouges, S.; Catroux, M.; Rigaux, J.C.; Laland, C.; Levêque, N.; Noury, U.; Larrieu, S.; Acef, S.; Habold, D.; et al. High impact of COVID-19 outbreak in a nursing home in the Nouvelle-Aquitaine region, France, March to April 2020. *BMC Infect. Dis.* **2021**, *21*, 198. [CrossRef]

27. Akinbami, L.J.; Chan, P.A.; Vuong, N.; Sami, S.; Lewis, D.; Sheridan, P.E.; Lukacs, S.L.; Mackey, L.; Grohskopf, L.A.; Patel, A.; et al. Severe Acute Respiratory Syndrome Coronavirus 2 Seropositivity among Healthcare Personnel in Hospitals and Nursing Homes, Rhode Island, USA, July-August 2020. *Emerg. Infect. Dis.* **2021**, *27*, 823–834. [CrossRef]

28. Focosi, D.; Maggi, F.; Mazzetti, P.; Pistello, M. Viral infection neutralization tests: A focus on severe acute respiratory syndrome-coronavirus-2 with implications for convalescent plasma therapy. *Rev. Med. Virol.* **2021**, *31*, e2170. [CrossRef]

29. Sethuraman, N.; Jeremiah, S.S.; Ryo, A. Interpreting Diagnostic Tests for SARS-CoV-2. *JAMA* **2020**, *323*, 2249–2251. [CrossRef]

30. Ripperger, T.J.; Uhrlaub, J.L.; Watanabe, M.; Wong, R.; Castaneda, Y.; Pizzato, H.A.; Thompson, M.R.; Bradshaw, C.; Weinkauf, C.C.; Bime, C.; et al. Orthogonal SARS-CoV-2 Serological Assays Enable Surveillance of Low-Prevalence Communities and Reveal Durable Humoral Immunity. *Immunity* **2020**, *53*, 925–933.e924. [CrossRef]

31. El-Khoury, J.M.; Schulz, W.L.; Durant, T.J.S. Longitudinal Assessment of SARS-CoV-2 Antinucleocapsid and Antispike-1-RBD Antibody Testing Following PCR-Detected SARS-CoV-2 Infection. *J. Appl. Lab. Med.* **2021**, *6*, 1005–1011. [CrossRef]

32. Nakano, Y.; Kurano, M.; Morita, Y.; Shimura, T.; Yokoyama, R.; Qian, C.; Xia, F.; He, F.; Kishi, Y.; Okada, J.; et al. Time course of the sensitivity and specificity of anti-SARS-CoV-2 IgM and IgG antibodies for symptomatic COVID-19 in Japan. *Sci. Rep.* **2021**, *11*, 2776. [CrossRef] [PubMed]

33. Cook, G.W.; Benton, M.G.; Akerley, W.; Mayhew, G.F.; Moehlenkamp, C.; Raterman, D.; Burgess, D.L.; Rowell, W.J.; Lambert, C.; Eng, K.; et al. Structural variation and its potential impact on genome instability: Novel discoveries in the EGFR landscape by long-read sequencing. *PLoS ONE* **2020**, *15*, e0226340. [CrossRef] [PubMed]

34. Meyerowitz, E.A.; Richterman, A.; Bogoch, I.I.; Low, N.; Cevik, M. Towards an accurate and systematic characterisation of persistently asymptomatic infection with SARS-CoV-2. *Lancet Infect. Dis.* **2021**, *21*, e163–e169. [CrossRef]

35. Jeffery-Smith, A.; Dun-Campbell, K.; Janarthanan, R.; Fok, J.; Crawley-Boevey, E.; Vusirikala, A.; Fernandez Ruiz De Olano, E.; Sanchez Perez, M.; Tang, S.; Rowland, T.A.; et al. Infection and transmission of SARS-CoV-2 in London care homes reporting no cases or outbreaks of COVID-19: Prospective observational cohort study, England 2020. *Lancet Reg. Health Eur.* **2021**, *3*, 100038. [CrossRef]

36. Aggarwal, D.; Myers, R.; Hamilton, W.L.; Bharucha, T.; Tumelty, N.M.; Brown, C.S.; Meader, E.J.; Connor, T.; Smith, D.L.; Bradley, D.T.; et al. The role of viral genomics in understanding COVID-19 outbreaks in long-term care facilities. *Lancet Microbe* **2022**, *3*, e151–e158. [CrossRef]

37. McMichael, T.M.; Currie, D.W.; Clark, S.; Pogosjans, S.; Kay, M.; Schwartz, N.G.; Lewis, J.; Baer, A.; Kawakami, V.; Lukoff, M.D.; et al. Epidemiology of Covid-19 in a Long-Term Care Facility in King County, Washington. *N. Engl. J. Med.* **2020**, *382*, 2005–2011. [CrossRef] [PubMed]

38. White, E.M.; Kosar, C.M.; Feifer, R.A.; Blackman, C.; Gravenstein, S.; Ouslander, J.; Mor, V. Variation in SARS-CoV-2 Prevalence in U.S. Skilled Nursing Facilities. *J. Am. Geriatr. Soc.* **2020**, *68*, 2167–2173. [CrossRef]

39. Arons, M.M.; Hatfield, K.M.; Reddy, S.C.; Kimball, A.; James, A.; Jacobs, J.R.; Taylor, J.; Spicer, K.; Bardossy, A.C.; Oakley, L.P.; et al. Presymptomatic SARS-CoV-2 Infections and Transmission in a Skilled Nursing Facility. *N. Engl. J. Med.* **2020**, *382*, 2081–2090. [CrossRef]

40. Patel, M.C.; Chaisson, L.H.; Borgetti, S.; Burdsall, D.; Chugh, R.K.; Hoff, C.R.; Murphy, E.B.; Murskyj, E.A.; Wilson, S.; Ramos, J.; et al. Asymptomatic SARS-CoV-2 Infection and COVID-19 Mortality During an Outbreak Investigation in a Skilled Nursing Facility. *Clin. Infect. Dis.* **2020**, *71*, 2920–2926. [CrossRef]

41. Janssens, H.; Heytens, S.; Meyers, E.; De Schepper, E.; De Sutter, A.; Devleesschauwer, B.; Formukong, A.; Keirse, S.; Padalko, E.; Geens, T.; et al. Pre-vaccination SARS-CoV-2 seroprevalence among staff and residents of nursing homes in Flanders (Belgium) in fall 2020. *Epidemiol. Infect.* **2022**, *150*, 1–25. [CrossRef]

42. Khoury, D.S.; Cromer, D.; Reynaldi, A.; Schlub, T.E.; Wheatley, A.K.; Juno, J.A.; Subbarao, K.; Kent, S.J.; Triccas, J.A.; Davenport, M.P. Neutralizing antibody levels are highly predictive of immune protection from symptomatic SARS-CoV-2 infection. *Nat. Med.* **2021**, *27*, 1205–1211. [CrossRef] [PubMed]

43. Ladhani, S.N.; Jeffery-Smith, A.; Patel, M.; Janarthanan, R.; Fok, J.; Crawley-Boevey, E.; Vusirikala, A.; Fernandez Ruiz De Olano, E.; Perez, M.S.; Tang, S.; et al. High prevalence of SARS-CoV-2 antibodies in care homes affected by COVID-19: Prospective cohort study, England. *EClinicalMedicine* **2020**, *28*, 100597. [CrossRef] [PubMed]

Article

Characterizing Longitudinal Antibody Responses in Recovered Individuals Following COVID-19 Infection and Single-Dose Vaccination: A Prospective Cohort Study

Andrea D. Olmstead [1,2,†], Aidan M. Nikiforuk [2,3,†], Sydney Schwartz [2], Ana Citlali Márquez [2], Tahereh Valadbeigy [2], Eri Flores [2], Monika Saran [1], David M. Goldfarb [1,4], Althea Hayden [5], Shazia Masud [1,6], Shannon L. Russell [1,2], Natalie Prystajecky [1,2], Agatha N. Jassem [1,2], Muhammad Morshed [1,2] and Inna Sekirov [1,2,*]

[1] Department of Pathology and Laboratory Medicine, University of British Columbia, 2211 Wesbrook Mall, Vancouver, BC V6T 1Z7, Canada
[2] British Columbia Centre for Disease Control Public Health Laboratory, Provincial Health Services Authority, 655 West 12th Ave, Vancouver, BC V5Z 4R4, Canada
[3] School of Population and Public Health, University of British Columbia, 2206 E Mall, Vancouver, BC V6T 1Z3, Canada
[4] Department of Pathology and Laboratory Medicine, British Columbia Children's and Women's Hospital, 4500 Oak Street, Vancouver, BC V6H 3N1, Canada
[5] Office of the Chief Medical Health Officer, Vancouver Coastal Health, Vancouver, BC V5Z 4C2, Canada
[6] Department of Pathology and Laboratory Medicine, Surrey Memorial Hospital, Surrey, BC V3V 1Z2, Canada
* Correspondence: inna.sekirov@bccdc.ca
† These authors contributed equally to this work.

Citation: Olmstead, A.D.; Nikiforuk, A.M.; Schwartz, S.; Márquez, A.C.; Valadbeigy, T.; Flores, E.; Saran, M.; Goldfarb, D.M.; Hayden, A.; Masud, S.; et al. Characterizing Longitudinal Antibody Responses in Recovered Individuals Following COVID-19 Infection and Single-Dose Vaccination: A Prospective Cohort Study. *Viruses* **2022**, *14*, 2416. https://doi.org/10.3390/v14112416

Academic Editor: Jason Yiu Wing KAM

Received: 20 September 2022
Accepted: 26 October 2022
Published: 31 October 2022

Publisher's Note: MDPI stays neutral with regard to jurisdictional claims in published maps and institutional affiliations.

Abstract: Background: Investigating antibody titers in individuals who have been both naturally infected with SARS-CoV-2 and vaccinated can provide insight into antibody dynamics and correlates of protection over time. Methods: Human coronavirus (HCoV) IgG antibodies were measured longitudinally in a prospective cohort of qPCR-confirmed, COVID-19 recovered individuals (k = 57) in British Columbia pre- and post-vaccination. SARS-CoV-2 and endemic HCoV antibodies were measured in serum collected between Nov. 2020 and Sept. 2021 (n = 341). Primary analysis used a linear mixed-effects model to understand the effect of single dose vaccination on antibody concentrations adjusting for biological sex, age, time from infection and vaccination. Secondary analysis investigated the cumulative incidence of high SARS-CoV-2 anti-spike IgG seroreactivity equal to or greater than 5.5 log10 AU/mL up to 105 days post-vaccination. No re-infections were detected in vaccinated participants, post-vaccination by qPCR performed on self-collected nasopharyngeal specimens. Results: Bivariate analysis (complete data for 42 participants, 270 samples over 472 days) found SARS-CoV-2 spike and RBD antibodies increased 14–56 days post-vaccination ($p < 0.001$) and vaccination prevented waning (regression coefficient, B = 1.66 [95%CI: 1.45–3.46]); while decline of nucleocapsid antibodies over time was observed (regression coefficient, B = −0.24 [95%CI: −1.2-(−0.12)]). A positive association was found between COVID-19 vaccination and endemic human β-coronavirus IgG titer 14–56 days post vaccination (OC43, $p = 0.02$ & HKU1, $p = 0.02$). On average, SARS-CoV-2 anti-spike IgG concentration increased in participants who received one vaccine dose by 2.06 log10 AU/mL (95%CI: 1.45–3.46) adjusting for age, biological sex, and time since infection. Cumulative incidence of high SARS-CoV-2 spike antibodies (>5.5 log10 AU/mL) was 83% greater in vaccinated compared to unvaccinated individuals. Conclusions: Our study confirms that vaccination post-SARS-CoV-2 infection provides multiple benefits, such as increasing anti-spike IgG titers and preventing decay up to 85 days post-vaccination.

Keywords: SARS-CoV-2; COVID-19; cohort study; antibody waning; seroreactivity; electrochemiluminescence assay; fixed-effect models

1. Introduction

The coronavirus disease 2019 (COVID-19) pandemic, caused by the novel beta (β)-coronavirus, severe acute respiratory syndrome coronavirus 2 (SARS-CoV-2), has resulted in significant morbidity, mortality, economic impact, and disruption of health care and societal systems. Prior to the emergence of COVID-19, four seasonal human coronaviruses (HCoV) were identified that typically cause self-limited respiratory infections with mild symptoms, i.e., the 'common cold' in otherwise healthy people [1]. Like SARS-CoV-2, HCoV-OC43, and HCoV-HKU1 are β-coronaviruses, while HCoV-229E and HCoV-NL63 are classified as alpha (α)-coronaviruses [2]. Coronavirus genera are separated by unique serological and genomic characteristics; viral species from the same genus share cross-neutralizing (non-specific) antibodies which arise from homology in viral genes and structural proteins [3].

In the province of British Columbia (BC), Canada, the first confirmed case of COVID-19 was reported on January 25, 2020; strict and swift public health measures were largely effective at controlling spread during the first wave, which peaked locally between the third week of March and late April in 2020 [4]. During the first epidemiological wave of the pandemic, little was known about antibody responses to SARS-CoV-2 infection and studies were needed to understand if and how quickly infected individuals develop a detectable, protective, and durable antibody-mediated immune response. Understanding the durability or waning of antibodies over time helps elucidate the risk of re-infection and inform vaccination schedules. Studies have shown that most SARS-CoV-2 infected individuals seroconvert within 14–28 days; the spike (S) and the nucleocapsid (N) proteins elicit the strongest humoral response [5,6]. Predictably, SARS-CoV-2 antibody concentrations wane over time; the rate of decline varies widely depending on various factors (e.g., age, biological sex, and disease severity) [7–9]. Neutralizing antibodies acquired naturally or from vaccination protect against infection and re-infection [10]. Several studies have shown a strong correlation between anti-S, anti-receptor binding domain (RBD) and neutralizing antibody titers, as such measuring anti-S and anti-RBD can be used as a proxy for antibody-mediated protection [8,11,12].

Despite the success of SARS-CoV-2 vaccines, many individuals are still hesitant to be immunized against COVID-19; supply shortages combined with social and economic inequity hamper global vaccination efforts [13–15]. The study of antibody dynamics following natural infection and the impact of vaccination on those who have been previously infected is needed, as novel SARS-CoV-2 variants with increasing capacity to escape pre-existing immunity continue to evolve and spread [16–18]. Observational studies agree that vaccination benefits those who have been previously infected, but the number of doses required for optimal protection remains unclear [19–21].

We describe a prospective cohort that was established to monitor antibody responses over three months in people that recovered from SARS-CoV-2 infection. Many participants were offered a single dose COVID-19 vaccine during the study; therefore, we expanded the aims to consider the dynamics of antibodies against both SARS-CoV-2, as well as endemic HCoVs, in recovered individuals pre- and post- vaccination against SARS-CoV-2, and investigated their relationship with age, biological sex, and time from qPCR diagnosis.

2. Materials and Methods

2.1. Study Design

A prospective observational cohort termed CARE (Characterizing the Antibody Response to Emerging COVID-19) was established from individuals who recovered from SARS-CoV-2 infection, for the purposes of investigating antibody responses against several SARS-CoV-2 epitopes (full spike (S), receptor binding domain (RBD) and nucleocapsid (N)), as well as against the S protein of endemic HCoVs (OC43, HKU1, NL63, 229E), at least 2 weeks post natural SARS-CoV-2 infection, with or without subsequent SARS-CoV-2 vaccination. Information on age, biological sex and date of real time PCR (qPCR) diagnosis was collected through medical records, while information on the duration of COVID-19 symptoms, hospitalization, and vaccination was collected using an online self-reporting survey. The date of SARS-CoV-2 qPCR diagnosis was used to estimate 'days post-infection'. Survey data and participant informed consent were collected and managed using REDCap electronic data capture tools hosted at BC Children's Hospital (Vancouver, BC). REDCap (Research Electronic Data Capture) is a secure, web-based application designed to support data capture for research studies [22]. Participants were enrolled in the cohort from 19 November 2020, to 7 September 2021. During this time the most prevalent SARS-CoV-2 variant in British Columbia transitioned between the Alpha, Beta, Gamma and Delta genotypes. The Beta variant was detected in late 2020 and January 2021, increase prevalence of the Alpha variant shortly followed and it remained dominant until June 2020. The Gamma variant was first detected in late February 2021, it's incidence surpassed Alpha in July 2021. Public health surveillance first recorded the Delta variant in March, and it was responsible for most sequenced cases over the summer from July to September 2021 [23]. Vaccinated CARE Study participants received any of the three SARS-CoV-2 glycoprotein-based vaccines approved by Health Canada during the study period: COMIRNATY (BioNTech (Mainz, Germany), Pfizer (New York, NY, USA)), Spikevax (Moderna, Cambridge, MA, USA) and Vaxzevria (Oxford, UK, Astra Zeneca) vaccine. All data analysis was performed in R version 4.0.4 using the packages: 'DataExplorer', 'survival', 'survminer', 'dplyr', 'ggfortify', 'tableone', 'naniar', 'RColorBrewer', 'lme4', 'mgcv', 'gam.check' and 'readr' [24].

2.2. Recruitment Criteria

Adults 18 years of age and older from the greater Vancouver metropolitan area were recruited if they had a confirmed real time PCR (qPCR)-positive SARS-CoV-2 infection and if they were no longer required to self-isolate per the BC provincial public health guidelines (i.e., tested positive for SARS-CoV-2 at least 14 days prior). Initial diagnostic qPCR testing was done in accordance with standard laboratory practices in British Columbia during the time of the study. Either nasopharyngeal swab or saline gargles were acceptable sample types. Diagnostic laboratories in the Lower Mainland area, where participants were tested, used a variety of both commercial (GeneXpert Cepheid, Sunnyvale, CA, USA; Cobas® Roche Diagnostics, Indiannapolis, IN, USA; Panther Fusion Hologic, Marlborough, MA, USA) and laboratory developed assay (E gene and RdRP gene) [25]. Only positive qPCR results were allowed as recruitment criterion (i.e., potential participants with indeterminate or invalid test results were excluded). The study protocol was approved by the University of British Columbia (UBC) Clinical Research Ethics Board (H20-01089).

2.3. Sample Collection and Processing

Participants were required to donate 10 mL blood samples collected by venipuncture (for serological testing) and concurrent self-collected saline gargle samples (for SARS-CoV-2 qPCR testing by a laboratory-developed assay targeting the E gene and RdRP genes) [25], every two weeks for 3 months post-recruitment (up to 7 collections total). Blood was drawn in gold-top serum separator tube with polymer gel (BD, cat# 367989); after at least 30 min of clotting at room temperature, the blood sample was then centrifuged at 1400 g by staff at the collection site and sent to the British Columbia Centre for Disease Control Public Health Laboratory (BCCDC PHL). At the BCCDC PHL the samples were divided into serum aliquots that were frozen at $-80°C$ within four hours of receipt. Blood collections occurred at four sites in the Greater Vancouver Area, British Columbia Canada: BC Children's Hospital, St. Paul's Hospital, Abbotsford Regional Hospital and Surrey Memorial Hospital. Saline gargle samples were self-collected by the participants at home, in accordance with well-validated instructions [26], on the day of blood collection and transported to the BCCDC PHL by the blood collection site. Self-collected saline gargle samples were tested for SARS-CoV-2 by qPCR.

SARS-CoV-2 whole genome sequencing was done on all available participants' SARS-CoV-2 positive diagnostic clinical specimens; detailed methods have been described elsewhere [27]. Samples were sequenced on an Illumina NextSeq instrument (San Diego, CA, USA) using a tiled 1200bp amplicon scheme and analyzed using a modified ARC-

TIC Nextflow pipeline (https://github.com/BCCDC-PHL/ncov2019-artic-nf, accessed on 15 September 2022). Called variants were kept if the variant allele frequency was above 0.25 with $\geq$10X coverage. Sequences passing QC (85% genome completeness, 10X depth of coverage and no quality flags) were included in the phylogenetic analysis. A phylogenetic tree was constructed using Fasttree [28] and visualized in Nextstrain [29] and lineage assignment was performed using the Phylogenetic Assignment of Named Global Outbreak Lineages tool (Pango/Usher Version 1.15.1) [30]. All molecular, genomic, and serological testing for participant specimens (described below) was conducted centrally at the BCCDC PHL, a College of American Pathologists accredited laboratory.

2.4. Measurements of Humoral Immunity

Serum samples were initially tested using a combination of three Health Canada approved chemiluminescent immunoassays: (1) total antibodies to SARS-CoV-2 RBD (Siemens SARS-CoV-2 Total Assay [COV2T], Munich, Germany), (2) total antibodies to SARS-CoV-2 S (Ortho VITROS^{TM} Anti-SARS-CoV-2 Total, Raritan, NJ, USA) and IgG anti-N antibodies (Abbott ARCHITECT^{TM} SARS-CoV-2 IgG, Abbott Park, IL, USA), as per manufacturer guidelines, with results interpreted as reactive or non-reactive using the manufacturer-recommended signal to cut-off ratios [31,32]. All available samples were then tested using the V-PLEX COVID-19 Coronavirus Panel 2 (IgG) (Mesoscale Diagnostics LLC (MSD): #K15369U, Rockville, MD, USA), the diagnostic accuracy of the MSD assay was previously validated through comparison with alternative Health Canada approved tests at the BCCDC PHL [33]. The MSD assay provides quantitative measures of IgG antibodies against RBD, S and N SARS-CoV-2 epitopes, as well as IgG antibodies against the glycoprotein (S) of the four seasonal endemic HCoVs. Serological specimens were processed as previously reported [33]. Quantitative antibody levels expressed as log10 antibody units (AU)/mL were recorded and evaluated for all tested samples. MSD results were interpreted as reactive or non-reactive using the MSD recommended signal thresholds for serum: SARS-CoV-2 anti-S IgG = 1960; anti-N IgG = 5000; anti-RBD IgG = 538. Cutoffs (derived at the BCCDC PHL) for seasonal HCoVs seropositive status are as follows: HCoV-229E anti-S IgG = 1700; HCoV-HKU1 anti-S IgG = 900; HCoV-NL63 anti-S IgG = 270; HCoV-OC43 anti-S IgG = 2000 [34]. Samples were stratified by collection time to <6 months and $\geq$6 months post infection and percent positivity was compared using a Chi-square test (χ^2 test).

2.5. Power Analysis for Investigating Association between IgG Concentration and Vaccination

A power calculation was conducted to determine the minimum number of paired participant samples needed to estimate at least a 70% association between COVID-19 vaccination and HCoV anti-IgG antibody concentration. Antibody concentrations were assumed to be normally distributed with a standard deviation of one. A significance level of 5% and two-sided alternative were used [35].

2.6. Analytic Data Selection

To analyze antibody dynamics, an analytic dataset was selected from the CARE COVID-19 cohort. At least k = 18 paired participant samples are required to estimate a 70% or greater association between COVID-19 vaccination and anti-HCoV IgG antibody concentration. Exclusion criteria were applied to select an analytic dataset from k = 57 participants with n = 341 observations. One participant had no follow up samples and was omitted from the analytic dataset (k = 1, n = 1). Six participants were excluded because they were vaccinated before collection of their baseline sample (k = 6, n = 37). Eight participants were removed from the analytic dataset because of missing data in their survey results (k = 8, n = 33). After applying the exclusion criteria, the analytic dataset contained k = 42 participants with n = 270 observations (Supplementary Figure S1). There were k = 41 participants with >1 pre-vaccine sample (n = 210 pre-vaccine observations) used for analysis of antibody waning pre-vaccination.

2.7. Bivariate Data Analysis

2.7.1. Antibody Waning

Waning of SARS-CoV-2 antibodies prior to vaccination was investigated in independent participant specimens measured at baseline using linear regression. HCoV anti-IgG antibody signals at baseline were compared to signals 14–56 days post-vaccination using a paired Wilcoxon signed-rank test in a sample of n = 21 participants who received a COVID-19 vaccine during the study. Waning of SARS-COV-2 specific IgG was measured prior to vaccination between participants using linear regression. Participant's baseline samples (defined as the first specimen taken after enrolling in the study) were plotted for anti-S and anti-N IgG over time.

2.7.2. Descriptive Statistics

Bivariate analysis was conducted between the exposure (a single dose of a Health Canada approved COVID-19 vaccine) and outcome (SARS-CoV-2 anti-S or anti-N IgG) of interest at baseline. Baseline represents the time of a participant's first blood draw after enrollment. The bivariate relationship between vaccine status and covariates was examined by t-test or Chi-square test (χ^2 test) depending on variable type. HCoV anti-IgG antibody signals were transformed to the logarithmic base ten scale for conformation to normality and ease of interpretation.

2.8. Primary Analysis

Primary analysis used a multivariable linear mixed-effects model to regress SARS-CoV-2 anti-IgG concentration on vaccine status adjusting for dependency within participant samples and covariates defined as potential confounders by the common cause criterion [36,37]. Separate models were fit for anti-S and anti-N IgG signals. Unconditional mean models were used to find the intraclass correlation coefficient (ICC) before covariates were added to build fixed effect models [38]. Effect modification terms were assessed by the Akaike information criterion and included in the fixed effect models to understand if time from infection influences SARS-CoV-2 antibody concentrations [39].

2.9. Secondary Analysis

Secondary analysis employed a Kaplan–Meier curve to estimate the cumulative incidence of seroreactivity stratified by vaccine status. The survival function was transformed to cumulative incidence by 1-S(τ) [40]. Seroreactivity was defined from the distribution of SARS-CoV-2 anti-S IgG concentration at first blood draw (baseline); the 95th percentile was chosen as the threshold (5.5 log10 AU/mL). Participants were censored if they were not seroreactive before loss to follow-up (right censoring). A log-rank test was used to test the hypothesis that the cumulative incidence of seroreactivity between unvaccinated and vaccinated persons, who have been previously naturally infected with SARS-CoV-2, does not differ [41].

3. Results

3.1. CARE COVID-19 Cohort

Fifty-seven individuals recovered from COVID-19 infection were recruited into the CARE COVID-19 Cohort. Recruited subjects (17 male, 40 female; 18 to 76 years old) represented a range of COVID-19 disease severity. Most subjects had a mild case of COVID-19, defined as not requiring hospitalization; 6 reported being asymptomatic and 12 reported experiencing fever. Only four of the recruited subjects (7%) reported being hospitalized for COVID-19; one required intensive care. The observed case severity distribution was consistent with the general distribution of COVID-19 disease severity in BC (~5% of diagnosed cases hospitalized as of April 2022) [42].

Participants were required to have recovered from COVID-19 (i.e., 14 days post-qPCR diagnosis) before providing their first blood and saline gargle sample. Collection dates ranged from 18–490 days (median 152 days) since a positive qPCR test (used as proxy for time since infection), with the baseline collection date ranging from 18–339 days (median 114 days). Participants submitted between 1 and 7 samples, with approximately 2 weeks (median 14 days; range 7–83 days) between each collection, with an average of 6 samples collected per participant and a total of 341 samples collected. No reinfections or persistent virus shedding were detected in self-collected saline gargle samples using qPCR (data not shown).

Virus whole genome sequencing was performed on all available primary diagnostic specimens obtained from recruited participants [27] to determine the SARS-CoV-2 variant responsible for infection. Forty-one sequences were obtained (28 from the analytic dataset) (Supplementary Tables S2 and S3). SARS-CoV-2 variants were classified by pangolin lineage (Supplementary Tables S2 and S3) and visualized as a phylogenetic tree (Supplementary Figure S2). Viral genomes detected in the study sample are representative of variants circulating at the time of respective participants' diagnoses [29]. Whole genome sequencing data was missing for ~33% of participants in the analytic dataset and, therefore, was not included as a covariate in the analysis. Multiple studies corroborate no significant difference in neutralising antibodies between the alpha variant and the ancestral isolate post mRNA vaccination with BNT162b2 or mRNA-1273. Noteworthy reduction of post-vaccination neutralising sera was observed for the beta variant in persons vaccinated with mRNA-1273 [43].

3.2. Comparison of Anti-SARS-CoV IgG Antibody Responses across Four Commercial Assays

All available samples (n = 340; 1 missing) were initially tested using a combination of three commercial serology assays supplied by Siemens (COV2T), Abbott (ARCHITECT™), or Ortho (VITROS™) clinical diagnostics. Of the n = 340 samples tested, n = 338 were classified as reactive using at least one assay (Supplementary Table S1). All available samples (n = 339) were subsequently tested using a highly sensitive and multiplex electrochemiluminescent assay offered by Meso Scale Diagnostics (MSD). Percent positivity differed across the platforms and by antigenic target. Overall detection of anti-S was more sensitive than anti-N SARS-CoV-2 IgG. Comparing anti-S results, the Ortho assay had the highest positivity rate (100%) followed by Siemens (95%) and MSD (89%) (Supplementary Table S1). For anti-N results MSD (58%) outperformed Abbott (47%) with a 11% increase in positivity (χ^2 test, $p = 0.01$). When samples were stratified by collection time to less than or greater than 6 months post-infection, the anti-N positivity rate decreased for both the Abbott (72% to 13%) and MSD, (76% to 33%) ($p < 0.001$). A 7% decline in positivity was observed for anti-S ($p = 0.06$) and 2% for anti-RBD ($p = 0.53$) when tested by MSD (Supplementary Table S1). Only antibody measurements from the MSD assay were used in the multivariable analysis as the anti-S IgG results compared well with Ortho and anti-N IgG results were superior to Abbott. Waning of anti-S and anti-N IgG concentrations over time were measured between participants using the first baseline observation for each of the k = 42 participants in the analytic dataset. Using linear regression analysis, overall waning was observed in both anti-N and anti-S and the slope did not differ significantly across the two measures ($p = 0.46$; Figure 1). On average SARS-CoV-2 antibodies wane at a rate of -0.0029 log10 AU/mL per day ($p < 0.001$) or ~4228 AU/mL per month. These results confirm waning of anti-SARS-CoV-2 antibodies over time in people who have recovered from natural SARS-CoV-2 infection before vaccination. Estimates of anti-SARS-CoV-2 IgG waning are calculated post-vaccination using a mixed-effects linear regression model and reported as the 'primary analysis'.

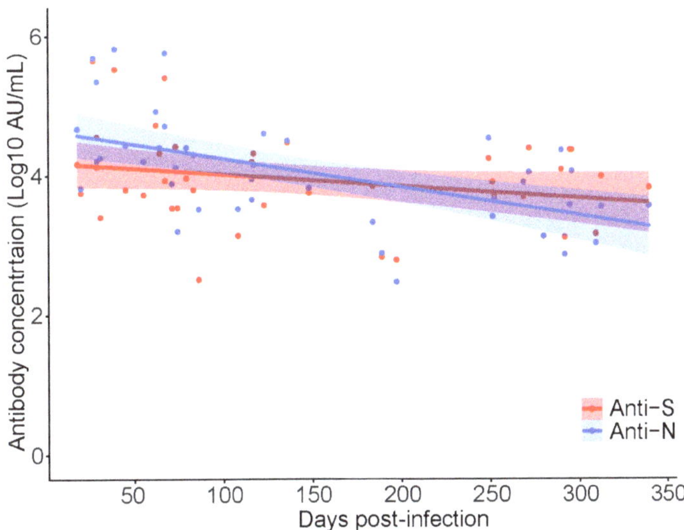

Figure 1. Longitudinal decay of SARS-CoV-2 anti-N and anti-S IgG concentration over time in natural SARS-CoV-2 infected CARE participants prior to vaccination (k = 42, n = 42 samples). Participant samples were restricted to the first collection date (baseline) and plotted independently. Linear regression was used to estimate the decrease in anti-S and anti-N titer over time since qPCR test result.

3.3. Serological Response to SARS-CoV-2 Vaccination

Bivariate analysis was conducted on the analytic dataset to compare participant antibody responses pre- and post-vaccination for COVID-19. Participant's serology results and survey responses are summarized and stratified at baseline by the exposure of interest, one dose of a COVID-19 vaccine (Table 1). No difference in the distribution of covariates between participants who received and did not receive a COVID-19 vaccine over the study period was observed for all variables except the number of participant visits. Though follow-up time did not significantly differ between the two groups, on average unvaccinated participants were observed 0.95 (approximately one) fewer times than those who received a COVID-19 vaccine ($p = 0.014$) (Table 1). Importantly, age, biological sex, days from positive qPCR test (diagnosis), symptom duration and endemic anti-coronavirus IgG signals did not differ by exposure at baseline; therefore, we expect limited confounding from these covariates when estimating the association between COVID-19 vaccination and anti-SARS-CoV-2 IgG signals. Covariates, which met the definition of a confounder by the common cause criteria, were adjusted for in the primary analysis using a linear mixed effects model.

In k = 21 paired participants, SARS-CoV-2 anti-S and anti-RBD IgG antibody concentrations increased post vaccination by 1.63 ($p \leq 0.001$) and 1.82 ($p \leq 0.001$) log10 AU/mL (Figure 2A,B). Anti-N antibody concentration continued to decrease post vaccination by -0.3 ($p = 0.03$) log10 AU/mL (Figure 2C), consistent with waning observed prior to vaccination. Most participants (>99%) were found to be seropositive for anti-S antibodies against the endemic HCoVs. Post vaccination, anti-S antibody concentrations for endemic human β-coronaviruses HCoV-HKU1 and HCoV-OC43 increased ($p = 0.02$ and $p = 0.02$) (Figure 3B,D). No increase in antibody concentration was observed for the endemic human α-coronaviruses HCoV-229E and HCoV-NL63 ($p = 0.15$ and $p = 0.25$) (Figure 3A,C).

Table 1. Descriptive statistics of study participants at the beginning of the study (baseline) in the analytic dataset with complete data (k = 42) [$].

Variable Name	Level	Total (n)	No	Yes	*p*-Value *
			Vaccinated During Study		
	-	42	21	21	–
Biological Sex (n [%])					
	Male	13	5 (23.8)	8 (38.1)	0.504
	Female	29	16 (76.2)	13 (61.9)	0.504
Age (mean [SD])		42	41.48 (11.66)	46.33 (11.91)	0.189
Days Since Positive qPCR Test (mean [SD])	-	42	127.62 (88.45)	165.33 (115.90)	0.243
Pre-Vaccine Sample (n [%])					
	True	42	21 (100)	21 (100)	–
Duration of COVID-19 Symptoms (n [%])					
	≤2 Weeks	26	13 (61.9)	13 (61.9)	1.00
	>2 Weeks	16	8 (38.1)	8 (38.1)	–
SARS-CoV-2 anti-Spike-IgG Log10 AU/mL	-	-	4.00 (0.82)	3.88 (0.50)	0.583
SARS-CoV-2 anti-RBD-IgG Log10 AU/mL	-	-	3.63 (0.81)	3.61 (0.49)	0.891
SARS-CoV-2 anti-Nucleocapsid-IgG Log10 AU/mL	-	-	4.09 (0.82)	4.02 (0.72)	0.755
229E-CoV anti-Spike-IgG Log10 AU/mL	-	-	4.33 (0.42)	4.32 (0.53)	0.971
HKU1-CoV anti-Spike-IgG Log10 AU/mL	-	-	4.14 (0.47)	4.19 (0.52)	0.779
NL63-CoV-2 anti-Spike-IgG Log10 AU/mL	-	-	3.60 (0.46)	3.62 (0.41)	0.891
OC43-CoV-2 anti-Spike-IgG Log10 AU/mL	-	-	4.75 (0.53)	4.68 (0.54)	0.642
Follow Up Time (median [SD])	-	-	85 (25.87)	84 (9.20)	0.435
Number of Follow Up Visits Per-Participant (mean [SD])	-	-	5.95 (1.60)	6.90 (0.3)	0.014

[$] Participants are stratified by vaccine status (primary exposure) throughout the study period, k = 21 participants were vaccinated while under observation. Bivariate associations at baseline were examined by testing for a difference in the distribution of covariates between participants who did or did not receive one dose of a COVID-19 vaccine over the study period. * *p*-values are reported for parametric tests used for continuous (*t*-test) and categorical variables (χ^2 test).

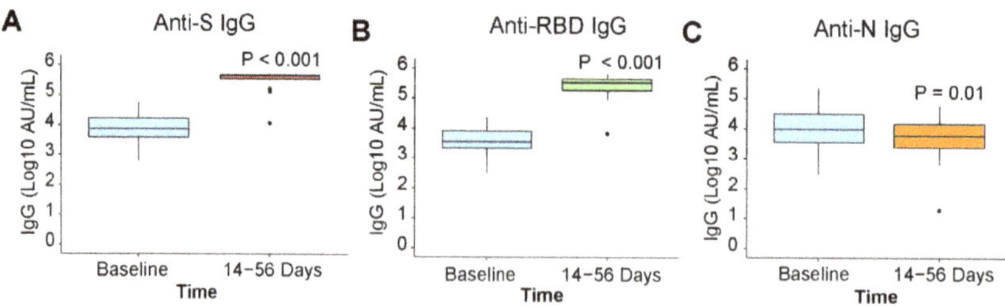

Figure 2. SARS-CoV-2 anti-IgG pre- and post-vaccination. Antibody signals in k = 21 paired participants, who received a COVID-19 vaccine during the study, at baseline and 14 to 56 days post vaccination, presented by individual SARS-CoV-2 antigen (k = 21): (**A**) anti-S, (**B**) anti-RBD, (**C**) anti-N. Differences in antibody signals were examined with a paired Wilcoxon signed rank test.

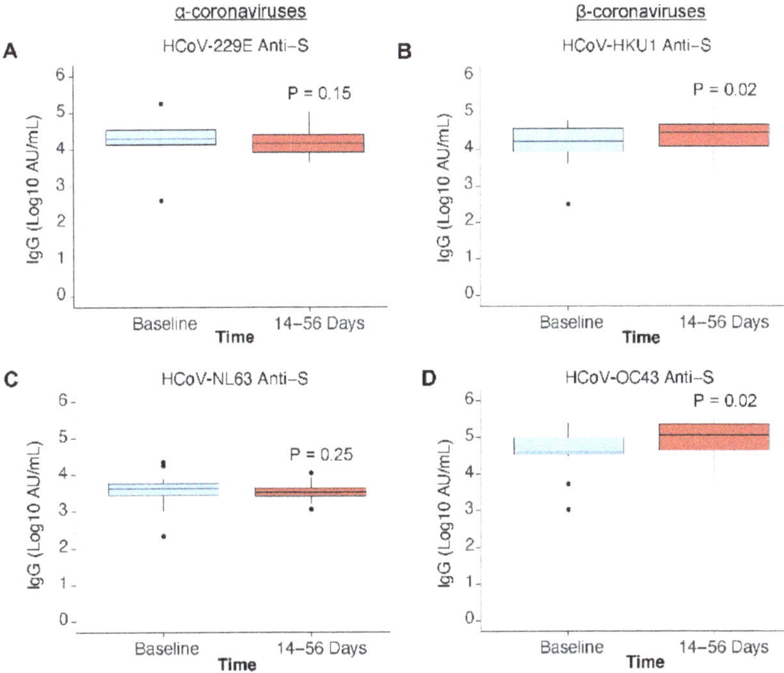

Figure 3. Endemic human coronavirus anti-S IgG antibody signals pre- and post-vaccination. HCoV antibody signals in n = 21 paired participants, who received a COVID-19 vaccine during the study, measured at baseline (be-fore vaccination) and 14 to 56 days post-vaccination, presented by HCoV species: (**A**) HCoV-229E anti-Spike (S), (**B**) HCoV-HKU1 anti-S, (**C**) HCoV-NL63 anti-S, (**D**) HCoV-OC43 anti-S. Difference in antibody signal was examined with a paired Wilcoxon signed rank test.

3.4. Primary Analysis

Linear mixed-effects regression models were used to estimate intraclass correlation within participant samples and the relationship between COVID-19 vaccination and SARS-CoV-2 anti-S or anti-N IgG antibody concentration. An unconditional mean model was fit to partition within participant variation from between participant variation (Table 2). The minority of variation in SARS-CoV-2 anti-S IgG concentration was attributable to differences between participants (ICC = 0.43) (Table 2). On average, anti-S IgG concentration increased over time in participants who received one dose of a COVID-19 vaccine during the study by 2.06 log10 AU/mL (95%CI: 1.45–3.46) adjusting for age, biological sex, days from positive qPCR test (time) and effect modification between COVID-19 vaccination and time (Table 2). In the adjusted model, the ICC increased to 0.89 indicating that between participant differences (e.g., COVID-19 vaccination) explains most of the variation in SASRS-CoV-2 anti-S IgG antibody concentration. COVID-19 vaccination has a positive association with SARS-CoV-2 anti-S IgG antibody concentration, which increases over time. Variation in anti-N IgG concentration was due to differences between participants in the unconditional mean model (ICC = 0.88). The average, anti-N IgG concentration decreased in vaccinated participants over time.

(−0.243 log10 AU/mL, 95%CI: −1.2–[0.12]) adjusting for age, biological sex, days from positive qPCR test (time) and effect modification between COVID-19 vaccination and time (Table 2). Variation in anti-N IgG concentration after fitting the adjusted model was explained by within participant variance (ICC = 30). Overall, these results indicate that waning of SARS-CoV-2 anti-N IgG is unaffected by COVID-19 vaccination. Anti- S IgG titers increase post vaccination; therefore, vaccination of recovered individuals benefits the durability of their humoral immune response.

Table 2. Summary of linear mixed effects models fit to examine the relationship between anti-S IgG log10 AU/mL (light grey) or anti-N IgG log10 AU/mL (dark grey) and COVID-19 vaccination status adjusting for: biological sex, age, and time from qPCR diagnosis.

Unconditional Mean Model (S)	Intraclass Correlation Coefficient		
Participant ID (n = 42)	0.434		
Residual	0.566		
Random Intercept Model	Variable	Fixed Effect Estimate	95%CI
Anti-Spike IgG	Intercept	4.84	3.27–6.39
	Vaccine-Yes	0.40	−0.41–1.20
	Biological Sex-Male	0.93	0.068–1.79
	Age (Years)	−0.029	−0.063–0.0057
	Time from +ve qPCR Test *	−0.20	−0.47–0.054
	Vaccine: Time *	1.86	1.39–2.21
Random Effects	Intraclass Correlation Coefficient		
	0.893		
Unconditional Mean Model (N)			
Participant ID (n = 42)	0.875		
Residual	0.125		
Random Intercept Model	Variable	Fixed Effect Estimate	95%CI
Anti-Nucleocapsid IgG	Intercept	3.14	2.48–3.79
	Vaccine-Yes	−0.080	−0.42–0.26
	Biological Sex-Male	0.27	−0.095–0.63
	Age (Years)	0.016	0.0017–0.03
	Time from +ve qPCR Test *	−0.40	−0.53–(−0.27)
	Vaccine: Time *	−0.077	-0.25–0.11
Random Effects	Intraclass Correlation Coefficient		
	0.30		

* An effect modification term was incorporated to explore how the effect of vaccination on antibody concentration differs by time since diagnosis with a qPCR test. Unconditional means models were fit to partition the variance by participant without inclusion of other exposure variables. Fixed effect models were built by applying the common cause criterion to select covariates which are a cause of the exposure, outcome, or both.

3.5. Secondary Analysis

Secondary analysis used the Kaplan–Meier method to estimate the cumulative incidence of seroreactivity above a defined threshold in vaccinated and unvaccinated participants over time. Seroreactive status was classified by the threshold of ≥5.5 log10 AU/mL SARS-CoV-2 anti-S IgG, as described in Methods. Participants with antibody measurements equal to or greater than the threshold were considered reactive. Over the 105 days follow up from baseline (first antibody measurement available for participants post-infection), 88% (95%CI: 42–98%) of vaccinated participants (n = 16) were seroreactive compared to 5% (95%CI: 0–14%) of unvaccinated participants (n = 1) (p = 0.03) (Figure 4). A single dose of COVID-19 vaccine increases the probability of a SARS-CoV-2 anti-S IgG antibody concentration ≥5.5 log10 AU/mL by 83%.

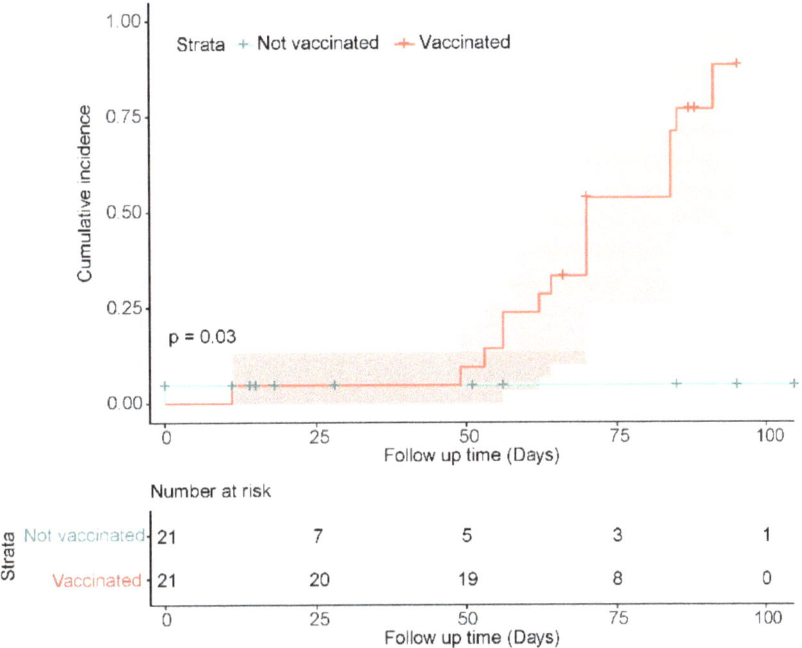

Figure 4. Cumulative incidence of seroreactivity ($\geq$5.5 SARS-CoV-2 anti-S IgG Log10 AU/mL) days from participant's first blood draw at baseline, stratified by vaccination status over the study period. Vaccinated participants achieved antibody titers not possible from natural infection alone (unvaccinated participants). Within 105 days of follow up, 88% (95%CI: 42–98%) of vaccinated participants were seropositive, an increase of 83% in comparison to the unvaccinated group ($p = 0.03$). In previously naturally infected individuals, COVID-19 vaccination increases SARS-CoV-2 anti-S IgG concentration over time to levels which are not attained by natural infection alone. No re-infections were detected by qPCR in the vaccinated or unvaccinated group during the study period, specimens were self-collected.

4. Discussion

4.1. Summary of Results

A prospective cohort study was carried out in Greater Vancouver, British Columbia to observe anti-SARS-CoV-2 and anti-endemic HCoV antibody dynamics in participants who were infected with SARS-CoV-2, a subset received the first dose of a Health Canada approved SARS-CoV-2 vaccine during the follow up. Several commercial serology assays were used to detect anti-coronavirus antibodies; detection of anti-SARS-CoV-2 antibodies was confirmed in all available samples, although both anti-S and anti-N antibodies decline over time post-infection. Bivariate analysis found that vaccination significantly increased the titer of SARS-CoV-2 anti-S IgG antibodies 14–56 days post vaccination. A positive association was found between SARS-CoV-2 vaccination and endemic human β-coronavirus (HCoV-OC43 and HCoV-HKU1) anti-S IgG antibodies. It cannot be ruled out that infection with an endemic HCoV is an alternative/competing cause of the observed increase; however, the incidence of influenza-like (syndromic) respiratory infections was low during the study period [44]. Vaccination was not observed to boost SARS-CoV-2 anti-N IgG titers, which waned overtime in both vaccinated and unvaccinated individuals. The rate of anti-N waning was approximately double that of anti-S. Secondary analysis used a Kaplan–Meier model to estimate the cumulative incidence of anti-S antibody titers equal to or above 5.5 log10 AU/mL ('seroreactivity threshold') in those vaccinated and unvaccinated. In the vaccinated group, 88% (95%CI: 42–98%) of participants had SARS-CoV-2 anti-S IgG titers greater than or equal to this threshold, while this level was achieved in only one unvaccinated participant measured twenty-seven days post infection. Asymptomatic and subclinical SARS-CoV-2 infections have been observed to prime the adaptive immune response and may explain observed increases in unvaccinated participants. In people with asymptomatic re-infections, anti-S IgG titers wane more slowly over time and anti-N may increase [45].

In unvaccinated participants, a few more substantial increases in anti-S titers were observed despite overall antibody waning. No reinfections were confirmed using qPCR in self-collected saline gargle samples throughout the study; however, one participant had a large (>8-fold) average increase in mean (anti-S, RBD and N) antibody levels seven months following initial SARS-CoV-2 diagnosis, which may be explained by a second exposure to SARS-CoV-2. A second participant had 6-fold increase of anti-S and anti-RBD levels, but not anti-N IgG levels, suggesting they may also have been re-exposed. Other detected increases in antibody levels were of much smaller magnitude and might be secondary to rising titers early in convalescence or be explained by technical variations rather than a biological mechanism.

4.2. Comparison with Literature

Previous studies have measured changes in SARS-CoV-2 antibody titers over time. Repeated exposure to SARS-CoV-2 antigens increases IgG titer, while antibodies generated from a single exposure wane overtime [46,47]. Following infection, SARS-CoV-2 specific antibody waning has been observed to decrease from the 8th to 9th week post symptoms onset, with detectable levels observed up to the end of the 12th week [48]. In those with multiple SARS-CoV-2 exposures or a hybrid immune response from infection and vaccination, decrease of antibody titers stops shortly after the secondary antigen exposure when stimulation of the memory B cell response produces additional antibodies [49]. A strong correlation between total lymphocyte count and SARS-CoV-2 anti-S IgG provides evidence that an ongoing/active immune response provides better protection than a dormant one [48]. Waning of SARS-CoV-2 antibodies differs by the SARS-CoV-2 epitope they target, anti-N IgG antibodies wane faster than anti-S. The difference in reactivity between anti-N to anti-S IgG was observed at the population level, anti-N seroprevalence underestimated the number of confirmed infections by 9–31% [50]. Vaccination post SARS-CoV-2 infection prevents waning of anti-S but has no effect on anti-N IgG [51]. Hybrid immunity also benefits the breadth of the antibody mediated response, increasing the probability that existing antibodies are effective against the novel variants. Persons who were infected prior to receiving one of two doses of a COVID-19 vaccine had more somatic mutations and antibody production from the IGHV2-5; IGHJ4-1 germline which was not active in the vaccinated but uninfected [52,53]. Additionally, hybrid immunity produces greater total and neutralizing anti-S titers than natural infection or vaccination alone [49]. Our study both supports and builds upon prior findings, as we show that SARS-CoV-2 IgG antibodies wane in SARS-CoV-2 infected people over time, with the rate of decline being greater for anti-N IgG than anti-S; vaccination post-infection boosts anti-S IgG titers and participants with hybrid immunity possess anti-S antibody levels which are not common in those infected but unvaccinated. We propose 5.5 log10 AU/mL anti-S IgG as a putative correlate of protection in persons convalescent for SARS-CoV-2 infection who have received a single dose of a Health Canada approved vaccine. Our calculated rates of antibody decline may be used to help estimate infection timing in seroprevalence studies.

4.3. Clinical and Epidemiological Interpretation

Our findings have important implications for clinical practice and public health guidelines as the pandemic progresses into its third year, novel viral variants continue to emerge, and vaccine doses are more widely distributed globally. Humoral immunity from natural infection wanes and vaccination with at least one dose of COVID-19 vaccine increases SARS-CoV-2 anti-S IgG titers immediately and over time. Therefore, we recommend that naturally infected individuals receive COVID-19 vaccination to increase protection from re-infection and severe disease and the duration of their humoral immune response against SARS-CoV-2. We demonstrate that a single dose of SARS-CoV-2 vaccine is effective in boosting anti-S antibody titers to high levels, which has implications in distribution of vaccine supplies in those countries with scarce access and low vaccination levels in the setting of high numbers of natural infection.

4.4. Strengths and Limitations

The strength of the described study stems from the prospective design, use of multiple serological tests, including the quantitative MSD option, and thorough analysis. A prospective cohort design offers several benefits, which allowed us to observe SARS-CoV-2 antibody dynamics over time with minimal bias. Recruiting participants post-infection but prior to vaccination delineated the sequence of temporal events, limiting the probability that any changes in antibody titers observed post-vaccination were due to causes other than the vaccine. Selection bias was minimized as the participants exposure (vaccination status) and outcome (IgG titer) were not known when they were recruited into the study. At the beginning of the study, the measured covariates were exchangeable between participants who were unvaccinated or vaccinated during follow-up. Balance of the covariates allowed for estimation of the relationship between vaccination and anti-SARS-CoV-2 IgG antibody titer with minimal bias from confounding. Utilizing multiple serological tests increased the rigor of our observations, limited instrument bias and allows for our findings to be generalized between different types of serological testing methods. Statistical power was optimized by analysis with a mixed effects linear regression model, which accommodated multiple repeated measures per participant.

Limitations of the work include differential loss to follow up in the vaccinated and unvaccinated groups, a small sample size, and incomplete/missing survey responses and incomplete whole genome sequencing data. Unvaccinated participants were observed to have approximately one fewer visit than those who received a COVID-19 vaccine. Vaccines were not an originally planned intervention in the study and were made available in British Columbia on a stage roll-out basis about half-way through the study period. The difference in visit numbers between the vaccinated and unvaccinated groups may be related to surveillance bias- those who receive a medical intervention are more open to clinical follow up than those who do not. Obtaining a larger sample size initially planned for the study was difficult due to low enrollment uptake, likely related to the social and economic stress of the pandemic on the public and geographic limitations on recruitment related to the availability of sample collection sites.

Despite the small sample size, we detected a significant increase in human β-coronavirus (HCoV-HKU1 and HCoV-OC43) anti-S titers following SARS-CoV-2 vaccination, while no difference was observed for the α-coronaviruses (HCoV-NL63 and HCoV-229E). Antigenic cross-reactivity between human β-coronaviruses may allow for 'back-boosting' a phenomenon which has been well described for Influenza A viruses [54]. Antibodies are 'back-boosted' when a secondary exposure to a novel viral strain generates new antibodies and increases the titer of antibodies against a previously encountered strain. As a result, SARS-CoV-2 vaccinations may provide non-specific protection in children who have a high incidence of endemic coronavirus infections. The study design may have underestimated any association between existing endemic coronavirus IgG titers and COVID-19 vaccination as the sample was restricted to persons previously infected with SARS-CoV-2. COVID-19 has been shown to affect endemic coronavirus antibody levels and as such, the effect of vaccination should be observed in a cohort of SARS-CoV-2 naive persons prior to vaccination [55–57]. The overall effect of SARS-CoV-2 vaccination and/or infection on the circulating antibodies against endemic HCoVs in the population may have implications for their seasonal epidemiology.

5. Conclusions

In summary, we report that single dose vaccination in a British Columbia-based cohort after natural infection significantly increases SARS-CoV-2 anti-S IgG titer by 1.63 $\log_{10}$ units and that vaccination increases the durability of high anti-S titers over time. Vaccination post-natural infection had no significant association with SARS-CoV-2 anti-N IgG titer; a significant trend towards higher anti-S IgG against the endemic human β-coronaviruses (HCoV-HKU1 and HCoV-OC43) was observed. Our results provide support that vaccination is beneficial for achieving higher and more persistent SARS-CoV-2 anti-S IgG titers. We also report an estimated rate of decay of anti-N antibodies, which may be useful for measuring ongoing population seroprevalence estimates. Future studies should examine the impact of infection following vaccination on antibody dynamics, as vaccine breakthrough infections with omicron or other variants of concern continue to occur.

Supplementary Materials: The following supporting information can be downloaded at: https://www.mdpi.com/article/10.3390/v14112416/s1, Table S1: Summary of serological test results; Table S2: Descriptive summary of variables not included in the analysis for n= 21 vaccinated participants; Table S3: Descriptive summary of variables not included in the analysis for n= 21 unvaccinated participants.; Figure S1: Exclusion criteria were applied to select an analytic data of n = 270 observations from k = 42 dependent participants (clusters); Figure S2: Phylogenetic analysis of CARE Study Participants' SARS-CoV-2 genomes.

Author Contributions: Conceptualization, A.D.O., A.N.J., M.M. and I.S.; methodology, A.M.N., A.C.M., E.F., M.S., A.H. and S.M.; software, A.D.O. and A.M.N.; formal analysis, A.D.O. and A.M.N.; investigation, A.D.O. and I.S.; resources, A.D.O., A.N.J., M.M., I.S., A.C.M., E.F., M.S., A.H., D.M.G. and S.M.; data curation, S.S., A.C.M. and T.V.; writing—original draft preparation, A.D.O. and A.M.N.; writing—review and editing, All Authors; visualization, A.D.O. and A.M.N.; supervision, A.N.J. and I.S.; project administration, A.C.M., E.F. and A.H.; genomic data provision and analysis S.L.R. and N.P.; funding acquisition, A.N.J., M.M. and I.S. All authors have read and agreed to the published version of the manuscript.

Funding: This study was funded by Genome BC's COVID-19 Rapid Response Funding Program (#COV-050).

Institutional Review Board Statement: The study was conducted in accordance with the Declaration of Helsinki, and approved by the Institutional Review Board (or Ethics Committee) of The University of British Columbia (H20-01089 on 4/21/2020).

Informed Consent Statement: Informed consent was obtained from all subjects involved in the study.

Data Availability Statement: The datasets generated during and/or analysed during the current study are available from the corresponding author on reasonable request.

Acknowledgments: We thank the dedicated staff at the BCCDC PHL for processing, testing, and sequencing clinical specimens. We also thank all the study participants who volunteered their time and specimens for this research. We also are thankful to all the staff who helped to collect and process samples for this study at Abbotsford Regional Hospital, Surrey Memorial Hospital and St. Paul's Hospital. Notably, we thank the staff at BC Children's Hospital for providing extra guidance and support to the researchers in conducting the study.

Conflicts of Interest: The authors declare no conflict of interest. The funders had no role in the design of the study; in the collection, analyses, or interpretation of data; in the writing of the manuscript; or in the decision to publish the results.

References

1. Fung, T.S.; Liu, D.X. Similarities and Dissimilarities of COVID-19 and Other Coronavirus Diseases. *Annu. Rev. Microbiol.* **2021**, *75*, 19–47. [CrossRef] [PubMed]
2. Fehr, A.R.; Perlman, S. Coronaviruses: An Overview of Their Replication and Pathogenesis. *Coronaviruses Methods Protoc.* **2015**, *1282*, 1–23. [CrossRef]
3. Huang, A.T.; Garcia-Carreras, B.; Hitchings, M.D.T.; Yang, B.; Katzelnick, L.C.; Rattigan, S.M.; Borgert, B.A.; Moreno, C.A.; Solomon, B.D.; Trimmer-Smith, L.; et al. A Systematic Review of Antibody Mediated Immunity to Coronaviruses: Kinetics, Correlates of Protection, and Association with Severity. *Nat. Commun.* **2020**, *11*, 4704. [CrossRef] [PubMed]
4. Skowronski, D.M.; Sekirov, I.; Sabaiduc, S.; Zou, M.; Morshed, M.; Lawrence, D.; Smolina, K.; Ahmed, M.A.; Galanis, E.; Fraser, M.N.; et al. Low SARS-CoV-2 Sero-Prevalence Based on Anonymized Residual Sero-Survey before and after First Wave Measures in British Columbia, Canada, March-May 2020. *medRxiv* 2020. [CrossRef]
5. Long, Q.X.; Liu, B.Z.; Deng, H.J.; Wu, G.C.; Deng, K.; Chen, Y.K.; Liao, P.; Qiu, J.F.; Lin, Y.; Cai, X.F.; et al. Antibody Responses to SARS-CoV-2 in Patients with COVID-19. *Nat. Med.* **2020**, *26*, 845–848. [CrossRef] [PubMed]
6. To, K.K.W.; Tsang, O.T.Y.; Leung, W.S.; Tam, A.R.; Wu, T.C.; Lung, D.C.; Yip, C.C.Y.; Cai, J.P.; Chan, J.M.C.; Chik, T.S.H.; et al. Temporal Profiles of Viral Load in Posterior Oropharyngeal Saliva Samples and Serum Antibody Responses during Infection by SARS-CoV-2: An Observational Cohort Study. *Lancet Infect. Dis.* **2020**, *20*, 565–574. [CrossRef]
7. Dorigatti, I.; Lavezzo, E.; Manuto, L.; Ciavarella, C.; Pacenti, M.; Boldrin, C.; Cattai, M.; Saluzzo, F.; Franchin, E.; Del Vecchio, C.; et al. SARS-CoV-2 Antibody Dynamics and Transmission from Community-Wide Serological Testing in the Italian Municipality of Vo'. *Nat. Commun.* **2021**, *12*, 4383. [CrossRef] [PubMed]
8. Garcia-Beltran, W.F.; Lam, E.C.; Astudillo, M.G.; Yang, D.; Miller, T.E.; Feldman, J.; Hauser, B.M.; Caradonna, T.M.; Clayton, K.L.; Nitido, A.D.; et al. COVID-19-Neutralizing Antibodies Predict Disease Severity and Survival. *Cell* **2021**, *184*, 476–488.e11. [CrossRef]
9. Noh, J.Y.; Kwak, J.E.; Yang, J.S.; Hwang, S.Y.; Yoon, J.G.; Seong, H.; Hyun, H.; Lim, C.S.; Yoon, S.Y.; Ryou, J.; et al. Longitudinal Assessment of Antisevere Acute Respiratory Syndrome Coronavirus 2 Immune Responses for Six Months Based on the Clinical Severity of Coronavirus Disease 2019. *J. Infect. Dis.* **2021**, *224*, 754–763. [CrossRef]
10. Khoury, D.S.; Cromer, D.; Reynaldi, A.; Schlub, T.E.; Wheatley, A.K.; Juno, J.A.; Subbarao, K.; Kent, S.J.; Triccas, J.A.; Davenport, M.P. Neutralizing Antibody Levels Are Highly Predictive of Immune Protection from Symptomatic SARS-CoV-2 Infection. *Nat. Med.* **2021**, *27*, 1205–1211. [CrossRef]
11. Dolscheid-Pommerich, R.; Bartok, E.; Renn, M.; Kümmerer, B.M.; Schulte, B.; Schmithausen, R.M.; Stoffel-Wagner, B.; Streeck, H.; Saschenbrecker, S.; Steinhagen, K.; et al. Correlation between a Quantitative Anti-SARS-CoV-2 IgG ELISA and Neutralization Activity. *J. Med. Virol.* **2022**, *94*, 388–392. [CrossRef] [PubMed]
12. Dogan, M.; Kozhaya, L.; Placek, L.; Gunter, C.; Yigit, M.; Hardy, R.; Plassmeyer, M.; Coatney, P.; Lillard, K.; Bukhari, Z.; et al. SARS-CoV-2 Specific Antibody and Neutralization Assays Reveal the Wide Range of the Humoral Immune Response to Virus. *Commun. Biol.* **2021**, *4*, 129. [CrossRef] [PubMed]
13. Solís Arce, J.S.; Warren, S.S.; Meriggi, N.F.; Scacco, A.; McMurry, N.; Voors, M.; Syunyaev, G.; Malik, A.A.; Aboutajdine, S.; Adeojo, O.; et al. COVID-19 Vaccine Acceptance and Hesitancy in Low- and Middle-Income Countries. *Nat. Med.* **2021**, *27*, 1385–1394. [CrossRef] [PubMed]
14. Padma, T.V. COVID Vaccines to Reach Poorest Countries in 2023-despite Recent Pledges. *Nature* **2021**, *595*, 342–343. [CrossRef] [PubMed]
15. Mathieu, E.; Ritchie, H.; Ortiz-Ospina, E.; Roser, M.; Hasell, J.; Appel, C.; Giattino, C.; Rodés-Guirao, L. A Global Database of COVID-19 Vaccinations. *Nat. Hum. Behav.* **2021**, *5*, 947–953. [CrossRef] [PubMed]
16. Tao, K.; Tzou, P.L.; Nouhin, J.; Gupta, R.K.; de Oliveira, T.; Kosakovsky Pond, S.L.; Fera, D.; Shafer, R.W. The Biological and Clinical Significance of Emerging SARS-CoV-2 Variants. *Nat. Rev. Genet.* **2021**, *22*, 757–773. [CrossRef] [PubMed]
17. Planas, D.; Saunders, N.; Maes, P.; Guivel-Benhassine, F.; Planchais, C.; Buchrieser, J.; Bolland, W.-H.; Porrot, F.; Staropoli, I.; Lemoine, F.; et al. Considerable Escape of SARS-CoV-2 Omicron to Antibody Neutralization. *Nature* **2021**, *602*, 671–675. [CrossRef]
18. Bernal, J.L.; Andrews, N.; Gower, C.; Gallagher, E.; Simmons, R.; Thelwall, S.; Stowe, J.; Tessier, E.; Groves, N.; Dabrera, G.; et al. Effectiveness of Covid-19 Vaccines against the B.1.617.2 (Delta) Variant. *NEJM* **2021**, *385*, 585–594. [CrossRef] [PubMed]
19. Sokal, A.; Barba-Spaeth, G.; Fernández, I.; Broketa, M.; Azzaoui, I.; de La Selle, A.; Vandenberghe, A.; Fourati, S.; Roeser, A.; Meola, A.; et al. MRNA Vaccination of Naive and COVID-19-Recovered Individuals Elicits Potent Memory B Cells That Recognize SARS-CoV-2 Variants. *Immunity* **2021**, *54*, 2893–2907.e5. [CrossRef] [PubMed]
20. Goel, R.R.; Painter, M.M.; Apostolidis, S.A.; Mathew, D.; Meng, W.; Rosenfeld, A.M.; Lundgreen, K.A.; Reynaldi, A.; Khoury, D.S.; Pattekar, A.; et al. MRNA Vaccines Induce Durable Immune Memory to SARS-CoV-2 and Variants of Concern. *Science (1979)* **2021**, *374*, abm0829. [CrossRef] [PubMed]
21. Stamatatos, L.; Czartoski, J.; Wan, Y.-H.; Homad, L.J.; Rubin, V.; Glantz, H.; Neradilek, M.; Seydoux, E.; Jennewein, M.F.; Maccamy, A.J.; et al. MRNA Vaccination Boosts Cross-Variant Neutralizing Antibodies Elicited by SARS-CoV-2 Infection. *Science (1979)* **2021**, *372*, 1413. [CrossRef]
22. Harris, P.A.; Taylor, R.; Thielke, R.; Payne, J.; Gonzalez, N.; Conde, J.G. Research Electronic Data Capture (REDCap)—A Metadata-Driven Methodology and Workflow Process for Providing Translational Research Informatics Support. *J. Biomed. Inf.* **2009**, *42*, 377–381. [CrossRef]

23. British Columbia Centre for Disease Control Weekly Update on Variants of Concern. Available online: http://www.bccdc.ca/Health-Info-Site/Documents/VoC/VoC_weekly_09172021.pdf (accessed on 15 September 2022).

24. R Core Team. *R: A Language and Environment for Statistical Computing*; R Foundation for Statistical Computing: Vienna, Austria, 2018. Available online: https://www.R-project.org/ (accessed on 15 September 2022).

25. LeBlanc, J.J.; Gubbay, J.B.; Li, Y.; Needle, R.; Arneson, S.R.; Marcino, D.; Charest, H.; Desnoyers, G.; Dust, K.; Fattouh, R.; et al. Real-Time PCR-Based SARS-CoV-2 Detection in Canadian Laboratories. *J. Clin. Virol.* **2020**, *128*, 104433. [CrossRef] [PubMed]

26. Goldfarb, D.M.; Tilley, P.; Al-Rawahi, G.N.; Srigley, J.A.; Ford, G.; Pedersen, H.; Pabbi, A.; Hannam-Clark, S.; Charles, M.; Dittrick, M.; et al. Self-Collected Saline Gargle Samples as an Alternative to Health Care Worker-Collected Nasopharyngeal Swabs for COVID-19 Diagnosis in Outpatients. *J. Clin. Microbiol.* **2021**, *59*, e02427-20. [CrossRef] [PubMed]

27. Hickman, R.; Nguyen, J.; Lee, T.D.; Tyson, J.R.; Azana, R.; Tsang, F.; Hoang, L.; Prystajecky, N. Rapid, High-Throughput, Cost Effective Whole Genome Sequencing of SARS-CoV-2 Using a Condensed One Hour Library Preparation of the Illumina DNA Prep Kit. *medRxiv* 2022. [CrossRef]

28. Price, M.N.; Dehal, P.S.; Arkin, A.P. FastTree 2–Approximately Maximum-Likelihood Trees for Large Alignments. *PLoS ONE* **2010**, *5*, e9490. [CrossRef]

29. Hadfield, J.; Megill, C.; Bell, S.M.; Huddleston, J.; Potter, B.; Callender, C.; Sagulenko, P.; Bedford, T.; Neher, R.A. Nextstrain: Real-Time Tracking of Pathogen Evolution. *Bioinformatics* **2018**, *34*, 4121–4123. [CrossRef] [PubMed]

30. O'Toole, Á.; Scher, E.; Underwood, A.; Jackson, B.; Hill, V.; McCrone, J.T.; Colquhoun, R.; Ruis, C.; Abu-Dahab, K.; Taylor, B.; et al. Assignment of Epidemiological Lineages in an Emerging Pandemic Using the Pangolin Tool. *Virus Evol.* **2021**, *7*, veab064. [CrossRef]

31. Sekirov, I.; Barakauskas, V.E.; Simons, J.; Cook, D.; Bates, B.; Burns, L.; Masud, S.; Charles, M.; McLennan, M.; Mak, A.; et al. SARS-CoV-2 Serology: Validation of High-Throughput Chemiluminescent Immunoassay (CLIA) Platforms and a Field Study in British Columbia. *J. Clin. Virol.* **2021**, *142*, 104914. [CrossRef]

32. Stein, D.R.; Osiowy, C.; Gretchen, A.; Thorlacius, L.; Fudge, D.; Lang, A.; Sekirov, I.; Morshed, M.; Levett, P.N.; Tran, V.; et al. Evaluation of Commercial SARS-CoV-2 Serological Assays in Canadian Public Health Laboratories. *Diagn. Microbiol. Infect. Dis.* **2021**, *101*, 115412. [CrossRef]

33. Li, F.F.; Liu, A.; Gibbs, E.; Tanunliong, G.; Marquez, A.C.; Gantt, S.; Frykman, H.; Krajden, M.; Morshed, M.; Prystajecky, N.A.; et al. A Novel Multiplex Electrochemiluminescent Immunoassay for Detection and Quantification of Anti-SARS-CoV-2 IgG and Anti-Seasonal Endemic Human Coronavirus IgG. *J. Clin. Virol.* **2022**, *146*, 105050. [CrossRef]

34. Tanunliong, G.; Liu, A.C.; Kaweski, S.; Irvine, M.; Reyes, R.C.; Purych, D.; Krajden, M.; Morshed, M.; Sekirov, I.; Gantt, S.; et al. Age-Associated Seroprevalence of Coronavirus Antibodies: Population-Based Serosurveys in 2013 and 2020, British Columbia, Canada. *Front. Immunol.* **2022**, *13*, 1189. [CrossRef]

35. Cohen, J. A Power Primer. *Psychol. Bull.* **1992**, *112*, 155–159. [CrossRef]

36. Vanderweele, T.J.; Shpitser, I. A New Criterion for Confounder Selection. *Biometrics* **2011**, *67*, 1406–1413. [CrossRef]

37. Gałecki, A.; Burzykowski, T. *Linear Mixed-Effects Models Using R*; Springer: New York, NY, USA, 2013; ISBN 978-1-4614-3899-1.

38. Shoukri, M.M.; Donner, A.; El-Dali, A. Covariate-Adjusted Confidence Interval for the Intraclass Correlation Coefficient. *Contemp. Clin. Trials* **2013**, *36*, 244–253. [CrossRef]

39. Bozdogan, H. Model Selection and Akaike's Information Criterion (AIC): The General Theory and Its Analytical Extensions. *Psychometrika* **1987**, *52*, 345–370. [CrossRef]

40. Satagopan, J.M.; Ben-Porat, L.; Berwick, M.; Robson, M.; Kutler, D.; Auerbach, A.D. A Note on Competing Risks in Survival Data Analysis. *Br. J. Cancer* **2004**, *91*, 1229–1235. [CrossRef]

41. Bland, J.M.; Altman, D.G. The Logrank Test. *BMJ* **2004**, *328*, 1073. [CrossRef]

42. BC Centre for Disease Control BCCDC COVID-19 Situational Report—Week 2: January 09–January 15, 2022. 2022. Available online: http://www.bccdc.ca/Health-Info-Site/Documents/COVID_sitrep/Week_2_2022_BC_COVID-19_Situation_Report.pdf (accessed on 15 September 2022).

43. Torbati, E.; Krause, K.L.; Ussher, J.E. The Immune Response to SARS-CoV-2 and Variants of Concern. *Viruses* **2021**, *13*, 1911. [CrossRef]

44. Public Health Agency of Canada FluWatch. Available online: https://www.canada.ca/en/public-health/services/publications/diseases-conditions/fluwatch/2020-2021/week-49-november-29-december-5-2020.html#a3 (accessed on 15 September 2022).

45. Boyton, R.J.; Altmann, D.M. The Immunology of Asymptomatic SARS-CoV-2 Infection: What Are the Key Questions? *Nat. Rev. Immunol.* **2021**, *21*, 762–768. [CrossRef]

46. Lau, C.S.; Phua, S.K.; Liang, Y.L.; Lin, M.; Oh, H.; Aw, T.C. SARS-CoV-2 Spike and Neutralizing Antibody Kinetics 90 Days after Three Doses of BNT162b2 MRNA COVID-19 Vaccine in Singapore. *Vaccines* **2022**, *10*, 331. [CrossRef] [PubMed]

47. Levin, E.G.; Lustig, Y.; Cohen, C.; Fluss, R.; Indenbaum, V.; Amit, S.; Doolman, R.; Asraf, K.; Mendelson, E.; Ziv, A.; et al. Waning Immune Humoral Response to BNT162b2 Covid-19 Vaccine over 6 Months. *N. Engl. J. Med.* **2021**, *385*, e84. [CrossRef]

48. Li, K.; Huang, B.; Wu, M.; Zhong, A.; Li, L.; Cai, Y.; Wang, Z.; Wu, L.; Zhu, M.; Li, J.; et al. Dynamic Changes in Anti-SARS-CoV-2 Antibodies during SARS-CoV-2 Infection and Recovery from COVID-19. *Nat. Commun.* **2020**, *11*, 6044. [CrossRef] [PubMed]

49. Bates, T.A.; McBride, S.K.; Leier, H.C.; Guzman, G.; Lyski, Z.L.; Schoen, D.; Winders, B.; Lee, J.-Y.; Lee, D.X.; Messer, W.B.; et al. Vaccination before or after SARS-CoV-2 Infection Leads to Robust Humoral Response and Antibodies That Effectively Neutralize Variants. *Sci. Immunol.* **2022**, *7*, eabn8014. [CrossRef]

50. Fenwick, C.; Croxatto, A.; Coste, A.T.; Pojer, F.; André, C.; Pellaton, C.; Farina, A.; Campos, J.; Hacker, D.; Lau, K.; et al. Changes in SARS-CoV-2 Spike versus Nucleoprotein Antibody Responses Impact the Estimates of Infections in Population-Based Seroprevalence Studies. *J. Virol.* **2021**, *95*, e01828-20. [CrossRef] [PubMed]

51. Jalkanen, P.; Kolehmainen, P.; Häkkinen, H.K.; Huttunen, M.; Tähtinen, P.A.; Lundberg, R.; Maljanen, S.; Reinholm, A.; Tauriainen, S.; Pakkanen, S.H.; et al. COVID-19 MRNA Vaccine Induced Antibody Responses against Three SARS-CoV-2 Variants. *Nat. Commun.* **2021**, *12*, 3991. [CrossRef] [PubMed]

52. Andreano, E.; Paciello, I.; Piccini, G.; Manganaro, N.; Pileri, P.; Hyseni, I.; Leonardi, M.; Pantano, E.; Abbiento, V.; Benincasa, L.; et al. Hybrid Immunity Improves B Cells and Antibodies against SARS-CoV-2 Variants. *Nature* **2021**, *600*, 530–535. [CrossRef]

53. Wang, Z.; Muecksch, F.; Schaefer-Babajew, D.; Finkin, S.; Viant, C.; Gaebler, C.; Hoffmann, H.-H.; Barnes, C.O.; Cipolla, M.; Ramos, V.; et al. Naturally Enhanced Neutralizing Breadth against SARS-CoV-2 One Year after Infection. *Nature* **2021**, *595*, 426–431. [CrossRef]

54. Skowronski, D.M.; Sabaiduc, S.; Leir, S.; Rose, C.; Zou, M.; Murti, M.; Dickinson, J.A.; Olsha, R.; Gubbay, J.B.; Croxen, M.A.; et al. Paradoxical Clade- and Age-Specific Vaccine Effectiveness during the 2018/19 Influenza A(H3N2) Epidemic in Canada: Potential Imprint-Regulated Effect of Vaccine (I-REV). *Eurosurveillance* **2019**, *24*, 1900585. [CrossRef]

55. Guo, L.; Wang, Y.; Kang, L.; Hu, Y.; Wang, L.; Zhong, J.; Chen, H.; Ren, L.; Gu, X.; Wang, G.; et al. Cross-Reactive Antibody against Human Coronavirus OC43 Spike Protein Correlates with Disease Severity in COVID-19 Patients: A Retrospective Study. *Emerg. Microbes Infect.* **2021**, *10*, 664. [CrossRef]

56. Lin, C.Y.; Wolf, J.; Brice, D.C.; Sun, Y.; Locke, M.; Cherry, S.; Castellaw, A.H.; Wehenkel, M.; Crawford, J.C.; Zarnitsyna, V.I.; et al. Pre-Existing Humoral Immunity to Human Common Cold Coronaviruses Negatively Impacts the Protective SARS-CoV-2 Antibody Response. *Cell Host Microbe* **2022**, *30*, 83–96.e4. [CrossRef] [PubMed]

57. Anderson, E.M.; Goodwin, E.C.; Verma, A.; Arevalo, C.P.; Bolton, M.J.; Weirick, M.E.; Gouma, S.; McAllister, C.M.; Christensen, S.R.; Weaver, J.E.; et al. Seasonal Human Coronavirus Antibodies Are Boosted upon SARS-CoV-2 Infection but Not Associated with Protection. *Cell* **2021**, *184*, 1858. [CrossRef] [PubMed]

viruses

MDPI

Communication

Enhancing the Immunogenicity of RBD Protein Variants through Amino Acid E484 Mutation in SARS-CoV-2

Zhikai Zhang [1,†], Xuan Wan [2,†], Xinyue Li [1], Shaoxi Cai [2,*] and Chengsong Wan [1,*]

[1] BSL-3 Laboratory (Guangdong), Guangdong Provincial Key Laboratory of Tropical Disease Research, School of Public Health, Southern Medical University, Guangzhou 510515, China
[2] Chronic Airways Diseases Laboratory, Department of Respiratory and Critical Care Medicine, Nanfang Hospital, Southern Medical University, Guangzhou 510515, China
*   Correspondence: hxkc@smu.edu.cn (S.C.); gzwcs@smu.edu.cn (C.W.)
†   These authors contributed equally to this work.

Abstract: In the context of the COVID-19 pandemic, conducting antibody testing and vaccination is critical. In particular, the continued evolution of SARS-CoV-2 raises concerns about the effectiveness of vaccines currently in use and the activity of neutralizing antibodies. Here, we used the *Escherichia coli* expression system to obtain nine different SARS-CoV-2 RBD protein variants, including six single-point mutants, one double-point mutant, and two three-point mutants. Western blotting results show that nine mutants of the RBD protein had strong antigenic activity in vitro. The immunogenicity of all RBD proteins was detected in mice to screen for protein mutants with high immunogenicity. The results show that the mutants E484K, E484Q, K417T-E484K-N501Y, and K417N-E484K-N501Y, especially the former two, had better immunogenicity than the wild type. This suggests that site E484 has a significant impact on the function of the RBD protein. Our results demonstrate that recombinant RBD protein expressed in *E. coli* can be an effective tool for the development of antibody detection methods and vaccines.

Keywords: COVID-19; SARS-CoV-2; RBD; immunogenicity; mutant

check for **updates**

Citation: Zhang, Z.; Wan, X.; Li, X.; Cai, S.; Wan, C. Enhancing the Immunogenicity of RBD Protein Variants through Amino Acid E484 Mutation in SARS-CoV-2. *Viruses* **2022**, *14*, 2020. https://doi.org/10.3390/v14092020

Academic Editor: Jason Yiu Wing KAM

Received: 16 August 2022
Accepted: 2 September 2022
Published: 13 September 2022

Publisher's Note: MDPI stays neutral with regard to jurisdictional claims in published maps and institutional affiliations.

1. Introduction

The beta-coronavirus SARS-CoV-2 has become the seventh discrete coronavirus species that is capable of causing human disease [1]. SARS-CoV-2 is easily transmitted and highly pathogenic [2,3]. There are now outbreaks in more than 216 countries, areas, and territories around the world. As of 27 July 2022, the total number of confirmed cases has exceeded 570 million, with more than 6.3 million deaths, and these numbers are increasing every day. Vaccination is a highly effective strategy to prevent and stop the spread of SARS-CoV-2 in light of its high pathogenicity and transmissible nature. Well-protected vaccines can significantly reduce the incidence and transmission of the virus and are of great significance to the prevention and treatment of COVID-19. Many different SARS-CoV-2 vaccines are being developed around the world. According to the technology, the current vaccines are divided into three main categories: novel coronavirus-inactivated vaccines, subunit recombinant protein vaccines, and nucleic acid vaccines [4,5].

Studies have shown that SARS-CoV-2 spike (S) protein can serve as a suitable antigen with strong immunogenicity that can effectively stimulate the host immune system to induce the production of neutralizing antibodies [5–7]. The receptor-binding domain (RBD) in the SARS-CoV-2 S protein mediates viral cell fusion to induce host infection through site-directed binding to the receptor protein angiotensin converting enzyme 2 (ACE2) [8,9]. Therefore, the S protein has become a priority target for the design of recombinant subunit vaccines [5,10,11]. SARS-CoV-2 was first detected in 2019 [12]. As the virus has spread globally, new variants of the virus have emerged in the past two years, some of which have significantly increased their infectivity, transmissibility, and immune escape potential compared to wild viruses [13–15]. In multiple variants of SARS-CoV-2 that have been reported so far, the S protein has been mutated, especially at some sites in the RBD, resulting in an increase in the binding affinity between the S protein and the receptor protein ACE2 [16,17]. For example, the RBD mutation N501Y appears in lineages B.1.1.7 (Alpha), B.1.351 (Beta), P.1 (Gamma), and B.1.1.529 (Omicron) [15,17]. The L452R mutation appears in lineages B.1.617.2 (Delta), B.1.429 (Epsilon), and B.1.617.1 (Kappa) [18]. The K417N mutation appears in lineage B.1.351 (Beta), while the K417T mutation appears in lineages P.1 (Gamma) and B.1.1.529 (Omicron) [15,17]. The E484K mutation is present in lineages B.1.351 (Beta), P.1 (Gamma), P.2 (Zeta), B.1.525 (Eta), and B.1.526 (Iota), the E484Q mutation is present in lineage B.1.617.1 (Kappa), and the E484A mutation is present in lineage B.1.1.529 (Omicron) [15,19]. Therefore, it can be inferred that K417, L452, E484, and N501 in the RBD are key sites affecting the function of the S protein and are prone to mutation. In particular, three different mutations have appeared in the E484 site.

Previous studies have shown that different SARS-CoV-2 mutants have different mutations in the RBD of the S protein, causing changes in viral pathogenicity and infectivity [17,20]. This means that some vaccines currently in use may be less protective against some of the currently existing variants, which could lead to the widespread transmission of the mutated virus in the population, making it more difficult to control future COVID-19 outbreaks. Therefore, there is an urgent need to develop vaccines that confer strong broad-spectrum protection against existing or emerging mutated viruses. In this study, different mutant RBD proteins were expressed in an *Escherichia coli* expression system, and their antigenicity and immunogenicity were compared and evaluated in mice in order to screen candidates for vaccine design and drug targets and to facilitate virus antibody detection kit development.

2. Materials and Methods

2.1. Bacterial Strains, Construction, and Growth Conditions

E. coli XL1-Blue was used as a host to express the RBD protein. The bacterial strains in our experiment were cultured in LB medium at 37 °C. Ampicillin was added as needed. The fragment of the SARS-CoV-2 S protein gene (GenBank: NC_045512.2) (991–1749 bp) corresponding to amino acids 331−583 of the SARS-CoV-2 S protein (GenBank: YP_009724390.1) was synthesized. The expression vector pQE30 was digested with *Bam*HI and *Hind*III restriction enzymes. The ClonExpress® Ultra One Step Cloning Kit (Vazyme, Nanjing, China) was used to ligate the target fragment into the vector. Primers used for cloning and mutant construction are shown in Table 1.

Table 1. List of primers used in this study.

Primers Name	Sequence (5′-3′)
RBD-F	TCGCATCACCATCACCATCACAATATTACAAACTTGTGCCCTTTTG
RBD-R	GAGTCCAAGCTCAGCTAATTTTACTCAAGTGTCTGTGGATCACGG
K417T-F	CAAACTGGAACCATTGCTGATTATAATTATAAATTACC
K417T-R	CAGCAATGGTTCCAGTTTGCCCTGGAGCGATTTGTC
K417N-F	CAAACTGGAAACATTGCTGATTATAATTATAAATTACC
K417N-R	CAGCAATGTTTCCAGTTTGCCCTGGAGCGATTTGTC
L452R-F	ATAATTACCGCTATAGATTGTTTAGGAAGTCTAATC
L452R-R	CAATCTATAGCGGTAATTATAATTACCACCAACCTTAG
E484K-F	AATGGTGTTAAGGGTTTTAATTGTTACTTTCCTTTAC
E484K-R	TTAAAACCCTTAACACCATTACAAGGTGTGCTACCG
E484Q-F	AATGGTGTTCAGGGTTTTAATTGTTACTTTCCTTTAC
E484Q-R	TTAAAACCCTGAACACCATTACAAGGTGTGCTACCG
N501Y-F	CCAACCCACTTACGGTGTTGGTTACCAACCATACAGAG
N501Y-R	CAACACCGTAAGTGGGTTGGAAACCATATGATTGT

2.2. Protein Expression and Purification

The recombinant strains were streaked on LB plates (containing 200 µg/mL ampicillin). Single colonies were inoculated in LB medium containing 100 µg/mL ampicillin and cultured overnight at 37 °C. Cultures were transferred to 1 L LB medium. When the OD600 value of the bacterial solution reached 0.5, IPTG was added at a final concentration of 0.5 mM and bacteria were cultured at 37 °C for 8–12 h. The cultured cells were harvested and resuspended in 15 mL lysis buffer (10 mM imidazole, 300 mM NaCl, 50 mM NaH_2PO_4, pH 8.0). Then, 0.5 mL lysozyme (50 mg/mL) was added and the lysate was placed on ice for 20 min. Ultrasonic disruption was performed with the following parameters: total time 16 min, working time 6 s, intermittent time 3 s, and power 300 W. After centrifugation for 20 min at 12,000 rpm and 4 °C, the supernatant was discarded, and the precipitate (inclusion body) was retained. Next, 20 mL of precipitation lysis buffer (8 M urea, 100 mM NaH_2PO_4, 100 mM Tris-HCl, pH 8.0) was added to the inclusion body. After sufficient oscillation, ultrasonication was carried out. The supernatant (containing protein after inclusion body dissolution) was collected by centrifugation for 20 min at 12,000 rpm and 4 °C, and the precipitate was discarded. The RBD protein was purified under denaturing conditions with HisSep Ni-NTA Agarose Resin. The inactive RBD protein was added to refolding buffer (100 mM Tris, 400 mM L-arginine, 2 mM EDTA, 5 mM GSH, 0.5 mM GSSG, 10% (*v*/*v*) glycerol, pH 8.0) and incubated for 12 h at 4 °C. Finally, the refolded RBD protein was concentrated and desalted by a 10 kDa ultrafiltration tube.

2.3. SDS-PAGE and Immunoblotting

The recombinant RBD protein was mixed with protein loading buffer, boiled for 5 min, and centrifuged for 10 min at 12,000 rpm. Next, proteins were separated by 12% SDS-PAGE (5 µL of supernatant per lane). For Coomassie Bright Blue staining, CBB Fast Staining Solution (Tiangen biotech, Beijing, China) was used. For Western blot (WB) analysis, the proteins were transferred from the gel to a nitrocellulose membrane by semi-dry membrane transfer (the membrane transfer parameters: 25 V, 1 A, and 25 min). The membranes were incubated in blocking buffer (5% BSA in TBST buffer (Tris 9.68 g/L, NaCl 32 g/L, 0.02% Tween-20)) at room temperature for 2 h. After washing the membrane with TBST buffer three times, the membrane was incubated with mouse monoclonal anti-His tag (1:5000, Clone: 9C11, Yeasen Biotechnology, Shanghai, China) or human monoclonal anti-RBD (1:1000, Clone: 7H6, KMD Bioscience, Tianjin, China) overnight at 4 °C. Following incubation, the membranes were washed and incubated with horseradish peroxidase (HRP)-conjugated goat anti-mouse (1:10,000) or HRP-conjugated goat anti-human for 1 h at room temperature. After washing, the membrane was immersed in a chromogenic solution and protein bands were imaged with a UVI gel imager (UVItec, Cambridge, UK).

2.4. Mouse Immunization

Ten RBD proteins were prepared and mixed with Freund's adjuvants at a ratio of 1:1. In this experiment, three mice were immunized in each experimental group. Female BALB/c mice aged 6–8 weeks were subcutaneously immunized with 500 µL of RBD protein solution (30 µg of protein/mouse) or PBS and boosted on days 7 and 14. Blood was sampled on days 0 (prevaccination), 13, and 25. After coagulation at room temperature for 1 h, sera were collected by centrifugation (5000 rpm, 30 min, 4 °C) and stored at −80 °C until use.

2.5. Enzyme-Linked Immunosorbent Assay (ELISA)

ELISA was used to analyze antibody responses in serum. Microplate wells were coated with RBD protein (100 ng/well) in PBS coating buffer and incubated overnight at 4 °C. Plates were washed three times using PBST containing 0.1% Tween 20 and blotted dry. The plates were sealed at room temperature with 5% BSA for 2 h. Continuously diluted serum with PBS buffer was added to the microplate (100 µL/well), which was incubated for 1 h at 37 °C. Then, the plate was washed. HRP-conjugated goat anti-mouse (100 µL/well, 1:2000 in PBS) was added to each well, followed by incubation for 1 h at room temperature. Next, the plate was washed six times and incubated with 100 µL 3,3′,5,5′-tetramethylbenzidine (TMB) solution for 15 min. To stop the reaction, 50 µL of stop solution (1 M H_2SO_4) was added to each well, and Infinite M200 fluorescent multifunctional enzyme marker (Tecan, Männedorf, Switzerland) was used to read the absorbance at 450 nm.

2.6. Statistical Analysis

Data are expressed as mean ± standard deviation. Significant differences were determined by a one-way analysis of variance followed by the Tukey's multiple comparison test. Data were analyzed and graphs were drawn using Origin software (version 8.0, OriginLab, Northampton, MA, USA) and GraphPad Prism software (version 5.0, GraphPad Software, San Diego, CA, USA). Each experiment was independently replicated three times and statistical significance was defined as $p < 0.05$.

3. Results

3.1. The Nine RBD Protein Mutants Were Obtained by Prokaryotic Expression and Affinity Chromatography

The expression of the recombinant SARS-CoV-2 RBD protein was induced in *E. coli* XL1-Blue cells and detected by SDS-PAGE. Although the recombinant RBD protein could be expressed in *E. coli* cells, the protein existed in the form of an inclusion body (Figure 1A). Therefore, denaturation and refolding were required. A large number of induced bacterial cells were collected, and the inclusion body protein was obtained after cell rupture. After dissolution with 8 M urea, a Ni-NTA agarose flow column was used for purification (Figure 1B). When the purified denatured RBD protein was refolded, a recombinant RBD protein with biological activity was obtained, and the molecular weight of ≈30 kDa was consistent with the expected theoretical value. In this study, nine RBD protein mutants were constructed by reverse PCR [21], including six single-point mutants, K417T, K417N, L452R, E484K, E484Q, and N501Y; a two-locus mutation, L452R-E484Q; and two three-site mutants, K417T-E484K-N501Y and K417N-E484K-N501Y. In this experiment, affinity chromatography and ultrafiltration were used to remove the endotoxin in the recombinant protein. After repeated operations, the endotoxin content in all mutant protein solutions was less than 2.0 eu/mL. The SDS-PAGE results after denaturation, refolding, and concentration of the nine RBD protein mutants and wild-type are shown (Figure 1C,D).

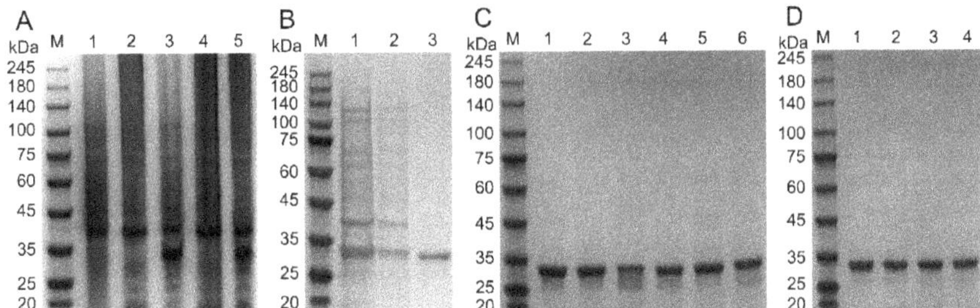

Figure 1. Expression and purification of the RBD protein mutants. (**A**) Lane 1: total protein of wild Enterobacter; Lane 2: total protein of uninduced recombinant strain; Lane 3: total protein of the recombinant strain after induction; Lane 4: supernatant protein after ultrasonic enucleation; Lane 5: precipitated protein after ultrasonic enucleation. (**B**) Lane 1: 8M urea dissolved total inclusion body protein; Lane 2: unbound protein collection solution; Lane 3: recombinant RBD protein collection solution. (**C,D**) SDS-PAGE analysis of denatured, renatured, and concentrated recombinant RBD protein and mutant protein. (**C**) Lane 1: mutant K417T; Lane 2: mutant K417N, Lane 3: mutant L452R; Lane 4: mutant E484K; Lane 5: mutant E484Q; Lane 6: mutant N501Y. (**D**) Lane 1: wild-type RBD protein; Lane 2: mutant L452R-E484Q; Lane 3: mutant K417T-E484K-N501Y; Lane 4: mutant K417N-E484K-N501Y.

3.2. The RBD Protein Mutants Possessed Antigenicity In Vitro

In order to ensure that the purified wild-type RBD protein and mutant proteins have a biological function, their specific antigenicity was analyzed using WB. Mouse monoclonal anti-His tag (Figure 2A) and human monoclonal anti-RBD (Figure 2B) were used as primary antibodies. The results show that a distinct single band is detected at approximately 30 kDa, which indicates that the wild RBD protein and nine purified mutant proteins had strong antigenicity and could be used as antigens for subsequent immune experiments.

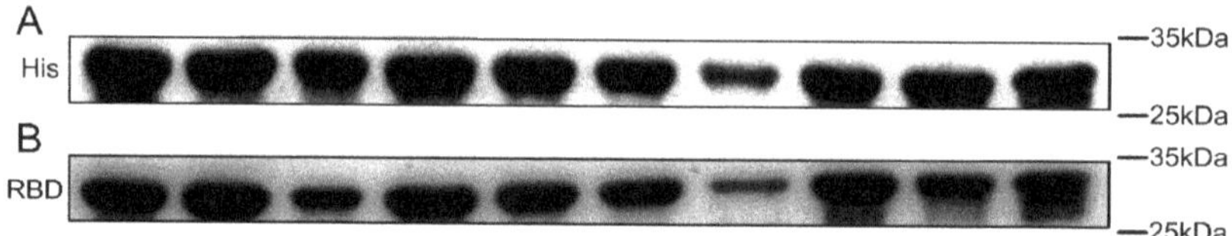

Figure 2. Western blotting analysis of RBD protein mutants. (**A**) His-tag mouse monoclonal antibody. (**B**) RBD human monoclonal antibody. The bands from left to right were wild RBD, K417T, K417N, L452R, E484K, E484Q, N501Y, L452R-E484Q, K417T-E484K-N501Y, and K417N-E484K-N501Y.

3.3. The Mutants E484K, E484Q, K417T-E484K-N501Y, and K417N-E484K-N501Y Displayed Excellent Immunogenicity in Mice

To evaluate the in vivo immunogenicity of the mutated RBD proteins, mice were immunized with both wild-type RBD and the nine RBD protein mutants, and blood samples were taken on days 0, 13, and 25. The total antibody titer against the SARS-CoV-2 RBD protein was evaluated by ELISA using HRP-conjugated goat anti-mouse as the secondary antibody. Before prevaccination, no anti-RBD was detected in the serum of all mice (Figure 3A). On day 13, anti-RBD was detected in mice immunized with both wild-type RBD and the nine RBD protein mutants (Figure 3B). On day 25, total antibody levels were significantly increased in all mice immunized with RBD and RBD mutant proteins (Figure 3C). After receiving three doses, the total resistance was greatly enhanced compared with the resistance levels after receiving only two doses. Interestingly, different RBD proteins yielded significantly different antibody titers after immunization (Figure 3D). On day 25, the antibody titers induced by mutants E484K, E484Q, K417T-E484K-N501Y, and K417N-E484K-N501Y were 4.7-, 4.8-, 6.8-, and 5.0-fold higher than those induced by wild-type RBD, respectively. These results indicate that the mutants E484K, E484Q, K417T-E484K-N501Y, and K417N-E484K-N501Y have excellent immunogenicity. In addition, the titers of mutants K417N, L452R, and L452R-E484Q were about 2-fold higher than that of wild-type RBD. However, surprisingly, the titers of mutants K417T and N501Y were approximately equal to that of wild-type RBD, suggesting that the single mutations at K417T and N501Y may not cause significant changes in the immunogenicity of RBD.

3.4. The Site E484 Has a Significant Impact on the Function of the RBD Protein

To further evaluate the extensive antigenicity of mutants E484K, E484Q, K417T-E484K-N501Y, and K417N-E484K-N501Y, we coated microplate wells with E484K, E484Q, K417T-E484K-N501Y, and K417N-E484K-N501Y proteins, respectively, and after blocking, antibody titers in mouse serum collected on day 25 were determined. When RBD mutant proteins of E484K, E484Q, K417T-E484K-N501Y, and K417N-E484K-N501Y were coated separately as antigens, higher antibody titers were detected in the sera of mice immunized with E484Q (Figure 4B). These results indicate that the mutant E484Q can induce high levels of antibodies, and the antibodies induced by the mutant have high antigenic binding ability to a wide spectrum of mutant RBD proteins. After coating with E484K protein, a high antibody titer was detected in the sera of all immunized mice (Figure 4A), indicating that E484K as an antigen has a high binding ability to antibodies induced by wild-type and other mutant RBD proteins. Although antibody titers against RBD were detected in the serum of all immunized mice after coating with K417T-E484K-N501Y (Figure 4C) and K417N-E484K-N501Y (Figure 4D) proteins alone, there was no significant difference between the antibody titer produced after the mice were immunized with other mutant proteins, except for the mutants E484K and E484Q. This means that the mutant proteins K417T-E484K-N501Y and K417N-E484K-N501Y may not be suitable as antigens in antibody detection. In conclusion, the above results indicate that the mutant proteins E484K, E484Q, K417T-E484K-N501Y, and K417N-E484K-N501Y all have good immunogenicity and can induce antibody production more strongly than wild-type RBD protein in mice.

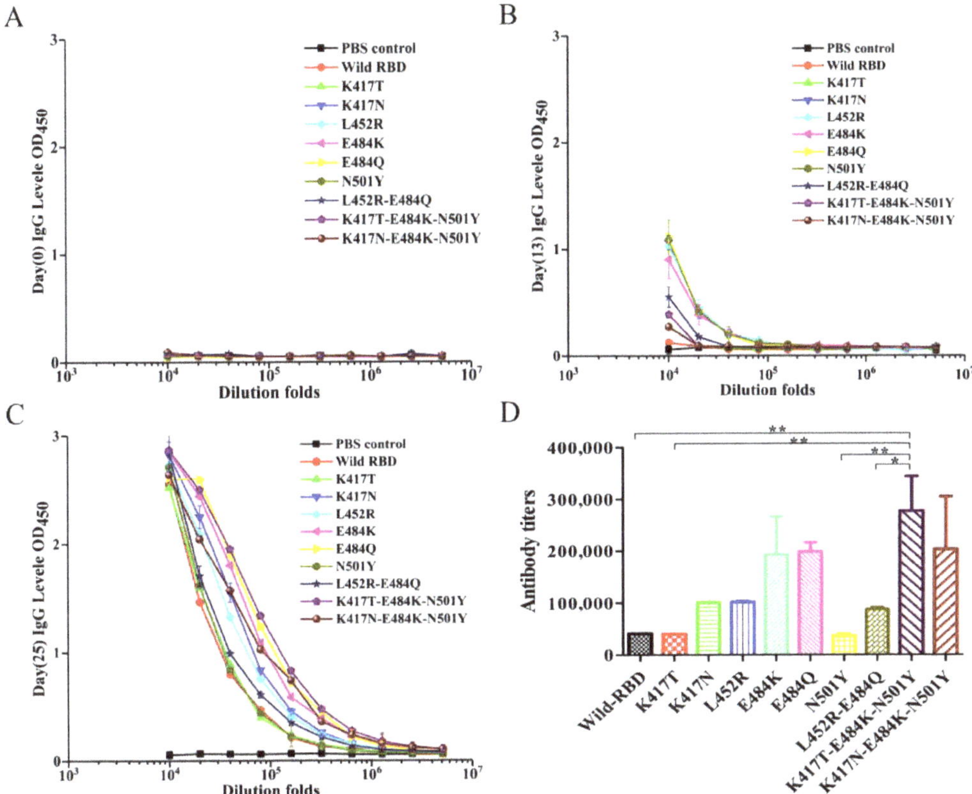

Figure 3. Anti-SARS-CoV-2 RBD antibody levels in mice by ELISA. It was coated with 100 ng of wild RBD protein, and after blocking with 2 times gradient diluted serum. (**A–C**) represent the antibody levels of RBD on days 0, 13, and 25, respectively. (**D**) The RBD antibody titers were expressed as the minimum concentration (maximum dilution) required for binding antigen. Statistical significance was defined as $p < 0.05$, and * $p < 0.05$, ** $p < 0.01$.

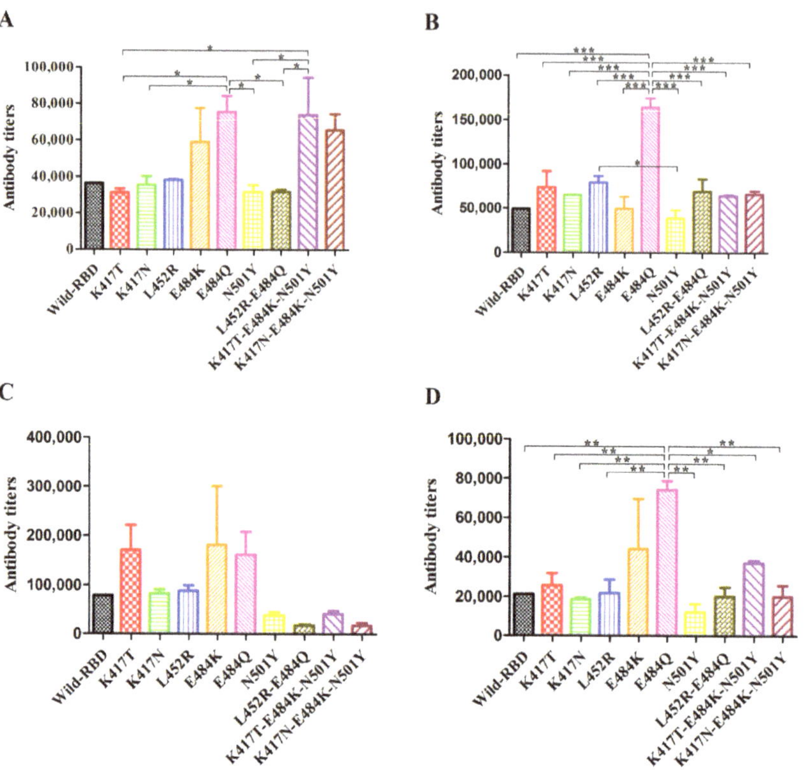

Figure 4. Extensive analysis of the antigenicity of RBD mutant protein. It was coated with 100 ng/well RBD mutant protein, and, after blocking, 2 times gradient diluted serum was added. (**A**) E484K protein. (**B**) E484Q protein. (**C**) K417T-E484K-N501Y protein. (**D**) K417N-E484K-N501Y protein. Statistical significance was defined as $p < 0.05$, and * $p < 0.05$, ** $p < 0.01$, *** $p < 0.001$.

4. Discussion

The cumulative multiple mutations of SARS-CoV-2 have formed highly infectious Delta and Omicron variants [22,23], which will trigger another outbreak of the COVID-19 epidemic around the world, exacerbating the global pandemic and threatening public health. The SARS-CoV-2 Omicron variant is of particular concern because of its increased transmissivity and the high number of mutations in the spike protein, which have the potential to evade neutralizing antibodies induced by the currently used COVID-19 vaccines [24,25]. In addition, it has been shown that the mechanical stability of SARS-CoV-2 RBD protein was 250 pN, while that of SARS-CoV RBD protein was 200 pN, which may play an important role in increasing transmissivity of the COVID-19 pandemic [26].

The spike protein of SARS-CoV-2 consists of two major functional domains containing a total of 1273 amino acids. Located at the N-terminus of the S protein is the S1 functional region containing the NTD and RBD domains. The remaining part is the S2 functional region containing two trimeric structures mediating membrane fusion [4]. Therefore, the NTD and RBD domains in the S1 functional region are candidates for the development of vaccines or superior antigens [27,28]. Several SARS-CoV-2 mutants are currently in focus, such as the Alpha (B.1.1.7), Beta (B.1.351), Gamma (P.1), Delta (B.1.617.2), Kappa (B.1.617.1), and Omicron (B.1.1.529) variants. The binding affinity of their corresponding mutant RBD protein to ACE2 also changes. For example, the Alpha (B.1.1.7), Beta (B.1.351, Gamma (P.1), Kappa (B.1.617.1), and Omicron (B.1.1.529) RBD proteins showed a higher affinity for ACE2 than wild-type RBD. Alpha (B.1.1.7) and Kappa (B.1.617.1) RBD proteins had a much higher affinity for ACE2 than wild-type RBD, while the Delta (B.1.617.2) RBD protein had a lower affinity for ACE2 than wild-type RBD [17,22]. In addition, it is interesting that K417T-E484K-N501Y, K417N-E484K-N501Y, and K417T-E484A-N501Y mutant RBD proteins had a lower affinity for ACE2 than the N501Y mutant RBD [16,29]. E484 and N501 are key sites to improve the binding affinity between RBD and ACE2 [17,22]. However, Koelher et al. investigated some RBD mutations, which described the effect of receptor binding energetics and neutralization of the SARS-CoV-2 variants by atomic force microscopy and molecular dynamics. They found that N501Y and E484Q mutations were particularly important for greater stability of the RBD-ACE2 complex, but the N501Y mutations were unlikely to significantly affect antibody neutralization [30].

Several RBD mutants selected in this study were present in the above SARS-CoV-2 variants. The recombinant RBD protein was expressed by *E. coli*, and after purification by affinity chromatography, BALB/c mice were immunized and antibodies against RBD were detected in serum (Figure 3B,C). The results show that RBD protein expressed by a prokaryotic expression system had good immunogenicity although it could not be modified by folding and glycosylation. The mutants E484K, E484Q, K417T-E484K-N501Y, and K417T-E484K-N501Y all showed higher antibody titers than wild-type RBD (Figure 3D), indicating that these sites can significantly enhance the immunogenicity of RBD protein to induce the production of high levels of neutralizing antibodies. However, the mutant N501Y showed a lower antibody titer than the wild type (Figure 3D). Although many studies have shown that the mutant N501Y can significantly increase the binding affinity between RBD and ACE2 [17,22,29,31], its immunogenicity to RBD in our study was not enhanced but decreased. Therefore, we assume that the binding affinity between RBD and ACE2 may not be directly related to the immunogenicity of the RBD protein.

To verify the above hypothesis, we used four RBD mutants, E484K, E484Q, K417T-E484K-N501Y, and K417T-E484K-N501Y, as antigens and detected the antibodies against RBD in the serum of all immunized mice. The results show that single mutants E484K and E484Q, when used as antigens, could not only detect high antibody titers in the sera of all immunized mice but also detect higher antibody titers in the sera of mice immunized with mutants E484K and E484Q as compared with mice immunized with wild-type and other mutant RBD proteins (Figure 4). It has been shown that intratype antigenic variation due to mutation(s) is widely considered the main hurdle to appropriate FMD vaccine development, such that two substitutions of distantly located aa at B-C (T48I) and G-H (A143V) loops, in combination, distorted the VP1 G-H loop, which leads to the variation of the antigen [32]. The study of Huang et al. showed that single mutations L452R and F490S reduce the antigenicity to neutralizing antibodies [33]. The mutant E484Q can also significantly enhance the binding affinity between RBD and ACE2 [17,30], but the results of our study indicate that the mutant E484Q can also significantly enhance the immunogenicity of RBD protein. In addition, surprisingly, although the mutant E484K could not significantly enhance the binding affinity between RBD and ACE2, the results of this study show that the mutant E484K could also enhance the immunogenicity of RBD protein. The above results verify our conjecture that the binding affinity between RBD mutants and ACE2 does not determine the immunogenicity of the mutant as an antigen.

5. Conclusions

In a pandemic where a virus frequently mutates, the broad-spectrum effectiveness of neutralizing antibodies and vaccines is crucial, so these studies are now the focus of many researchers. In the present study, different RBD mutants were selected as immunogens to investigate the differences in the levels of antibodies induced by them, and two significant mutants, E484K and E484Q, were found. Therefore, it can be inferred that the E484 amino acid residue on RBD not only significantly affects the binding affinity with the receptor ACE2, but also has a significant impact on the immunogenicity of the RBD protein, and it may even affect the transmissibility and pathogenicity of the mutant virus.

Author Contributions: Conceptualization, S.C. and C.W.; investigation, Z.Z. and X.W.; resources, X.W. and X.L.; software and data curation, X.W. and X.L.; writing—original draft preparation, Z.Z.; writing—review and editing, S.C. and C.W.; funding acquisition, Z.Z. and C.W. All authors have read and agreed to the published version of the manuscript.

Funding: This research was funded by the China Postdoctoral Science Foundation (2021M701594), Guangdong Science and Technology Program Key Projects (No.2021B1212030014).

Institutional Review Board Statement: The animal study was reviewed and approved by Animal Ethics Committee of Guangdong Medical Laboratory Animal Center (C202107-11).

Informed Consent Statement: Not applicable.

Data Availability Statement: Not applicable.

Acknowledgments: We thank Yushan Jiang and Chenguang Shen for their help in animal experimentation techniques.

Conflicts of Interest: The authors declare no conflict of interest.

References

1. Wu, J.T.; Leung, K.; Leung, G.M. Nowcasting and forecasting the potential domestic and international spread of the 2019-nCoV outbreak originating in Wuhan, China: A modelling study. *Lancet* **2020**, *395*, 689–697. [CrossRef]
2. Sanche, S.; Lin, Y.T.; Xu, C.; Romero-Severson, E.; Hengartner, N.; Ke, R. High Contagiousness and Rapid Spread of Severe Acute Respiratory Syndrome Coronavirus 2. *Emerg. Infect. Dis.* **2020**, *26*, 1470–1477. [CrossRef] [PubMed]
3. Harrison, A.G.; Lin, T.; Wang, P. Mechanisms of SARS-CoV-2 Transmission and Pathogenesis. *Trends Immunol.* **2020**, *41*, 1100–1115. [CrossRef] [PubMed]
4. Hu, B.; Guo, H.; Zhou, P.; Shi, Z. Characteristics of SARS-CoV-2 and COVID-19. *Nat. Rev. Microbiol.* **2021**, *19*, 141–154. [CrossRef]
5. Dai, L.; Gao, G.F. Viral targets for vaccines against COVID-19. *Nat. Rev. Immunol.* **2021**, *21*, 73–82. [CrossRef]
6. Keech, C.; Albert, G.; Cho, I.; Robertson, A.; Reed, P.; Neal, S.; Plested, J.S.; Zhu, M.; Cloney-Clark, S.; Zhou, H.; et al. Phase 1–2 Trial of a SARS-CoV-2 Recombinant Spike Protein Nanoparticle Vaccine. *N. Engl. J. Med.* **2020**, *383*, 2320–2332. [CrossRef]
7. Liu, X.; Drelich, A.; Li, W.; Chen, C.; Sun, Z.; Shi, M.; Adams, C.; Mellors, J.W.; Tseng, C.; Dimitrov, D.S. Enhanced elicitation of potent neutralizing antibodies by the SARS-CoV-2 spike receptor binding domain Fc fusion protein in mice. *Vaccine* **2020**, *38*, 7205–7212. [CrossRef]
8. Lan, J.; Ge, J.; Yu, J.; Shan, S.; Zhou, H.; Fan, S.; Zhang, Q.; Shi, X.; Wang, Q.; Zhang, L.; et al. Structure of the SARS-CoV-2 spike receptor-binding domain bound to the ACE2 receptor. *Nature* **2020**, *581*, 215–220. [CrossRef]
9. Xu, C.; Wang, Y.; Liu, C.; Zhang, C.; Han, W.; Hong, X.; Wang, Y.; Hong, Q.; Wang, S.; Zhao, Q.; et al. Conformational dynamics of SARS-CoV-2 trimeric spike glycoprotein in complex with receptor ACE2 revealed by cryo-EM. *Sci. Adv.* **2021**, *7*, eabe5575. [CrossRef]
10. Yang, S.; Li, Y.; Dai, L.; Wang, J.; He, P.; Li, C.; Fang, X.; Wang, C.; Zhao, X.; Huang, E.; et al. Safety and immunogenicity of a recombinant tandem-repeat dimeric RBD-based protein subunit vaccine (ZF2001) against COVID-19 in adults: Two randomised, double-blind, placebo-controlled, phase 1 and 2 trials. *Lancet Infect. Dis.* **2021**, *21*, 1107–1119. [CrossRef]
11. Gao, F.; An, C.; Bian, L.; Wang, Y.; Zhang, J.; Cui, B.; He, Q.; Yuan, Y.; Song, L.; Yang, J.; et al. Establishment of the first Chinese national standard for protein subunit SARS-CoV-2 vaccine. *Vaccine* **2022**, *40*, 2233–2239. [CrossRef]
12. Lu, R.; Zhao, X.; Li, J.; Niu, P.; Yang, B.; Wu, H.; Wang, W.; Song, H.; Huang, B.; Zhu, N.; et al. Genomic characterisation and epidemiology of 2019 novel coronavirus: Implications for virus origins and receptor binding. *Lancet* **2020**, *395*, 565–574. [CrossRef]
13. Weissman, D.; Alameh, M.; de Silva, T.; Collini, P.; Hornsby, H.; Brown, R.; LaBranche, C.C.; Edwards, R.J.; Sutherland, L.; Santra, S.; et al. D614G Spike Mutation Increases SARS CoV-2 Susceptibility to Neutralization. *Cell Host Microbe* **2021**, *29*, 23–31. [CrossRef]
14. David, D. What scientists know about new, fast-spreading coronavirus variants. *Nature* **2021**, *594*, 19–20.
15. Kannan, S.R.; Spratt, A.N.; Sharma, K.; Chand, H.S.; Byrareddy, S.N.; Singh, K. Omicron SARS-CoV-2 variant: Unique features and their impact on pre-existing antibodies. *J. Autoimmun.* **2022**, *126*, 102779. [CrossRef]
16. Laffeber, C.; de Koning, K.; Kanaar, R.; Lebbink, J.H.G. Experimental Evidence for Enhanced Receptor Binding by Rapidly Spreading SARS-CoV-2 Variants. *J. Mol. Biol.* **2021**, *433*, 167058. [CrossRef]
17. Kumar, V.; Singh, J.; Hasnain, S.E.; Sundar, D. Possible Link between Higher Transmissibility of Alpha, Kappa and Delta Variants of SARS-CoV-2 and Increased Structural Stability of Its Spike Protein and hACE2 Affinity. *Int. J. Mol. Sci.* **2021**, *22*, 9131. [CrossRef]
18. Escalera, A.; Gonzalez Reiche, A.S.; Aslam, S.; Mena, I.; Laporte, M.; Pearl, R.L.; Fossati, A.; Rathnasinghe, R.; Alshammary, H.; van de Guchte, A.; et al. Mutations in SARS-CoV-2 variants of concern link to increased spike cleavage and virus transmission. *Cell Host Microbe* **2022**, *30*, 373–387. [CrossRef]
19. Biswas, S.; Dey, S.; Chatterjee, S.; Nandy, A. Combatting future variants of SARS-CoV-2 using an in-silico peptide vaccine approach by targeting the spike protein. *Med. Hypotheses* **2022**, *161*, 110810. [CrossRef]

20. Chen, L.; Lu, L.; Choi, C.Y.; Cai, J.; Tsoi, H.; Chu, A.W.; Ip, J.D.; Chan, W.; Zhang, R.R.; Zhang, X.; et al. Impact of Severe Acute Respiratory Syndrome Coronavirus 2 (SARS-CoV-2) Variant-Associated Receptor Binding Domain (RBD) Mutations on the Susceptibility to Serum Antibodies Elicited by Coronavirus Disease 2019 (COVID-19) Infection or Vaccination. *Clin. Infect. Dis.* **2022**, *74*, 1623–1630. [CrossRef]

21. Li, C.; Wen, A.; Shen, B.; Lu, J.; Huang, Y.; Chang, Y. FastCloning: A highly simplified, purification-free, sequence- and ligation-independent PCR cloning method. *BMC Biotechnol.* **2011**, *11*, 92. [CrossRef] [PubMed]

22. Han, P.; Li, L.; Liu, S.; Wang, Q.; Zhang, D.; Xu, Z.; Han, P.; Li, X.; Peng, Q.; Su, C.; et al. Receptor binding and complex structures of human ACE2 to spike RBD from omicron and delta SARS-CoV-2. *Cell* **2022**, *185*, 630–640. [CrossRef]

23. Zhang, Y.; Wei, M.; Wu, Y.; Wang, J.; Hong, Y.; Huang, Y.; Yuan, L.; Ma, J.; Wang, K.; Wang, S.; et al. Cross-species tropism and antigenic landscapes of circulating SARS-CoV-2 variants. *Cell Rep.* **2022**, *38*, 110558. [CrossRef]

24. Pérez-Then, E.; Lucas, C.; Monteiro, V.S.; Miric, M.; Brache, V.; Cochon, L.; Vogels, C.B.F.; Malik, A.A.; De la Cruz, E.; Jorge, A.; et al. Neutralizing antibodies against the SARS-CoV-2 Delta and Omicron variants following heterologous CoronaVac plus BNT162b2 booster vaccination. *Nat. Med.* **2022**, *28*, 481–485. [CrossRef]

25. Guo, H.; Gao, Y.; Li, T.; Li, T.; Lu, Y.; Zheng, L.; Liu, Y.; Yang, T.; Luo, F.; Song, S.; et al. Structures of Omicron Spike complexes and implications for neutralizing antibody development. *Cell Rep.* **2022**, *39*, 110770. [CrossRef]

26. Moreira, R.A.; Chwastyk, M.; Baker, J.L.; Guzman, H.V.; Poma, A.B. Quantitative determination of mechanical stability in the novel coronavirus spike protein. *Nanoscale* **2020**, *12*, 16409–16413. [CrossRef]

27. Ren, W.; Sun, H.; Gao, G.F.; Chen, J.; Sun, S.; Zhao, R.; Gao, G.; Hu, Y.; Zhao, G.; Chen, Y.; et al. Recombinant SARS-CoV-2 spike S1-Fc fusion protein induced high levels of neutralizing responses in nonhuman primates. *Vaccine* **2020**, *38*, 5653–5658. [CrossRef]

28. Su, Q.; Zou, Y.; Yi, Y.; Shen, L.; Ye, X.; Zhang, Y.; Wang, H.; Ke, H.; Song, J.; Hu, K.; et al. Recombinant SARS-CoV-2 RBD with a built in T helper epitope induces strong neutralization antibody response. *Vaccine* **2021**, *39*, 1241–1247. [CrossRef]

29. Liu, H.; Wei, P.; Zhang, Q.; Chen, Z.; Aviszus, K.; Downing, W.; Peterson, S.; Reynoso, L.; Downey, G.P.; Frankel, S.K.; et al. 501Y.V2 and 501Y.V3 variants of SARS-CoV-2 lose binding to bamlanivimab in vitro. *mAbs* **2021**, *13*, 1919285. [CrossRef]

30. Koehler, M.; Ray, A.; Moreira, R.A.; Juniku, B.; Poma, A.B.; Alsteens, D. Molecular insights into receptor binding energetics and neutralization of SARS-CoV-2 variants. *Nat. Commun.* **2021**, *12*, 6977. [CrossRef]

31. Liu, H.; Zhang, Q.; Wei, P.; Chen, Z.; Aviszus, K.; Yang, J.; Downing, W.; Jiang, C.; Liang, B.; Reynoso, L.; et al. The basis of a more contagious 501Y.V1 variant of SARS-CoV-2. *Cell Res.* **2021**, *31*, 720–722. [CrossRef] [PubMed]

32. Islam, M.R.; Rahman, M.S.; Amin, M.A.; Alam, A.; Siddique, M.A.; Sultana, M.; Hossain, M.A. Evidence of combined effect of amino acid substitutions within G-H and B-C loops of VP1 conferring serological heterogeneity in foot-and-mouth disease virus serotype A. *Transbound. Emerg. Dis.* **2021**, *68*, 375–384. [CrossRef]

33. Wang, M.; Zhang, L.; Li, Q.; Wang, B.; Liang, Z.; Sun, Y.; Nie, J.; Wu, J.; Su, X.; Qu, X.; et al. Reduced sensitivity of the SARS-CoV-2 Lambda variant to monoclonal antibodies and neutralizing antibodies induced by infection and vaccination. *Emerg. Microbes Infect.* **2022**, *11*, 18–29. [CrossRef] [PubMed]

Article

Chikungunya Virus E2 Structural Protein B-Cell Epitopes Analysis

João Paulo da Cruz Silva, Marielton dos Passos Cunha, Shahab Zaki Pour, Vitor Renaux Hering, Daniel Ferreira de Lima Neto and Paolo Marinho de Andrade Zanotto *

Laboratory of Molecular Evolution and Bioinformatics, Department of Microbiology, Biomedical Sciences Institute, University of São Paulo, São Paulo 05508-000, Brazil
* Correspondence: pzanotto@usp.br

Abstract: The *Togaviridae* family comprises a large and diverse group of viruses responsible for recurrent outbreaks in humans. Within this family, the Chikungunya virus (CHIKV) is an important *Alphavirus* in terms of morbidity, mortality, and economic impact on humans in different regions of the world. The objective of this study was to perform an IgG epitope recognition of the CHIKV's structural proteins E2 and E3 using linear synthetic peptides recognized by serum from patients in the convalescence phase of infection. The serum samples used were collected in the state of Sergipe, Brazil in 2016. Based on the results obtained using immunoinformatic predictions, synthetic B-cell peptides corresponding to the epitopes of structural proteins E2 and E3 of the CHIKV were analyzed by the indirect peptide ELISA technique. Protein E2 was the main target of the immune response, and three conserved peptides, corresponding to peptides P3 and P4 located at Domain A and P5 at the end of Domain B, were identified. The peptides P4 and P5 were the most reactive and specific among the 11 epitopes analyzed and showed potential for use in serological diagnostic trials and development and/or improvement of the Chikungunya virus diagnosis and vaccine design.

Keywords: Chikungunya virus; immunoinformatics; B-cell epitopes; peptides; ELISA

Citation: Silva, J.P.d.C.; Cunha, M.d.P.; Pour, S.Z.; Hering, V.R.; Neto, D.F.d.L.; Zanotto, P.M.d.A. Chikungunya Virus E2 Structural Protein B-Cell Epitopes Analysis. *Viruses* **2022**, *14*, 1839. https://doi.org/10.3390/v14081839

Academic Editor: Jason Yiu Wing KAM

Received: 26 July 2022
Accepted: 15 August 2022
Published: 22 August 2022

Publisher's Note: MDPI stays neutral with regard to jurisdictional claims in published maps and institutional affiliations.

1. Introduction

The Chikungunya virus (CHIKV) (family: *Togaviridae*; genera: *Alphavirus*) [1–3] is a viral species widely distributed in the world. CHIKV outbreaks impact, in terms of morbidity and socioeconomic problems, areas with circulation [2,4–6]. Currently, three genotypes have been identified based on the genealogical relationships within the CHIKV: (i) East-Central-South Africa (ECSA), (ii) West African (WA), and the (iii) Asian genotype. Within the ECSA genotype, there is a lineage that is considered very important due to its biological characteristics: the Indian Ocean Lineage (IOL). CHIKV outbreaks are aggravated by a simultaneous co-circulation with other arboviruses, including the Dengue, Zika, and Mayaro viruses, which share symptoms in acute symptomatic individuals, which is a limitation for the clinical diagnosis and treatment [6–10].

The development of methods for synthetic immunogen synthesis has become increasingly applied and explored in combination with the development and improvement of bioinformatics algorithms for predicting immunogenic targets and identifying potential immune molecules of interest [11–16]. Epitopes can be linear or conformational, being linear arranged in a protein (the sequence of amino acids of the primary structure), while the conformational are discontinuous in the amino acids sequence and are linked to the secondary, tertiary, and quaternary structures of the proteins. Specific antibodies secreted by B lymphocytes can recognize and bind to both the linear and conformational epitopes of the antigens [11–13,17,18]. The mapping epitopes technique has numerous and important applications, including the development of new vaccines and the improvement of existing vaccines. This process is fast, efficient, and key to understanding the immune system for any type of protein [11,12,14,15,18,19]. Antigenic recognition is based on determining the recognized CHIKV immunogenic proteins by specific antibodies generated in the convalescent phase of infection. During the CHIKV outbreak in Singapore, occurring in 2008, it was observed that response antibodies against most of these protein determinants were elevated in 2 to 3 months but progressively decreased, as occurs in other infections. The only response against E2 glycoprotein was still detectable at 21 months post-infection. This long-standing response against the N-term region of glycoprotein E2 (the amino acids' positions between 1 and 18) makes it an important candidate for specific serological diagnostic tools [12,13,19].

There have been no studies on the specific targets of the anti-CHIKV antibody-mediated immune response of infected patients in Brazil. At the moment, there is no approved vaccine, and no effective antiviral agents have been made available so far. Treatment for the virus infection is often limited to symptomatic treatment due to problems in drug specificity and effectiveness [20–24]. However, recent epidemiological data reveal increasing evidence of the importance of antibody-mediated protection against the CHIKV, where it is important to highlight the possibility of using anti-CHIKV antibodies in therapeutic or prophylactic treatment [12,13,19,20,25].

Within this context, the epitopes' validation promotes a greater understanding of the interaction of peptides and the activation of the immune response. Together, these observations strongly imply the importance of the interactions between the specific antibodies and the peptides corresponding to the amino acid level. To improve key information in the development of vaccines and diagnostic tests for the CHIKV, considering specific amino acid residues important in the recognition of anti-CHIKV antibodies [11–13,19,20], we aim to: (i) map the epitopes of the CHIKV structural glycoproteins (E2 and E3) to identify specific IgG class antibody binding sites of the CHIKV' previously infected patients and identify the reactive epitopes of anti-CHIKV IgG antibodies induced by primary infection in humans infected with the CHIKV ECSA genotype; (ii) identify antigenic signatures that may be used in the development of serological diagnostic trials and/or vaccine development. While this topic has already been addressed by researchers in other regions, in this study, we aim to bring a novel perspective on the still underexplored topics of CHIKV infection in South America [3,6,12,13,16,19,26].

2. Material and Methods

2.1. Ethics Committee

The serum samples for the present study were collected in Sergipe [27] following a protocol approved by the Research Ethics Committee of the Institute of Biomedical Sciences of the University of São Paulo (protocol CEPSH-ICB n° 1406/17). All consenting adults signed a written informed form, and for children, the signature was made by the parents or guardians with the written informed consent on their behalf. These samples were collected during a period of intense circulation of the Chikungunya virus in the northeast region of Brazil [27–29].

2.2. Phylogenetic Reconstruction

To choose representative sequences of all CHIKV genotypes and lineages, we performed an analysis of the phylogenetic reconstruction using 63 sequences previously analyzed by our group [28,30] and available in Genbank (http://www.ncbi.nlm.nih.gov) (accessed on 1 October 2020), referring to the complete genomes representing all genotypes. The phylogeny was performed with the maximum likelihood analysis using IQ-TREE software (James Barbetti, Camberra, Australia) (http://www.iqtree.org/) (accessed on 1 October 2020) using the standard criteria [31]. All sequences of this study are presented in the tree in the format: genotype/access number/country of isolation/year of isolation [31]. To predict the antigenicity and linear and conformational epitopes, we selected the Chikungunya virus' complete genome sequences representing the four CHIKV genotypes (Asian,

ECSA, IOL, and WA), which were used in the computational modeling as described in the next section (GenBank accession numbers: KP164572, KY055011, HM045817, and KJ796852).

2.3. Computational Modeling

Structural protein modeling of the CHIKV was performed using the homology strategy based on crystallographic structures available at the Protein Data Bank (PDB: 3N43) and the Chikungunya virus' complete genome sequences (GenBank accession numbers: KP164572, KY055011, HM045817, and KJ796852). These genome sequences were translated into amino acid sequences, and the corresponding regions of structural proteins E3 and E2 were separated and modeled individually and validated. This step was performed using I-TASSER v5.1 software, using the default settings and the gnu parallel option. Furthermore, the species that had their structures identified in crystals were aligned to our models, and the RMSD calculations were made [32]. The models were validated using the MolProbity website for flips and stereochemical corrections, and on the ProSA website, Ramachandram plots were generated to validate the models [33,34]. Using the obtained structures in the modeling, the antigenicity and conformational and linear epitopes were predicted. The antigenicity score was obtained by the Kolaskar and Tongaonkar antigenicity scale, a semi-empirical method that makes use of the physicochemical properties of amino acid residues and their frequencies of occurrence in experimentally known segmental epitopes were developed to predict antigenic determinants on the proteins [35].

The conformational and linear epitopes' predictions were performed based on the alignment of the sequences of different CHIKV isolates (access numbers: KP164572 (Asian), KJ796852 (ECSA-IOL), KY055011 (ECSA), and HM045817 (WA)) and using the IEDB's program package by National Institute of Allergy and Infectious Diseases(Bethesda, MD, USA) (https://www.iedb.org/) (accessed on 1 October 2020) [11]. Conformational epitopes were obtained after the analysis of the structures modeled in the Discotope 2.0 prediction algorithm (Haste Andersen P, Bethesda, MD, USA) (http://tools.iedb.org/discotope) (accessed on 1 October 2020), which integrates the combination of two scores linearly; one based on the hydrophobicity/hydrophilicity scale and a score of the epitopes' propensity, which is based on the calculation of the surface accessibility and residue contact area [11,17]. The linear epitopes were obtained after analysis of the structures modeled in the ElliPro prediction algorithm (Ponomarenko, Bethesda, MD, USA)(http://tools.iedb.org/ellipro/) (accessed on 1 October 2020) [11,17], which is based on the residue protrusion index. The Pymol (The PyMOL Molecular Graphics System, Version 1.8 (Warren Lyford DeLano, New York, NY, USA) developed by Schrödinger, LLC.) (https://pymol.org/2/) (accessed on 1 October 2020) and Jalview Version 2.11.2.4 (Andrew Waterhouse, Dundee, Scotland) (www.jalview.org) (accessed on 1 October 2020) software were used for editing and indicating the conformational and linear epitopes in the modeled proteins and the aligned sequences, respectively, for each of the CHIKV structural proteins. Using the results obtained in the combined linear and conformational epitopes' predictions with the previously obtained information in the literature [12,19] and focusing on epitopes conserved by all genotypes, a panel of 11 linear peptides (P1–P9 corresponding to the E2 protein and P10–P11 corresponding to the E3 protein) were selected for synthesis (Proteimax, São Paulo, Brazil) and epitope validation was obtained by Peptide-based indirect ELISA.

2.4. Enzyme Immunoassay Qualitative and Quantitative for IgG Anti-CHIKV Antibodies

To identify and quantify positive samples, we used the protocol for the IgG antibody developed by EuroImmun (Euroimmun, Lübeck, Germany). In the first step of the qualitative assay, the samples, calibrator 2, and the positive and negative controls were diluted 1:101 using the sample buffer and added to the microplate wells covered with CHIKV antigens. In the first step of the quantitative assay, the samples, calibrators 1, 2, and 3, and the positive and negative controls were diluted 1:101 and added to the microplate wells covered with the same CHIKV antigens. The samples were added to the plate, which was incubated for 1 h at 37 °C. In the second step, the plate was washed 3 times with 300 µL of a wash solution in each well and, after washing, 100 µL of enzymatic conjugate (anti-human IgG marked with peroxidase) was added to each well, and the plate was incubated at room temperature (18 to 25 °C) for 30 min. In the third step, the plate was washed 3 times with 300 µL of a wash solution and, after washing, 100 µL of a substrate/chromogenic solution was added to each microplate well following incubation at room temperature (18 to 25 °C) for 15 min. In the fourth step, 100 µL of stop solution was added to each well of the microplate, and the optical density (O.D.) was measured in an Epoch Microplate Spectrophotometer (BioTek, Winooski, VT, USA). The results were interpreted following the manufacturer's instructions.

2.5. Peptides Screening Using Peptide-Based Indirect ELISA

To identify the target epitopes of CHIKV-specific IgG antibodies, 18 sera samples from positive-IgG CHIKV patients, collected in the state of Sergipe in 2016, were characterized with a quantitative immunoenzyme assay developed by EuroImmun. The sera samples were analyzed by the Euroimunne test with a mean titration of 181 UR/mL and standard deviation of 0.1 UR/mL and subsequently used in the standardization assay. All samples used in this study were tested negative for the presence of CHIKV antibodies IgM by the EuroImmun CHIKV IgM qualitative test. For indirect ELISA standardization using synthetic peptides as antigens, we briefly performed the following: in the first step of the experiment, a pool of 18 Chikungunya positive IgG samples identified by the EuroImmun test was used, and a commercial negative IgG serum with an unspecific reaction for the Chikungunya virus and identified by the EuroImmun test was used as a negative control (Catalog number: S7023; Sigma Aldrich, St Luis, MO, USA). The peptides P1 to P9 corresponding to the E2 protein and the peptides P10 and P11 corresponding to the E3 protein were diluted in a filtered PBS buffer (1 M pH 7.4) at a final concentration of 200 µg/mL; then, serial dilutions of factor 100× to a concentration of 0.2 fg/mL were performed to obtain a dilution curve of the antigen concentration. The peptide dilutions were used to coat the polystyrene Nunc 96 microplate wells ELISA plate, 50 µL of solution per well were added, and the plates were incubated at room temperature until the wells dried. For background reference, peptide-free wells were used solely with the blocking solution. The blocking step was done with 5% albumin, 1% powdered milk solution, and 0.05% Tween20, diluted in PBS 1× (pH 7.4), followed by the addition of 50 µL per well and overnight incubation at 4 °C or 37 °C for two hours. After the blocking solution was removed, 50 µL of the diluted serum (1:100) was added to the PBS 1× solution (pH 7.4) and incubated for 1 h at 37 °C. Then, the wells were emptied, and the plate was washed 3 times with 200 µL per well of a wash solution. After the washing step, 50 µL of an enzymatic conjugate (Ig anti-human marked with peroxidase 1:12,000) was added to the plate wells and incubated for 1 h at 37 °C. The plate was washed 4 times with 200 µL of a wash solution after incubation, and 100 µL of a substrate/chromogenic solution was added to each microplate well, followed by incubation at room temperature (18 to 25 °C) for 15 min. In the last step, 100 µL of a stop solution was added to each well of the microplate, and then the photometric measurement at 450 nm and 620 nm was performed, and the results of optical density were analyzed. All trials were performed in sample technical triplicate and using individual patients' sera.

2.6. Validation of the Peptides Using Peptide-Based Indirect ELISA

Based on the previous screening results, the 5 peptides (P1–P5) that demonstrated the potential to respond differentially were selected for individual patient sera analysis, including 10 Chikungunya virus-positive IgG sera and, as the negative control, 8 Chikungunya virus-negative IgG sera samples previously characterized by the EuroImmun test. The P1 to P5 peptides corresponding to the E2 protein were diluted in a filtered PBS buffer (1M pH 7.4) at a final concentration of 20 µg/mL followed by the addition of 50 µL of a solution per well to coat the polystyrene Nunc 96 microplate wells; the plates were incubated at room temperature until the wells dried. For background reference, peptide-free wells were used solely with the blocking solution. The blocking step was done with 5% albumin, 1% powdered milk solution, and 0.05% Tween20, diluted in PBS 1× (pH 7.4), followed by the addition of 50 µL per well and overnight incubation at 4 °C or 37 °C for two hours. After the blocking solution was removed, 50 µL of the diluted serum (1:100) was added to the PBS 1× solution (pH 7.4) and incubated for 1 h, at 37 °C. Then the wells were emptied, and the plate was washed 3 times with 200 µL per well of a wash solution. After the washing step, 50 µL of enzymatic conjugate (Ig anti-human marked with peroxidase 1:12,000) was added to the plate wells and incubated for 1 h at 37 °C. The plate was washed 4 times with 200 µL of a wash solution after incubation, and 100 µL of a substrate/chromogenic solution was added to each microplate well, followed by incubation at room temperature (18 to 25 °C) for 15 min. In the last step, 100 µL of a stop solution was added to each well of the microplate, and then the photometric measurement at 450 nm and 620 nm was performed, and the results of the optical density were analyzed. All trials were performed in sample technical triplicate and using individual patients' sera.

2.7. Data Analysis

All experiments were performed in sample triplicate. All data were presented as means and standard deviations. The differences in responses between the analyzed group and the control group were analyzed using appropriate statistical tests. The Graphpad software was used for the analysis of classifier performance, and a Receiver Operating Characteristic curve (ROC curve) was performed for each one of the best peptides (P1–P5) with a 95% confidence interval; the sensitivity corresponds to the rate of true positives and the specificity to the rate of true negatives. Most of the graphs presented here were performed using R scripts and are available upon request.

3. Results

To select representative amino acid sequences from all genotypes and lineages spread worldwide causing the human CHIKV disease, we relied on phylogenetic reconstruction and tree topology, where four representative sequences were chosen for each of the Asian, ECSA, and WA genotypes and the IOL lineage (GenBank accession numbers: KP164572, KY055011, and HM045817 KJ796852) (Figure 1). Based on the obtained structures in the modeling step, the conformational and linear epitopes were predicted (Figures 2 and 3, Supplementary Figures S1–S3).

Based on the prediction results, the E2 protein (Figure 2, Supplementary Figures S1 and S2) presented the highest immunogenic potential compared with the E3 protein (Figure 3, Supplementary Figure S3) and demonstrated both conformational and linear epitopes in the conserved regions among the genotypes and/or lineages with both predictions demonstrating the E2 protein as the main target of the immune response (Figure 2, Supplementary Figures S1 and S2). In the epitopes' prediction for the E3 protein (Figure 3, Supplementary Figure S3), only one potential epitope was identified at positions 57 to 61 in the C-terminal region in both the linear and conformational predictions. Using the results obtained in the combined linear and conformational predictions with the literature information previously obtained [12,19] and focusing on the epitopes preserved among all the genotypes and/or lineages, a panel of 11 linear peptides was selected for epitope validation. For validating the epitopes and checking the serological reactivity, a group of CHIKV-positive IgG patient samples was tested and selected (Table 1) for peptide validation. The 11 selected peptides were synthesized and visually located in the E2 and E3 proteins crystallography structure (Figures 4 and 5) and then tested in the peptide-based indirect immunoenzyme assay.

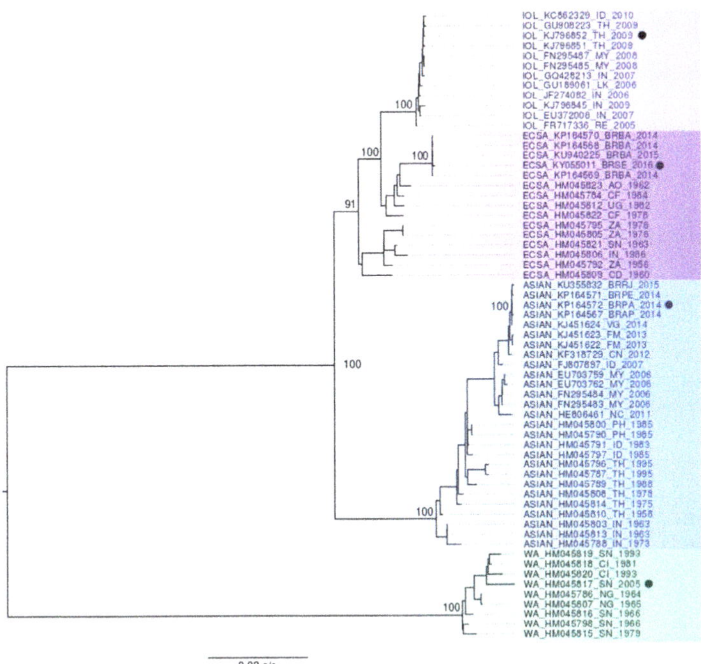

Figure 1. Phylogenetic tree representing the genotypes of the Chikungunya virus. The Chikungunya virus phylogenetic reconstruction, with 63 complete genome sequences. The green color is the WA genotype (West African), the blue color is the Asian genotype, the violet color is the ECSA genotype (East-Central-South African), and the gray color is the IOL lineage. The present values close to the main knots represent the bootstrap values.

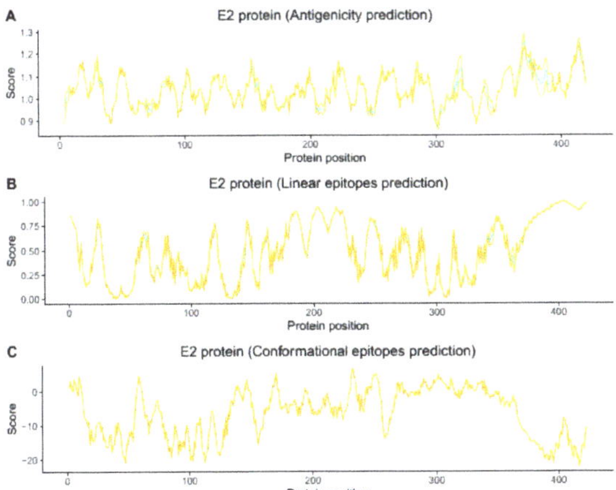

Figure 2. Chikungunya virus E2 Protein antigenicity (**A**) and linear (**B**) and conformational (**C**) epitopes' prediction. The blue line represents the mean value, and the yellow line represents the standard deviation value. In general, the places where only yellow appears are regions with a very small standard deviation and probably conserved regions among the CHIKV genotypes, and where blue appears, they correspond to regions with a higher standard deviation and probably variable or semi-conserved regions among all four CHIKV genotypes.

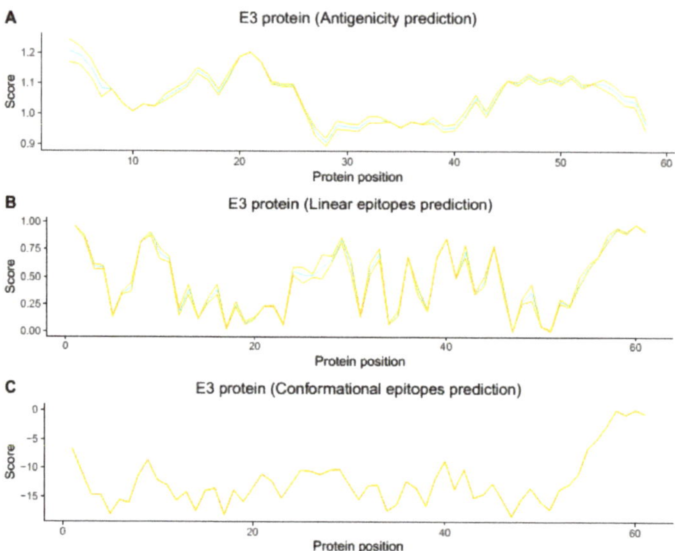

Figure 3. Chikungunya virus E3 Protein antigenicity (**A**) and linear (**B**) and conformational (**C**) epitopes' prediction. The blue line represents the mean value, and the yellow line represents the standard deviation value. In general, the places where only yellow appears are regions with a very small standard deviation and probably conserved regions among the CHIKV genotypes, and where blue appears, they correspond to regions with a higher standard deviation and probably variable or semi-conserved regions among all four CHIKV genotypes.

Table 1. Summary of E2 protein synthetic peptides' P1–P5 reactivity analyzed by IgG positive CHIKV individual serum samples and the Area Under the Curve (AUC) analysis.

Peptides	Control Samples	D.O Mean Control Samples	D.O Mean Control Samples Standard Deviation	Threshold (Control Samples Mean + 2SD)	CHIKV IgG Positive Samples	D.O mean CHIKV IgG Positivie Samples	D.O Mean CHIKV IgG Positive Samples Standard Deviation	Area under the Curve (AUC)	AUC *p*-Value
Peptide1	8	0.147	0.057	0.259	10	0.186	0.079	0.6932	0.1604
Peptide2	8	0.141	0.055	0.251	10	0.182	0.065	0.675	0.2135
Peptide3	8	0.085	0.083	0.229	10	0.244	0.095	0.8625	0.0032
Peptide4	8	0.074	0.054	0.176	10	0.219	0.073	0.9875	0.0005
Peptide5	8	0.084	0.032	0.146	10	0.152	0.040	0.9	0.0045

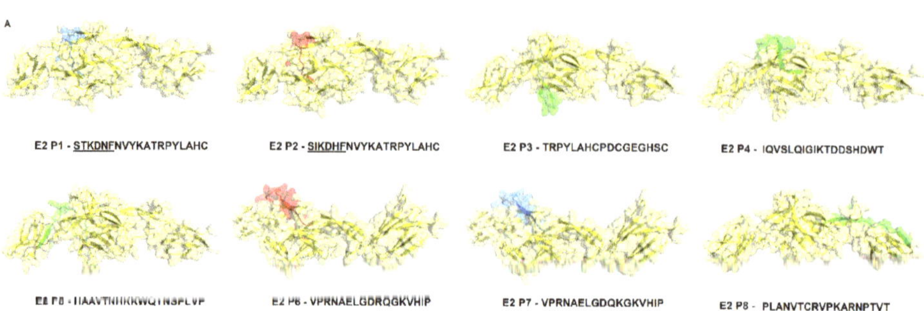

Figure 4. *Cont.*

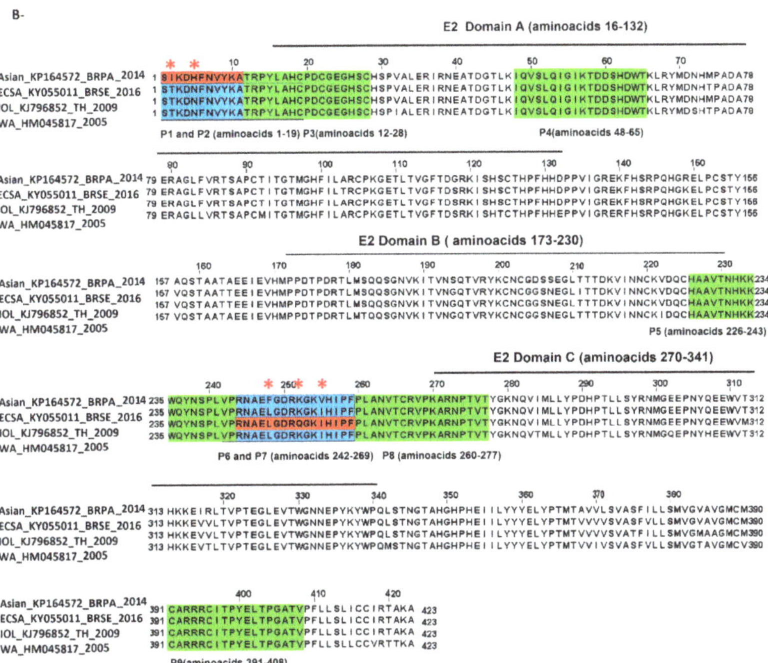

Figure 4. Synthetic peptides are highlighted on the E2 Protein structure and multiple sequence alignment. On top (**A**), predicted epitopes are highlighted on the structure of the protein E2. (**B**), Multiple alignment of E2 protein sequences corresponding to the four genotypes with highlighted peptides. In green, epitopes are conserved among all genotypes. In blue, epitopes are common to more than one genotype. In red are the single epitopes of each genotype. *—Amino acid changes in the regions predicted as epitopes. Underlined amino acids in peptides' sequences corresponding to regions not resolved in the crystallography structure.

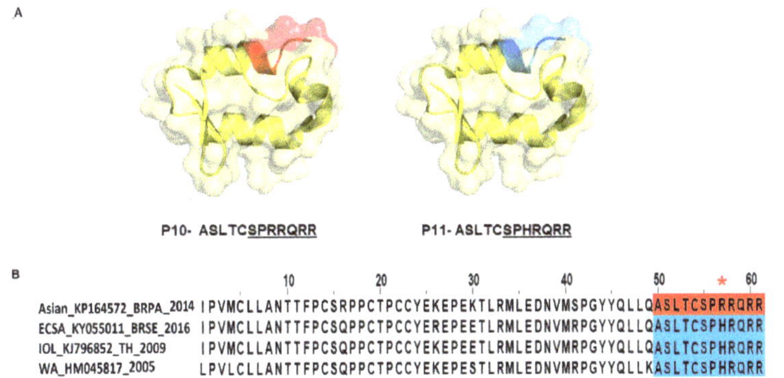

Figure 5. Synthetic peptides are highlighted on the E3 Protein structure and multiple sequence alignment. On top, (**A**) the predicted epitopes are highlighted on the structure of protein E3. (**B**), The multiple alignment of the E3 protein sequences corresponding to the four genotypes with highlighted peptides. Blue epitopes are common to more than one genotype. In red are the single epitope for individual genotype. *—Amino acid changes in the regions predicted as epitopes. Underlined amino acids in the peptides' sequences corresponding to regions not resolved in the crystallography structure.

The peptides P1 to P9 were synthesized with a size of 18 amino acids corresponding to protein E2 (Figure 4, Table 2), and the peptides P10 and P11 were synthesized with a size of 12 amino acids corresponding to the E3 protein (Figure 4, Table 2). The peptides P3, P4, P5, P8, and P9 are conserved among all the genotypes (Figure 4). The peptides P1, P6, and P10 correspond to the genotypes and/or lineage WA, ECSA, and IOL sequences (semi-conserved), while the P2 and P11 peptides are variants of these peptides, corresponding to the Asian genotype sequence exclusively, and P7 corresponding to the IOL genotype sequence exclusively (Figure 4, Table 2).

Table 2. Synthetic linear oligopeptides corresponding to the immunogenic regions of the E3 and E2 structural proteins of the Chikungunya virus.

Protein	Position	Sequence	Peptide
E3	2788–2799	ASLTCSPRRQRR	P11
	2788–2799	ASLTCSPHRQRR	P10
E2	2800–2818	STKDNFVYKATRPYLAHC	P1
	2800–2818	SIKDHFNVYKATRPYLAHC	P2
	2812–2828	TRPYLAHCPDCGEGHSC	P3
	2848–2865	IQVSLQIGIKTDDSHDWT	P4
	3026–3043	HAAVTNHKKWQYNSPLVP	P5
	3042–3059	VPRNAELGDRKGKIHIPF	P6
	3042–3059	VPRNAEFGDRQGKVHIPF	P7
	3060–3077	PLANVTCRVPKARNPTVT	P8
	3190–3208	CARRRCITPYELTPGATV	P9

To identify the target epitopes of CHIKV-specific IgG antibodies, a pool of 18 positive patient sera (IgG + CHIKV) and a non-specific human IgG serum as the control were used to screen for the epitopes that best react to previously characterized antibodies. The results indicated that the maximum peak achieved at a 1 ng concentration per well was obtained as the best reactivity and also the best difference between CHIKV IgG positive and the human IgG non-specific serum (Figure 6). Later, this concentration was used as a standard in the following assays with individual samples. In the peptide standardization assay, the P4 and P5 peptides were found to be the most reactive, respectively, and were selected to be evaluated with the individual samples from the CHIKV IgG positive and CHIKV IgG negative patients (Figure 7, Table 1). The previous epitopes, as described by Kam et al. [13,26], and corresponding to peptides P1–P3, were also included for the analysis with the individual samples. In this analysis, 10 positive IgG CHIKV sera samples with a mean titration of 160 UR/mL and a standard deviation of 46.87 UR/mL, characterized by the EuroImmun assay, were used. As controls, eight negative IgG CHIKV sera samples were used (Figure 6).

The following cut-off value was established as the comparison threshold, the average of the triplicate two standard deviations (the mean plus two SDs). The peptides P1 and P2 did not obtain a reactivity above the established cut-off value and demonstrated no potential for discrimination among the groups evaluated (Figure 7, Table 1). The peptides P3, P4, and P5 reacted above the established cut-off value and showed a discriminant potential among the groups evaluated (Figure 7, Table 1). Peptides P3 and P4 showed a stronger reactivity in the recognition among the peptides that reacted above the cut-off value. With the results obtained in this analysis, the peptides P1 to P5 were also analyzed for sensitivity and specificity through the analysis of the ROC (Receiver Operating Characteristic curve) (a 95% confidence interval) to verify the potential for the correct discrimination among the analyzed sample groups (Figure 8, Table 1). The P1 and P2 peptides demonstrated a low potential for classification among the analyzed groups and generated an AUC (Area under the curve) of below 0.7. The P3, P4, and P5 peptides, in addition to being reactive, demonstrated potential as discriminator classifiers, with AUC values of 0.86, 0.99, and 0.9, respectively (Figure 8, Table 1).

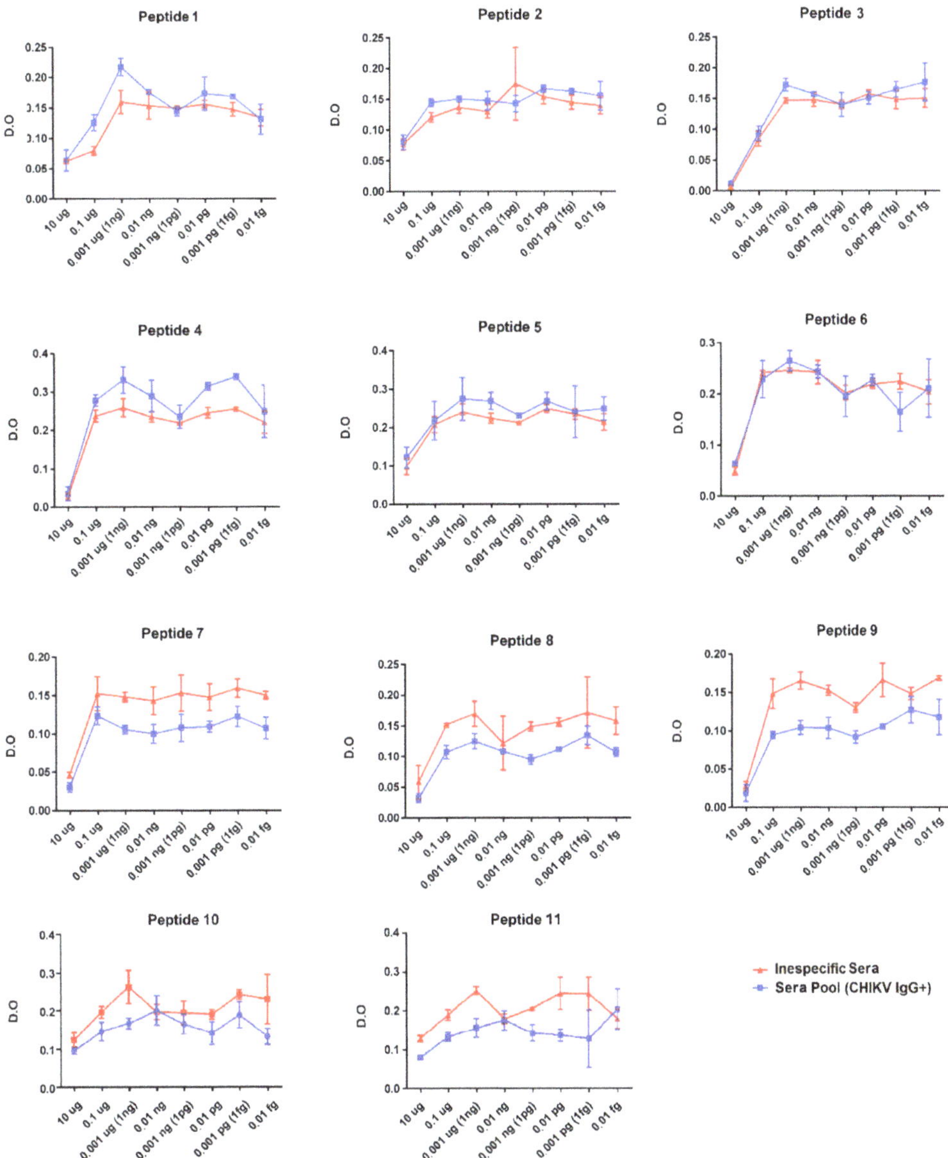

Figure 6. Standardization results of the indirect immunoenzyme peptide assay. A—The peptides P1–P9 correspond to the E2 protein. The peptides P10 and P11 correspond to the E3 protein. On the Y-axis, the optical density (O.D.) measurements for the positive IgG CHIKV patient sera pool (blue line) and, as the control, a non-specific IgG human serum (red line). On the X-axis the antigen dilutions. All assays were performed in sample triplicates.

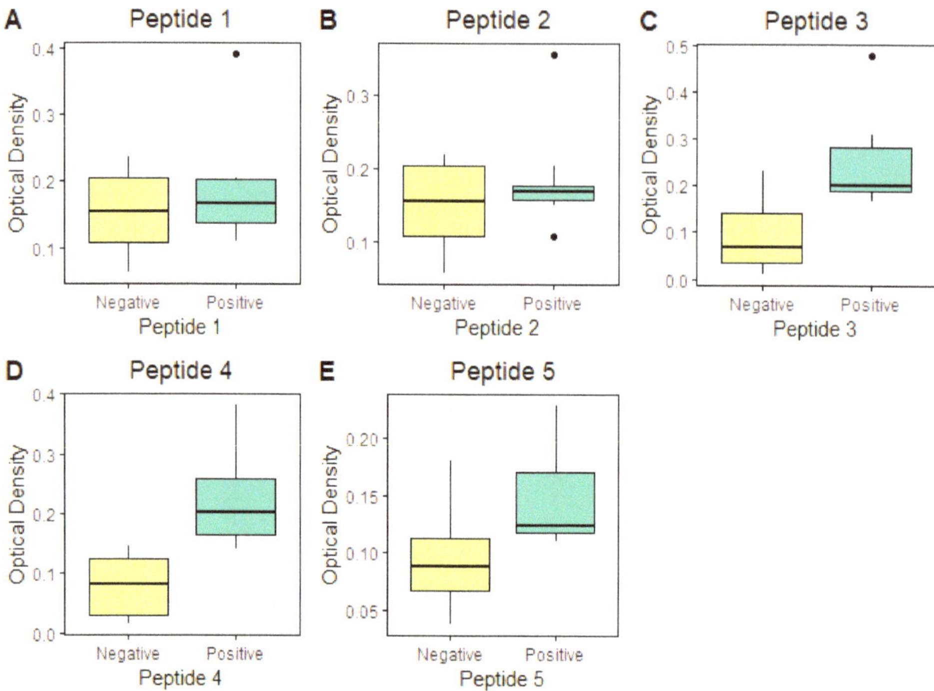

Figure 7. Protein E2 peptides' P1 (**A**), P2 (**B**), P3 (**C**), P4 (**D**), and P5 (**E**) reactivity using individual sera samples. Boxplots in yellow represent the CHIKV IgG negative sera, and green boxplots represent the CHIKV positive-IgG patients. The experiments were all performed in sample triplicates. The following cut-off value was established as the comparison threshold; mean of the triplicates +2x the standard deviation.

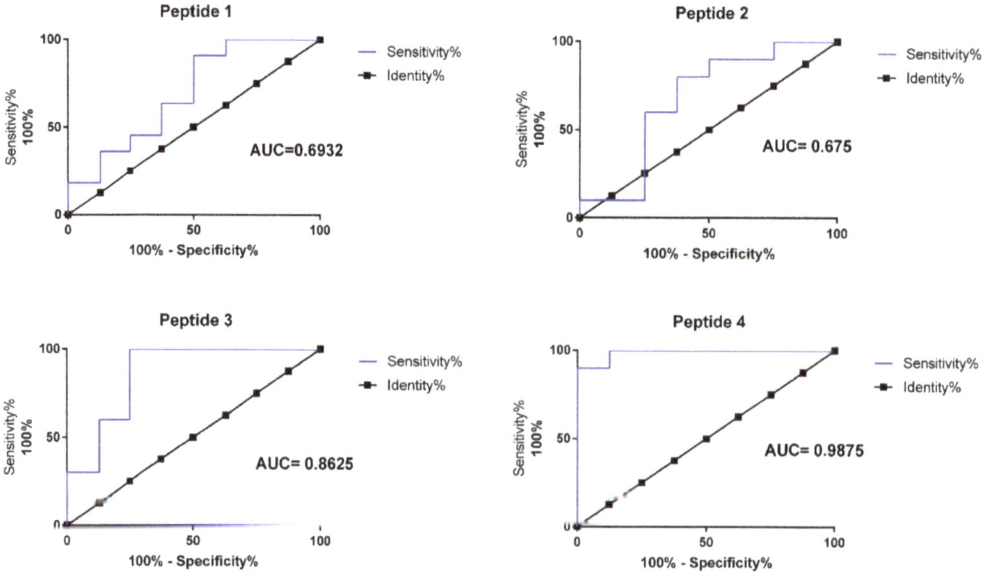

Figure 8. *Cont.*

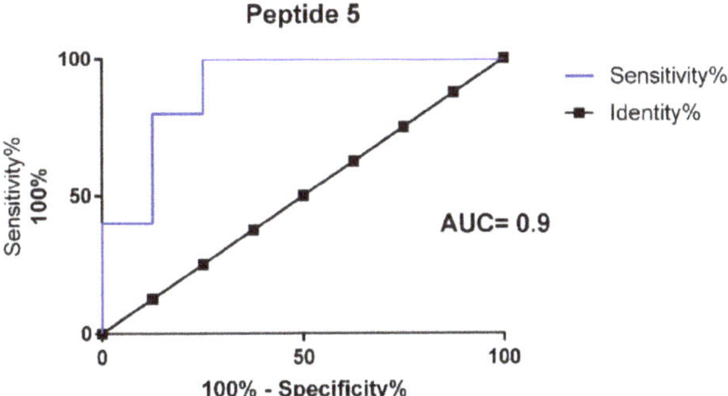

Figure 8. ROC curve for E2 protein synthetic peptides P1 to P5. 95% confidence interval. The peptides', P1 to P5, potential as classifiers was analyzed by the Receiving Operating Curve analysis. On the x-axis specificity score and the y-axis sensitivity score, the black dotted line represents the identity score, and the blue line represents the sensitivity score. The AUC (Area Under the Curve) score is shown for the individual peptides.

4. Discussion

In general, the surface proteins of the viruses are targets for neutralizing antibodies. The E3 and E2 CHIKV proteins have the function of binding to cell receptors, and the E1 protein is the protein responsible for membrane fusion in the endosomal vesicle protected by the E2 glycoprotein and is structurally less accessible to B lymphocyte antibodies, consequently with limited and low immunogenic potential. Our study focused on predicting the B-cell epitopes of the CHIKV structural proteins, E2 and E3, verifying the reactivity of synthetic linear peptides corresponding to the epitopes mapped by utilizing the CHIKV IgG-positive serum samples collected in the state of Sergipe, Brazil, in 2016. Screening studies using different patient cohorts, different geographic regions, and different genetic backgrounds help to identify and define usable signatures for the development of serological assays and vaccines [12,13,19]. So far, there are no studies on the target of the immune response of the Chikungunya virus previously infecting patients in Brazil and South America.

E2EP3, corresponding to peptide P1, was described by Kam et al., 2012 [12,13,19] as a marker epitope of the early convalescent phase of the CHIKV infection studying a cohort of patients on the Asian continent. In this study, Kam et al., 2012 [12,13,19] suggest that an ELISA based on the epitope E2EP3 may be useful to study CHIKV infections and to determine the magnitude of outbreaks. An analysis of this epitope with the sampled patient cohort showed no strong reactivity and potential as a marker for the CHIKV infection. These results may be explained by not using an ELISA protocol employing detection amplifiers as performed in the study by Kam et al., 2012 [12,13,19]. Another reason might be the sample collection period is not nearly the early stage of convalescence, considering the response pattern varies over time, and the sampled groups under analysis have different genetic backgrounds [12,19]. The P2 peptide with two amino acid changes lost recognition by the serum from patients infected with the ECSA genotype (Figure 6). This difference in reactivity requires further investigation and preferably by comparing well-characterized convalescent-phase sera from patients infected with the different genotypes.

Other significant immunodominant epitopes found by Kam et al., 2012 [12,13,19] are located among the amino acids 3025–3058 (corresponding to peptides P5, P6, and P7) in the E2 protein between Domain B and C in a linker region. This region is associated with early protection from the convalescent to the recovery phase, being considered an epitope with potential vaccine application. In this study, Kam et al., 2012 [12,13,19] also concluded that a single substitution in the amino acid K252Q in glycoprotein E2 has an important effect in decreasing the antibody binding capacity. The P6 and P7 peptides are variants of the same peptide; P7 has a lysine at position 252, previously identified as an escape mutation and was poorly recognized by the pool of the sera in the standardization test compared to the P6 peptide according to Kam et al. from their 2012 [12,13,19] findings (Figure 5). Both P6 and P7 peptides are adjacent to the P5 peptide (Figure 4). The P5 peptide was the most reactive and was selected to be evaluated with the individual samples (Figures 6–8, Table 1).

The peptides P7, P8, and P9 were weakly recognized by the specific CHIKV positive-IgG antibodies, similar to the findings described by Kam et al., 2012 [12,13,19] (Figure 6). These peptides correspond to the linker regions between Domains B and C (peptides P7 and P8), and after Domain C (peptide P9), they are accessible; however, they demonstrated a low potential for antigenic recognition (Figure 6; Supplementary Figures S1 and S2). The peptides P10 and P11, corresponding to the C-terminal region of the E3 protein, were weakly recognized by the specific CHIKV IgG antibodies and demonstrated a low antigenic recognition potential (Figure 6; Supplementary Figure S3).

The peptides P3, P4, and P5 are conserved among all the genotypes and are located in Domain A of the E2 protein (Figure 4, Table 2) and proved to be the most reactive peptides (Figure 6), showing the potential for antigenic recognition, being the major targets of the antibodies in the group of the patients sampled (Figures 6–8, Table 1). The P3 peptide is adjacent to the epitopes described by Kam et al. in 2012 [12,13,19] (the P1 peptide), overlapping three amino acids and confirming the N-terminal region of the E2 protein as one of the major immunogenic targets of specific CHIKV IgG antibodies (Figures 2 and 6–8; Supplementary Figures S1 and S2). The P4 and P5 peptides are also located in immunogenic regions, as previously described, withP4 in Domain A and P5 at the end of Domain B, in a region of arc accessible to the antibodies between Domains B and C (Figures 2, 3, 5 and 6; Supplementary Figures S1 and S2). Additional peptides in Domain B would also be of interest to test. The peptides P1, P2, P3, and P4, corresponding to Domain A and peptide P5 at the end of Domain B, demonstrate this region as the main antigenic and immunogenic target of the E2 structural protein and are expected to be present in different antigen and vaccine designs [3,13,26].

Additionally, the peptides P4 and P5 are reactive and have the potential as markers for serological tests, demonstrated by the performance analysis of the classifiers observed in the ROC curve graphs (Figure 8; Table 1). The performance of these peptides requires further confirmation by analyzing a larger number of samples and, preferably, collected in different regions. This peptide panel is useful for identifying the target regions of the immune response in Chikungunya virus-infected patients. Finally, the most reactive peptides, P1, P3, P4, P5, and P6, were located in the Chikungunya virus' glycoprotein envelope structural organization, resolved by x-ray crystallography (PDB:3N43), identifying the main antigenic and immunogenic epitopes analyzed in this study (Figure 9). A methodology with a higher detection capacity is also interesting and desirable, considering that the indirect peptide-based ELISA technique used in this study has limitations and is useful in exploratory studies. The potential biological activity of these peptides may also be analyzed in plate reduction neutralization assays to identify epitopes presenting a neutralizing potential, as well as for immunization with peptide-content vaccines [12,16,19]. Epitopes-based vaccines present an alternative capable of inducing a protective immune response with no adverse effects in vivo. Epitope mapping and validation studies are important for the discovery of B-cell epitopes that can be targeted by neutralizing antibodies and also for identifying T-cell epitopes that can be targeted by protective cytotoxic T-cells (CD8+ T-cells) and, finally, epitopes recognized by CD4 + auxiliary T-cells for the optimization of anti-CHIKV-specific antibody generation. The potential advantages of this strategy include safety, the ability to design and manufacture epitopes to increase vaccination efficiency, and the opportunity to design vaccines with a higher population coverage. Peptide vaccines containing sequences covering all genotypes have the potential to generate protection against CHIKV infections globally and, also, to cross-protect against the closely related Alphavirus [12,13,16,19].

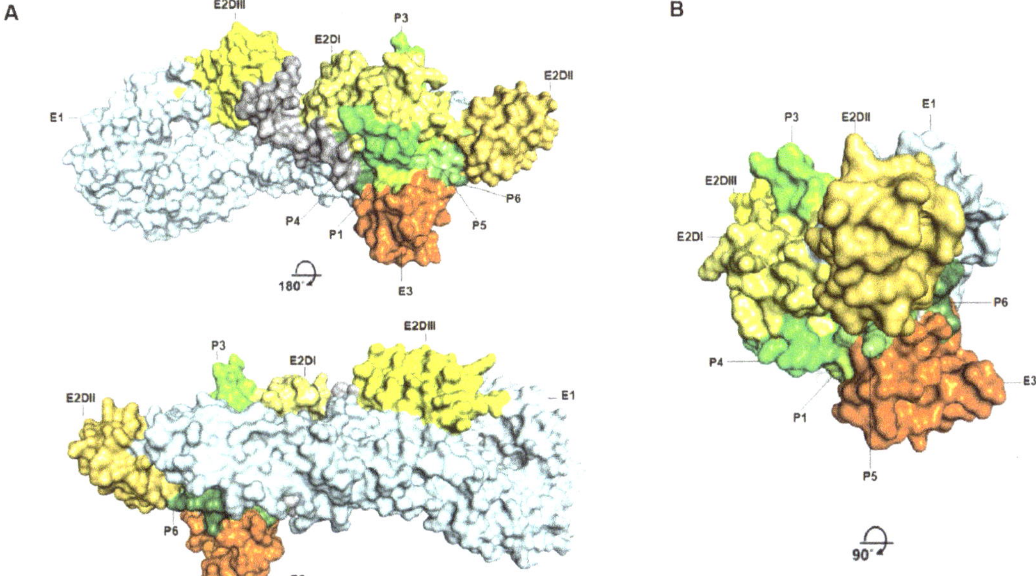

Figure 9. Antigenic and immunogenic B-cell epitopes are highlighted on the Chikungunya virus' glycoprotein envelope structural organization. (**A**) The Chikungunya virus' glycoprotein envelope complex structure's surface (PDB:3N43); protein E1 is in light blue; protein E2, Domains I, II, and III are in pale yellow, yellow-orange, and light yellow, respectively; and protein E3 is in orange. Epitopes corresponding to peptides P1, P3, P4, P5, and P6 in protein E2 are highlighted in green tones. (**B**) The upper view from (**A**).

Supplementary Materials: The supporting information can be downloaded at: https://www.mdpi.com/article/10.3390/v14081839/s1. Supplementary Figure S1. Conformational epitopes prediction for Chikungunya virus E2 Protein. Multiple alignment of E2 protein sequences corresponding to the four genotypes with highlighted epitopes conformational prediction. On green epitopes are conserved among all genotypes. Blue epitopes are common to more than one genotype. On red single epitopes of each genotype; *- Amino acid changes in the regions predicted as epitopes. *, ** CHIKV epitopes already identified and described. Supplementary Figure S2. Linear epitopes prediction for Chikungunya virus E2 Protein. Multiple alignment of E2 protein sequences corresponding to the four genotypes with highlighted epitopes linear prediction. On green epitopes are conserved among all genotypes. Blue epitopes are common to more than one genotype. On red single epitopes of each genotype; *- Amino acid changes in the regions predicted as epitopes. *, ** CHIKV epitopes already identified and described. Supplementary Figure S3. Conformational and Linear epitopes prediction for Chikungunya virus E3 Protein. Multiple alignment of E3 protein sequences corresponding to the four genotypes with highlighted conformational (A) and linear (B) epitopes predictions. Blue epitopes are common to more than one genotype. On red single epitopes of each genotype; *—Amino acid changes in the regions predicted as epitopes.

Author Contributions: J.P.d.C.S., M.d.P.C., D.F.d.L.N. and P.M.d.A.Z.: conceptualization. J.P.d.C.S., M.d.P.C., D.F.d.L.N. and P.M.d.A.Z.: data curation. J.P.d.C.S., M.d.P.C., S.Z.P., D.F.d.L.N. and P.M.d.A.Z.: formal analysis. P.M.d.A.Z.: funding acquisition. J.P.d.C.S. and M.d.P.C.: project administration. J.P.d.C.S., M.d.P.C., S.Z.P., V.R.H., D.F.d.L.N. and P.M.d.A.Z.: investigation, methodology, resources, and writing—review and editing. J.P.d.C.S. and M.d.P.C.: software. M.d.P.C., D.F.d.L.N. and P.M.d.A.Z.: supervision. J.P.d.C.S., M.d.P.C., S.Z.P., V.R.H., D.F.d.L.N. and P.M.d.A.Z.: validation and writing—original draft. J.P.d.C.S., M.d.P.C. and D.F.d.L.N.: visualization. All authors have read and agreed to the published version of the manuscript.

Funding: This work was supported by the Brazilian National Council of Scientific and Technological Development (CNPq) (process no. 441105/2016-5) and by the São Paulo Research Foundation (FAPESP) (process no. 2017/23281-6). Our MPC received a FAPESP fellowship (grant 2016/08204-2).

Institutional Review Board Statement: Not applicable.

Informed Consent Statement: Informed consent was obtained from all subjects involved in the study.

Data Availability Statement: Not applicable.

Acknowledgments: We thank all the professionals from the Fundação de Saúde Parreiras Horta/LACEN from Sergipe, Brazil, for helping to organize the human sample collection used in this study.

Conflicts of Interest: Not applicable.

References

1. Powers, A.N.N.M.; Brault, A.C.; Shirako, Y.; Strauss, E.G.; Kang, W.; Strauss, J.H.; Weaver, S.C. Evolutionary Relationships and Systematics of the Alphaviruses. *Am. Soc. Microbiol.* **2001**, *75*, 10118–10131. [CrossRef] [PubMed]
2. Weaver, S.C.; Forrester, N.L. Chikungunya: Evolutionary history and recent epidemic spread. *Antivir. Res.* **2015**, *120*, 32–39. [CrossRef] [PubMed]
3. Development, C.V. A Review on Chikungunya Virus Epidemiology, Pathogenesis and Current Vaccine Development. *Viruses* **2022**, *14*, 969.
4. Weaver, S.C. Arrival of Chikungunya Virus in the New World: Prospects for Spread and Impact on Public Health. *PLoS Negl. Trop. Dis.* **2014**, *8*, e2921. [CrossRef]
5. Nunes, M.R.T.; Faria, N.R.; De Vasconcelos, J.M.; Golding, N.; Kraemer, M.U.G.; de Oliveira, L.F.; Azevedo, S.; Coelho, G.E.; Cecília, A.; Cruz, R.; et al. Emergence and potential for spread of Chikungunya virus in Brazil Emergence and potential for spread of Chikungunya virus in Brazil. *BMC Med.* **2015**, *13*, 102. [CrossRef] [PubMed]
6. Cunha, P.; Alves, C.; Ferreira, D.; Neto, D.L.; Simone, M.; De Souza, F.; Duarte, D.; Marinho, P.; Zanotto, D.A. Outbreak of chikungunya virus in a vulnerable population of Sergipe, Brazil—A molecular and serological survey. *J. Clin. Virol.* **2017**, *97*, 44–49. [CrossRef]
7. Aparecido, N.; Raquel, N.; Douglas, F.; Morais, O. Chikungunya, Zika, Mayaro, and Equine Encephalitis virus detection in adult Culicinae from South Central Mato Grosso, Brazil, during the rainy season of 2018. *Braz. J. Microbiol.* **2022**, *53*, 63–70. [CrossRef]
8. Carvalho, V.L.; Azevedo, R.S.S.; Carvalho, L.; Azevedo, R.S.; Henriques, D.F.; Cecilia, A.; Cruz, R.; Vasconcelos, P.F.C.; Martins, L.C. Arbovirus outbreak in a rural region of the Brazilian Amazon. *J. Clin. Virol.* **2022**, *151*, 6–9. [CrossRef]
9. Gonzalez-escobar, G.; Churaman, C.; Rampersad, C.; Singh, R.; Nathaniel, S. Mayaro virus detection in patients from rural and urban areas in Trinidad and Tobago during the Chikungunya and Zika virus outbreaks. *Pathog. Glob. Health* **2021**, *115*, 188–195. [CrossRef]
10. Bogoch, I.I.; Brady, O.J.; Kraemer, M.U.G.; German, M.; Creatore, M.I.; Kulkarni, M.A.; Brownstein, J.S.; Mekaru, S.R.; Hay, S.I.; Groot, E.; et al. Anticipating the international spread of Zika virus from Brazil. *Lancet* **2016**, *387*, 335–336. [CrossRef]
11. Zhang, Q.; Wang, P.; Kim, Y.; Haste-andersen, P.; Beaver, J.; Bourne, P.E.; Bui, H.; Buus, S.; Frankild, S.; Greenbaum, J.; et al. Immune epitope database analysis resource. *Nucleic Acids Res.* **2008**, *36*, 513–518. [CrossRef] [PubMed]
12. Kam, Y.; Lum, F.; Teo, T.; Lee, W.W.L.; Simarmata, D.; Harjanto, S.; Chua, C.; Chan, Y.; Wee, J.; Chow, A.; et al. Early neutralizing IgG response to Chikungunya virus in infected patients targets a dominant linear epitope on the E2 glycoprotein. *EMBO Mol. Med.* **2012**, *4*, 330–343. [CrossRef] [PubMed]
13. Kam, Y.; Lee, W.W.L.; Simarmata, D.; Le Grand, R.; Tolou, H.; Merits, A.; Roques, P.; Ng, L.F.P.; Elisa, V. Unique Epitopes Recognized by Antibodies Induced in Chikungunya Virus-Infected Non-Human Primates: Implications for the Study of Immunopathology and Vaccine Development. *PLoS ONE* **2014**, *9*, e95647. [CrossRef]
14. Frank, R. The SPOT-synthesis technique Synthetic peptide arrays on membrane supports—Principles and applications. *J. Immunol. Methods* **2002**, *267*, 13–26. [CrossRef]
15. Volkmer, R. Synthesis and application of peptide arrays: Quo vadis SPOT technology. *ChemBioChem* **2009**, *10*, 1431–1442. [CrossRef]
16. Teng, T.; Kam, Y. Host response to Chikungunya virus and perspectives for immune-based therapies. *Future Virol.* **2011**, *6*, 975–984. [CrossRef]
17. Mukonyora, M. A Review of Important Discontinuous B-Cell Epitope Prediction Tools. *J. Clin. Cell. Immunol.* **2015**, *6*, 1–5. [CrossRef]
18. Freire, M.C.L.C.; Pol-Fachin, L.; Coêlho, D.F.; Viana, I.F.T.; Magalhães, T.; Cordeiro, M.T.; Fischer, N.; Loeffler, F.F.; Jaenisch, T.; Franca, R.F.; et al. Mapping Putative B-Cell Zika Virus NS1 Epitopes Provides Molecular Basis for Anti-NS1 Antibody Discrimination between Zika and Dengue Viruses. *ACS Omega* **2017**, *2*, 3913–3920. [CrossRef]
19. Kam, Y.; Lee, W.W.L.; Simarmata, D.; Harjanto, S.; Teng, T.; Tolou, H.; Chow, A.; Lin, R.T.P.; Leo, Y.; Rénia, L.; et al. Longitudinal Analysis of the Human Antibody Response to Chikungunya Virus Infection: Implications for Serodiagnosis and Vaccine Development. *J. Virol.* **2012**, *86*, 13005–13015. [CrossRef]

20. KAM, Y.-W.; Ong, E.K.S.; Rénia, L.; Tong, J.-C.; Ng, L.F.P. Immuno-biology of Chikungunya and implications for disease intervention. *Microbes Infect.* **2009**, *11*, 1186–1196. [CrossRef]

21. LEMANT, J.; Boisson, V.; Winer, A.; Thibault, L.; André, H.; Tixier, F.; Lemercier, M.; Antok, E.; Cresta, M.P.; Grivard, P.; et al. Serious acute chikungunya virus infection requiring intensive care during the reunion island outbreak in 2005–2006. *Crit. Care Med.* **2008**, *36*, 2536–2541. [CrossRef] [PubMed]

22. Tesh, R.B. Arthritides caused by mosquito-borne viruses. *Annu. Rev. Med.* **1982**, *33*, 31–40. [CrossRef] [PubMed]

23. Sissoko, D.; Malvy, D.; Ezzedine, K.; Renault, P.; Moscetti, F.; Pierre, V. Post-Epidemic Chikungunya Disease on Reunion Island: Course of Rheumatic Manifestations and Associated Factors over a 15-Month Period. *PLoS Negleted Trop. Dis.* **2009**, *3*, e389. [CrossRef] [PubMed]

24. Borgherini, G.; Poubeau, P.; Jossaume, A.; Gouix, A.; Cotte, L.; Michault, A.; Arvin-berod, C.; Paganin, F. Persistent Arthralgia Associated with Chikungunya Virus: A Study of 88 Adult Patients on Reunion Island. *Clin. Infect. Dis.* **2008**, *47*, 469–475. [CrossRef] [PubMed]

25. Kam, Y.; Simarmata, D.; Chow, A.; Her, Z.; Teng, T.; Ong, E.K.S.; Leo, Y.; Ng, L.F.P. Early Appearance of Neutralizing Immunoglobulin G3 Antibodies Is Associated With Chikungunya Virus Clearance and Long-term Clinical Protection. *J. Infect. Dis.* **2012**, *205*, 1147–1154. [CrossRef]

26. Costa-da-Silva, A.L.; Ioshino, R.S.; Petersen, V.; Lima, A.F.; Cunha, M.D.P.; Wiley, M.R.; Ladner, J.T.; Prieto, K.; Palacios, G.; Costa, D.D.; et al. First report of naturally infected Aedes aegypti with chikungunya virus genotype ECSA in the Americas. *PLoS Negl. Trop. Dis.* **2017**, *11*, e0005630. [CrossRef]

27. Soares-Schanoski, A.; Baptista-Cruz, N.; Castro-Jorge, L.A.; Carvalho, R.V.; Santos, C.A.; Ros, N.; Oliveira, U.; Costa, D.D.; Santos, C.L.; Cunha, M.P.; et al. Systems Analysis of Subjects Acutely Infected with Chikungunya Virus. *PLoS Pathog.* **2019**, *15*, e1007880. [CrossRef]

28. Trifinopoulos, J.; Nguyen, L.; Von Haeseler, A.; Minh, B.Q. W-IQ-TREE: A fast online phylogenetic tool for maximum likelihood analysis. *Nucleic Acids Res.* **2016**, *44*, 232–235. [CrossRef]

29. Yang, J.; Yan, R.; Roy, A.; Xu, D.; Poisson, J.; Zhang, Y. The I-TASSER Suite: Protein structure and function prediction. *Nat. Methods* **2015**, *12*, 7–8. [CrossRef]

30. Williams, C.J.; Headd, J.J.; Moriarty, N.W.; Prisant, M.G.; Videau, L.L.; Deis, L.N.; Verma, V.; Keedy, D.A.; Hintze, B.J.; Chen, V.B.; et al. MolProbity: More and better reference data for improved all-atom structure validation. *Protein Sci.* **2017**, *27*, 293–315. [CrossRef]

31. Wiederstein, M.; Sippl, M.J. ProSA-web: Interactive web service for the recognition of errors in three-dimensional structures of proteins. *Nucleic Acids Res.* **2007**, *35*, 407–410. [CrossRef] [PubMed]

32. Kolaskar, A.S.; Tongaonkar, P.C. A semi-empirical method for prediction of antigenic deteinants protein antigens. *FEBS Lett.* **1990**, *276*, 172–174. [CrossRef]

33. Voss, J.E.; Vaney, M.C.; Duquerroy, S.; Vonrhein, C.; Girard-Blanc, C.; Crublet, E.; Thompson, A.; Bricogne, G.; Rey, F.A. Glycoprotein organization of Chikungunya virus particles revealed by X-ray crystallography. *Nature* **2010**, *468*, 709–712. [CrossRef] [PubMed]

34. Venceslau-carvalho, A.A.; Teixeira, M.; Favaro, D.P.; Pereira, L.R.; Rodrigues-jesus, M.J.; Pereira, S.S.; Andreata-santos, R.; Prince, R.; Castro-amarante, M.F.; Rodrigues, K.B.; et al. Nano-multilamellar lipid vesicles loaded with a recombinant form of the chikungunya virus E2 protein improve the induction of virus-neutralizing antibodies. *Nanomed. Nanotechnol. Biol. Med.* **2021**, *37*, 102445. [CrossRef]

35. Fumagalli, M.J.; De Souza, M.; de Castro-jorge, L.A.; Homem, V. Chikungunya Virus Exposure Partially Cross-Protects against Mayaro Virus Infection in Mice. *J. Virol.* **2021**, *95*, e01122-21. [CrossRef]

Communication

SARS-CoV-2 Omicron Variant Neutralization after Third Dose Vaccination in PLWH

Alessandra Vergori [1,*], Alessandro Cozzi-Lepri [2], Giulia Matusali [3], Francesca Colavita [3], Stefania Cicalini [1], Paola Gallì [4], Anna Rosa Garbuglia [3], Marisa Fusto [1], Vincenzo Puro [5], Fabrizio Maggi [3], Enrico Girardi [6], Francesco Vaia [7] and Andrea Antinori [1] on behalf of the HIV-VAC Study Group

[1] HIV/AIDS Unit, National Institute for Infectious Diseases Lazzaro Spallanzani IRCCS, 00149 Rome, Italy; stefania.cicalini@inmi.it (S.C.); marisa.fusto@inmi.it (M.F.); andrea.antinori@inmi.it (A.A.)
[2] Centre for Clinical Research, Epidemiology, Modelling and Evaluation (CREME), Institute for Global Health, UCL, London WC1E 6BT, UK; a.cozzi-lepri@ucl.ac.uk
[3] Laboratory of Virology, National Institute for Infectious Diseases Lazzaro Spallanzani IRCCS, 00149 Rome, Italy; giulia.matusali@inmi.it (G.M.); francesca.colavita@inmi.it (F.C.); annarosa.garbuglia@inmi.it (A.R.G.); fabrizio.maggi@inmi.it (F.M.)
[4] Health Direction, National Institute for Infectious Diseases Lazzaro Spallanzani IRCCS, 00149 Rome, Italy; paola.galli@inmi.it
[5] Risk Management Unit, National Institute for Infectious Diseases Lazzaro Spallanzani IRCCS, 00149 Rome, Italy; vincenzo.puro@inmi.it
[6] Scientific Direction, National Institute for Infectious Diseases Lazzaro Spallanzani IRCCS, 00149 Rome, Italy; enrico.girardi@inmi.it
[7] General Direction, National Institute for Infectious Diseases Lazzaro Spallanzani IRCCS, 00149 Rome, Italy; francesco.vaia@inmi.it
* Correspondence: alessandra.vergori@inmi.it

Citation: Vergori, A.; Cozzi-Lepri, A.; Matusali, G.; Colavita, F.; Cicalini, S.; Gallì, P.; Garbuglia, A.R.; Fusto, M.; Puro, V.; Maggi, F.; et al. SARS-CoV-2 Omicron Variant Neutralization after Third Dose Vaccination in PLWH. *Viruses* **2022**, *14*, 1710. https://doi.org/10.3390/v14081710

Academic Editor: Jason Yiu Wing KAM

Received: 9 June 2022
Accepted: 1 August 2022
Published: 3 August 2022

Publisher's Note: MDPI stays neutral with regard to jurisdictional claims in published maps and institutional affiliations.

Abstract: The aim was to measure neutralizing antibody levels against the SARS-CoV-2 Omicron (BA.1) variant in serum samples obtained from vaccinated PLWH and healthcare workers (HCW) and compare them with those against the Wuhan-D614G (W-D614G) strain, before and after the third dose of a mRNA vaccine. We included 106 PLWH and 28 HCWs, for a total of 134 participants. Before the third dose, the proportion of participants with undetectable nAbsT against BA.1 was 88% in the PLWH low CD4 nadir group, 80% in the high nadir group and 100% in the HCW. Before the third dose, the proportion of participants with detectable nAbsT against BA.1 was 12% in the PLWH low nadir group, 20% in the high nadir group and 0% in HCW, respectively. After 2 weeks from the third dose, 89% of the PLWH in the low nadir group, 100% in the high nadir group and 96% of HCW elicited detectable nAbsT against BA.1. After the third dose, the mean log2 nAbsT against BA.1 in the HCW and PLWH with a high nadir group was lower than that seen against W-D614G (6.1 log2 ($\pm$1.8) vs. 7.9 ($\pm$1.1) and 6.4 ($\pm$1.3) vs. 8.6 ($\pm$0.8)), respectively. We found no evidence of a different level of nAbsT neutralization by BA.1 vs. W-D614G between PLWH with a high CD4 nadir and HCW (0.40 ($-$1.64, 2.43); p = 0.703). Interestingly, in PLWH with a low CD4 nadir, the mean log2 difference between nAbsT against BA.1 and W-D614G was smaller in those with current CD4 counts 201–500 vs. those with CD4 counts < 200 cells/mm^3 ($-$0.80 ($-$1.52, $-$0.08); p = 0.029), suggesting that in this target population with a low CD4 nadir, current CD4 count might play a role in diversifying the level of SARS-CoV-2 neutralization.

Keywords: SARS-CoV-2; HIV/AIDS; SARS-CoV-2 vaccine; Omicron variant; neutralization titers; third dose vaccine

1. Introduction

Persons living with HIV (PLWH) might have an increased risk of adverse outcomes following COVID-19 infection [1]. The high contagiousness and spread of the Omicron (BA.1 and BA.2) variant of SARS-CoV-2 and its ability to evade immunity elicited by vaccination or infection are of increasing concern. We previously showed that, in PLWH, after three doses of the COVID-19 mRNA vaccine, the humoral response elicited was strong and higher than that achieved with the second dose (>2 $\log_2$ difference), and neutralizing antibodies (nAbs) against the Whuan-D614G strain SARS-CoV-2 increased in most of the participants regardless of their CD4 count at the time of first dose vaccination [2]. In this work, we aimed to evaluate the potential susceptibility of the Omicron BA.1 variant to the mRNA vaccine and to compare neutralization titers in PLWH with those observed in a control sample of health care workers.

2. Materials and Methods

In a subset of vaccinated PLWH participating in the HIV-VAC study (details described elsewhere) [3], we measured levels of neutralization to BA.1 and Wuhan-D614G (W-D614G) in stored serum samples collected before and after the third dose. An external control group of HIV-negative health care workers (HCWs) vaccinated with the third booster dose (BD) was also included for comparison. The study was approved by the Scientific Committee of the Italian Drug Agency (AIFA) and by the Ethical Committee of the Lazzaro Spallanzani Institute, as the National Review Board for COVID-19 pandemic in Italy (approval number 323/2021). PLWH were firstly stratified according to their CD4 count nadir (<350 (low nadir group) vs. >350 cells/mm^3 (high nadir group)). In addition, the group with low nadir CD4 count (<350/mm^3) was further stratified according to the CD4 count at the time of the booster vaccine dose (>200/mm^3, 201–500/mm^3 and >500 cells/mm^3), which had increased as a result of treatment with ART. All participants received either an additional 3rd dose (full dose at least 28 days after the 2nd, PLWH, low nadir group) or a booster dose of vaccine (booster at least 5 months after the 2nd, high nadir and HCW groups). Neutralizing antibody titers (nAbsT) were measured by micro-neutralization assay based on live SARS-CoV-2 virus (described elsewhere [4]) for W-D614G (Ref-SKU: 008V–04005, from EVAg portal), and BA.1 (GISAID accession ID EPI_ISL_7716384), before and after the 3rd dose (after 15 days in PLWH and 30 days in HCW). The highest serum dilution inhibiting at least 90% of the cytopathic effect on Vero E6 cells was defined as neutralizing, and nAbs were categorized as undetectable if titers were <1:10. Proportions with detectable responses were compared using the McNemar's test for paired data and chi-square and Fisher exact test for unpaired data. Mean levels of nAbsT to BA.1 vs. W-D614G (in the log2 scale) were compared within groups using a paired t-test and across groups using a truncated linear regression model (to account for censored response data) after controlling for gender, age and time elapsed since the end of the primary vaccination cycle.

3. Results

We included 106 PLWH, of whom there were 81 in the low CD4 nadir group (27 (33%) with a CD4 count < 200/mm^3, 29 (36%) with a CD4 count 201–500/mm^3 and 25 (31%) with a CD4 count > 500/mm^3), and 25 in the high nadir group and 28 HCWs, for a total of 134 participants (characteristics shown in Supplementary Table S1).

At the time of receiving the third dose, the proportion of participants with detectable nAbsT against Whuan-D614G and BA.1 was 59 % and 12 % in the PLWH low nadir group ($p < 0.0001$) after a median of 156 days (IQR 152–159, min 4.4 months) after completion of the primary cycle (after the two-dose vaccine cycle), 80% and 20% in the high nadir group (after a median of 182 (177–186) days, min 4.9 months) ($p = 0.0001$) and 64% and 0% in the HCW (after a median of 283 (278–28) days, min 8.9 months), respectively.

After 2 weeks from receiving the third dose, 89% of PLWH in the low nadir group, 100% in the high nadir group and 96% of HCW elicited detectable nAbsT against BA.1 ($p = 0.123$). At the same time, the proportion of the participants with detectable nABsT against Whuan-D614G was 94%, 100% and 100%, respectively ($p = 0.182$).

After the third dose, the mean log2 nAbsT against BA.1 in the HCW and PLWH with a high nadir was lower than that seen against W-D614G (6.1 $\log_2$ (±1.8) vs. 7.9 (±1.1) and 6.4 (±1.3) vs. 8.6 (±0.8), respectively) ($p = < 0.0001$, Figure 1A). However, from fitting a truncated linear regression, we found no evidence of a different level of nAbsT neutralization by BA.1 vs. W-D614G between PLWH with a high CD4 nadir and HCW (0.40 (−1.64, 2.43); $p = 0.70$, Supplementary Table S2, panel A).

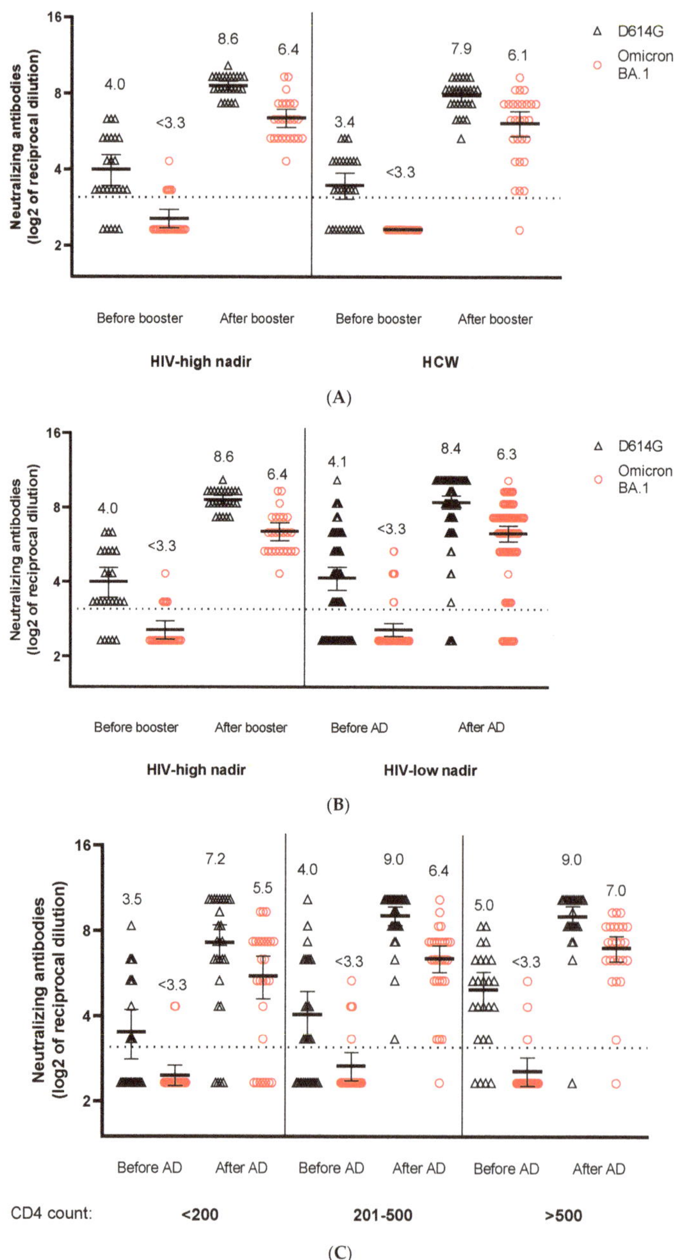

Figure 1. Log$_2$ mean titers of neutralizing antibodies against Wuhan-D614G and BA.1 in (**A**) PLWH with high CD4 count/high nadir and health care workers (HIV-negative); (**B**) PLWH with low and high CD4 nadir; (**C**) PLWH with low CD4 nadir and with CD4 count < 200/mm^3, 201–500/mm^3, >501/mm^3 at the time of third vaccine dose.

As well, the mean $\log_2$ nAbsT against BA.1 (vs. D614G) in PLWH with a high nadir and those with low nadir was 6.4 $\log_2$ (±1.3) vs. 8.6 (±0.8) and 6.3 (±2.1) vs. 8.4 (±2.4), respectively, (Figure 1B). From fitting a truncated linear regression, we found no evidence of a different level of nAbsT neutralization by BA.1 vs. W-D614G between PLWH with high and low CD4 nadir (0.23 (-0.44, 0.91); $p = 0.497$, Supplementary Table S2, panel B).

In contrast, even after controlling for potential confounding bias and censored responses by regression adjustment, among PLWH with a low CD4 nadir, the mean $\log_2$ difference between nAbsT against BA.1 and WD614G was smaller in those with a current CD4 count 201–500 vs. those with a CD4 count < 200 cells/mm^3 (-0.80 (-1.52, -0.08); $p = 0.03$) (Figure 1B, Supplementary Table S2, panel C).

4. Discussion

Our data show that neutralizing activity against BA.1 strongly increased after a third dose of a mRNA vaccine in all participants but was poorer than that seen against the original W-D614G strain, regardless of HIV status. This is likely due to the fact that the original mRNA vaccines were developed against the original strain, so they are less protective against current circulating variants [4]. There are limitations to this analysis. First, we only evaluated the short-term response (2 weeks on average) to the third dose and, therefore, we are unable to provide estimates of the durability and waning of nAbsT. Second, our analysis was not powered to establish whether the nAbsT recovering after a third dose was also associated with a reduced incidence of infection or severe disease. Third, nAbs activity against the past (e.g., Delta) and most recent variants (e.g., BA.1.1, BA.2) was not measured, although it has been recently demonstrated that the immune escape exhibited by all Omicron sublineages (BA.1.1 and BA.2) seems largely overcome by the booster dose compared with the two-dose cycle, suggesting a certain degree of cross-reactive immunity [5]. Last but not least, our results are only valid under the usual strong assumptions of a correctly specified model (liner regression predictor), the inclusion of all key potential confounding factors (age, gender and time since completion of primary cycle) and no unmeasured confounding factors being present. In particular, the fact that vaccination kinetics in the low CD4 count group vs. HCW was largely different, as per AIFA recommendations (although down to a maximum of one month in our sample) might have introduced residual confounding bias. Moreover, reasons for not seeing a clear dose-response relationship with CD4 count level (as there was no evidence of a difference between participants with a CD4 count < 200 and those with a CD4 count > 500 and the HCW) are unclear.

5. Conclusions

In conclusion, our results show that after at least 5 months from a primary two-dose vaccination cycle and shortly after the third dose, both with one additional and a booster, neutralizing activity against BA.1 strongly increased in all participants but was poorer than that seen against the original W-D614G strain, regardless of HIV status. Although, we found a signal for an association between CD4 count level and the extent of neutralization, and this should be further evaluated in future studies. These results are useful to appropriately define future boosting vaccination strategies in PLWH, accounting for the currently circulating SARS-CoV-2 variant.

Supplementary Materials: The following supporting information can be downloaded at: https://www.mdpi.com/article/10.3390/v14081710/s1, Table S1: Main descriptive characteristics of study population (n = 134); Table S2: Contrast high nadir vs. HCW from fitting a truncated regression in log2 scale (Panel A); Contrast by CD4 group in low nadir (Panel B).

Author Contributions: A.A. and A.V. conceptualized, designed the study and wrote the protocol. A.V. wrote the first draft of the manuscript and referred to appropriate literature. A.C.-L. was also the main person responsible for formal data analysis. A.A., A.C.-L., A.V., F.V. and G.M. conceived, supervised the study and contributed to data interpretation. S.C. and M.F. were responsible for data curation; V.P. is the main person responsible for data on immunogenicity in HCWs and revised the manuscript; V.P., F.M., A.R.G. and E.G. revised the manuscript content, reviewed and edited the manuscript.; G.M. and F.C. performed all the serology tests and neutralization assays; F.V. and P.G. supervised the anti-SARS-CoV-2 vaccination campaign at INMI for HIV-positive individuals. All authors have read and agreed to the published version of the manuscript.

Funding: The study was performed in the framework of the SARS-CoV-2 surveillance and response program implemented by the Lazio Region Health Authority. This study was supported by funds to the Istituto Nazionale per le Malattie Infettive Lazzaro Spallanzani IRCCS, Rome (Italy) from Ministero della Salute (Programma CCM 2020; Ricerca Corrente—Linea 1 e Linea 2; COVID-2020-12371675). EuCARE project funded by the EU under the HORIZON Europe programme, GA n. 6944397 received by Alessandro Cozzi Lepri, University College of London.

Institutional Review Board Statement: The study was approved by the Scientific Committee of the Italian Drug Agency (AIFA) and by the Ethical Committee of the Lazzaro Spallanzani Institute, as the National Review Board for COVID-19 pandemic in Italy (approval number 323/2021).

Informed Consent Statement: Informed consent was obtained from all subjects involved in the study.

Data Availability Statement: Anonymized participant data will be made available upon reasonable requests directed to the corresponding author. Proposals will be reviewed and approved by the investigator and collaborators on the basis of scientific merit. After approval of a proposal, data can be shared through a secure online platform after signing a data access agreement.

Acknowledgments: The authors gratefully acknowledge nursing staff, all the patients and all members of the HIV-VAC Study Group: Chiara Agrati, Alessandra Amendola, Andrea Antinori, Francesco Baldini, Rita Bellagamba, Aurora Bettini, Licia Bordi, Veronica Bordoni, Marta Camici, Rita Casetti, Concetta Castilletti, Stefania Cicalini, Francesca Colavita, Sarah Costantini, Flavia Cristofanelli, Alessandro Cozzi Lepri, Claudia D'Alessio, Veronica D'Aquila, Alessia De Angelis, Federico De Zottis, Lydia de Pascale, Massimo Francalancia, Marisa Fusto, Roberta Gagliardini, Paola Galli, Enrico Girardi, Giulia Gramigna, Germana Grassi, Elisabetta Grilli, Susanna Grisetti, Denise Iafrate, Roberta Iannazzo, Simone Lanini, Daniele Lapa, Patrizia Lorenzini, Alessandra Marani, Erminia Masone, Ilaria Mastrorosa, Davide Mariotti, Stefano Marongiu, Giulia Matusali, Valentina Mazzotta, Silvia Meschi, Annalisa Mondi, Stefania Notari, Sandrine Ottou, Jessica Paulicelli Luca Pellegrino, Carmela Pinnetti, Maria Maddalena Plazzi, Adriano Possi, Vincenzo Puro, Alessandra Sacchi, Eleonora Tartaglia, Francesco Vaia, Alessandra Vergori.

Conflicts of Interest: The authors declare no conflict of interest.

References

1. Danwang, C.; Noubiap, J.J.; Robert, A.; Yombi, J.C. Outcomes of patients with HIV and COVID-19 co-infection: A systematic review and meta-analysis. *AIDS Res Ther.* **2022**, *19*, 3. [CrossRef] [PubMed]
2. Vergori, A.; Cicalini, S.; Cozzi Lepri, A.; Matusali, G.; Bordoni, V.; Lanini, S.; Colavita, F.; Cimini, E.; Iannazzo, R.; De Pascale, L.; et al. Immunogenicity and Reactogenicity to COVID-19 mRNA Vaccine Additional Dose in PLWH. In Proceedings of the Conference on Retroviruses and Opportunistic Infections, Virtual, 12–16 February 2022.
3. Antinori, A.; Cicalini, S.; Meschi, S.; Bordoni, V.; Lorenzini, P.; Vergori, A.; Lanini, S.; De Pascale LMatusali, G.; Mariotti, D.; Cozzi Lepri, A.; et al. HIV-VAC Study Group. Humoral and cellular immune response elicited by mRNA vaccination against SARS-CoV-2 in people living with HIV (PLWH) receiving antiretroviral therapy (ART) according with current CD4 T-lymphocyte count. *Clin. Infect. Dis.* **2022**, ciac238. [CrossRef] [PubMed]
4. Tuekprakhon, A.; Huo, J.; Nutalai, R.; Dijokaite-Guraliuc, A.; Zhou, D.; Ginn, H.M.; Sekvaraj, M.; Liu, C.; Mentzer, A.J.; Supasa, P.; et al. Further antibody escape by Omicron BA.4 and BA.5 from vaccine and BA.1 serum. *bioRxiv* **2002**. [CrossRef]
5. Meschi, S.; Matusali, G.; Colavita, F.; Lapa, D.; Bordi, L.; Puro, V.; Leoni, B.D.; Galli, C.; Capobianchi, M.R.; Castilletti, C.; et al. Predicting the protective humoral response to a SARS-CoV-2 mRNA vaccine. *Clin. Chem. Lab. Med.* **2021**, *59*, 2010–2018. [CrossRef] [PubMed]

MDPI

St. Alban-Anlage 66

4052 Basel

Switzerland

www.mdpi.com

Viruses Editorial Office

E-mail: viruses@mdpi.com

www.mdpi.com/journal/viruses